Level 3

INSTALLING ELECTROTECHNICAL SYSTEMS & EQUIPMENT

Book B
NVQ/SVQ AND DIPLOMA

www.pearsonschoolsandfecolleges.co.uk

✓ Free online support
✓ Useful weblinks
✓ 24 hour online ordering

0845 630 44 44

Part of Pearson

Heinemann is an imprint of Pearson Education Limited, Edinburgh Gate, Harlow, Essex, CM20 2JE.

www.pearsonschoolsandfecolleges.co.uk

Heinemann is a registered trademark of Pearson Education Limited

Text © JTL 2011
Typeset by Saxon Graphics Ltd
Original illustrations © Pearson Education Ltd 2011
Illustrated by Oxford Designers
Cover design by Wooden Ark
Cover photo/illustration © Corbis: Tim Parnell

The right of JTL to be identified as author of this work has been asserted by them in accordance with the Copyright, Designs and Patents Act 1988

First published 2011

17 16 15 14
10 9 8 7 6 5 4 3

British Library Cataloguing in Publication Data
A catalogue record for this book is available from the British Library

ISBN 978 0 435 031275

Copyright notice
All rights reserved. No part of this publication may be reproduced in any form or by any means (including photocopying or storing it in any medium by electronic means and whether or not transiently or incidentally to some other use of this publication) without the written permission of the copyright owner, except in accordance with the provisions of the Copyright, Designs and Patents Act 1988 or under the terms of a licence issued by the Copyright Licensing Agency, Saffron House, 6–10 Kirby Street, London EC1N 8TS (www.cla.co.uk). Applications for the copyright owner's written permission should be addressed to the publisher.

Printed and bound in Slovakia by Neografia

Websites
Pearson Education Limited is not responsible for the content of any external internet sites. It is essential for tutors to preview each website before using it in class so as to ensure that the URL is still accurate, relevant and appropriate. We suggest that tutors bookmark useful websites and consider enabling students to access them through the school/college intranet.

The information and activities in this book have been prepared according to the standards reasonably to be expected of a competent trainer in the relevant subject matter. However, you should be aware that errors and omissions can be made and that different employers may adopt different standards and practices over time. Before doing any practical activity, you should always carry out your own risk assessment and make your own enquiries and investigations into appropriate standards and practices to be observed.

Acknowledgements

Every effort has been made to contact copyright holders of material reproduced in this book. Any omissions will be rectified in subsequent printings if notice is given to the publishers.

The author and publisher would like to thank the following individuals and organisations for permission to reproduce photographs:

(Key: b-bottom; c-centre; l-left; r-right; t-top)

Alamy Images: Art Directors and TRIP 449, 470, David J. Green 382, 434, 453bc, 468, imagebroker 33cr, Krys Bailey 294, Phil Degginger 290, Roberto Orecchia 416, sciencephotos 199cr, stu49 458, 478; **Corbis:** Martial Trezzini 260; **Dave Allen:** 372, 380; **Robert Harding World Imagery:** Josef Beck 18; **iStockphoto:** BartCo 163, stu49 484; **Megger Limited:** 144t; **Pearson Education Ltd:** Gareth Boden 31, 32t, 32b, 33cl, 176, 179cr, 179br, 182cl, 182bl, 183, 184cl, 184bl, 185, 186, 189cr, 189br, 191bl, 191br, 192, 351, 359, 391, 405tr, 433, 448cl, 448bl, 465; **Pearson Education Ltd:** Naki Photography 36, 37, 119, 142, 143t, 143b, 144b, 145t, 145b, 360, 363, 397; **Photolibrary.com:** originals imagebroker.net 75; **Science Photo Library Ltd:** Andrew Lambert Photography 370, GIPhotoStock 452tl, 452bl, 453cl, John McLean 1; **Shutterstock.com:** DavidXu 199tr, Evgeny Korshenhov 373, HomeStudio 406c, Ingvar Bjork 419, Ivaschenko Roman 405b, Jakub Pavlinec 211, Jiri Hera 405cl, Jouke van Keulen 406t, Piotr Wardynski 417, ra3rn 34, tr3gin 404

All other images © Pearson Education Ltd.

Every effort has been made to trace the copyright holders and we apologise in advance for any unintentional omissions. We would be pleased to insert the appropriate acknowledgement in any subsequent edition of this publication.

Contents

	Introduction	iv
	Features of this book	v
	Acknowledgements	vi
ELTK 04a	Understanding the principles of planning and selection for the installation of electrotechnical equipment and systems in buildings, structures and the environment	1
ELTK 06	Understanding principles, practices and legislation for the inspection, testing, commissioning and certification of electrotechnical systems and equipment in buildings, structures and the environment	75
ELTK 07	Understanding the principles, practices and legislation for diagnosing and correcting electrical faults in electrotechnical systems and equipment in buildings, structures and the environment	163
ELTK 08	Understanding the electrical principles with the design, building, installation and maintenance of electrical equipment and systems	211
	Index	501

Introduction

This book is designed to support the new NVQ Level 3 Diploma in Installing Electrotechnical Systems and Equipment. This new Diploma has been prepared by SummitSkills (the SSC) in consultation with employers, the main exam boards and training providers. This means that all exam boards offering this qualification will have the same unit structure and assessment strategy. Exam boards will then design their own assessment content.

This book is designed to support the following qualifications:

- 2357 City and Guilds Level 3 NVQ Diploma in Installing Electrotechnical Systems and Equipment
- 501/1605/8 EAL Level 3 NVQ Diploma in Installing Electrotechnical Systems and Equipment (Buildings, Structures and the Environment)

These qualifications are approved on the Qualifications and Credit Framework (QCF). The QCF is a new government framework which regulates all vocational qualifications to ensure they are structured and titled consistently and quality assured. SummitSkills has developed the new Diploma qualification with the awarding bodies.

Who the qualification is aimed at

The new diploma qualification is aimed at both new entrants (such as apprentices or adult career changers) as well as the existing workforce (so those looking to upskill). It is intended to train and assess candidates so that they can be recognised as occupationally competent by the industry in installing electrotechnical systems. Learners should gain the skills to:

- work as a competent electrician
- achieve a qualification recognised by the Joint Industry Board (JIB) for professional grading to the industry
- complete an essential part of the SummitSkills Advanced Apprenticeship.

About this book

This book supports the last four units of the Level 3 Diploma. In combination with Book A, which covers the remaining five units (01, 02, 03, 04 and 05), it is designed to cover all the information you will need to attain your Level 3 qualification in Installing Electrotechncial Systems and Equipment.

Each unit of this book relates to a particular unit of the Diploma and provides the information needed to gain the required knowledge and understanding of that area.

This book has been prepared by expert JTL trainers, who have many years of experience of training learners and delivering electrical qualifications. The content of each unit will underpin the various topics which you will be assessed on by your exam board.

Each unit has knowledge checks throughout, as well as a set of multiple choice questions at its conclusion, to allow you to measure your knowledge and understanding.

This book will also be a useful reference tool for you in your professional life once you have gained your qualifications and are a practising electrician.

Using this book

It is important to note that this book is intended to be used for training. It should not be regarded as being relevant to an actual installation. You should always make specific reference to the British Standards or manufacturer's data when designing electrical installations.

Features of this book

This book has been fully illustrated with artworks and photographs. These will help to give you more information about a concept or a procedure, as well helping you to follow a step-by-step procedure or identify a particular tool or material.

This book also contains a number of different features to help your learning and development.

Key term
These are new or difficult words. They are picked out in **bold** in the text and then defined in the margin.

Remember
This highlights key facts or concepts, sometimes from earlier in the text, to remind you of important things you will need to think about.

Safety tip
This feature gives you guidance for working safely on the tasks in this book.

Did you know?
This feature gives you interesting facts about the building services trade.

Find out
These are short activities and research opportunities, designed to help you gain further information about, and understanding of, a topic area.

Working life
This feature gives you a chance to read about and debate a real life work scenario or problem. Why has the situation occurred? What would you do?

Progress check
These are a series of short questions, usually appearing at the end of each learning outcome, which gives you the opportunity to check and revise your knowledge.

Getting ready for assessment

This feature provides guidance for preparing for the practical assessment. It will give you advice on using the theory you have learnt about in a practical way.

CHECK YOUR KNOWLEDGE

This is a series of multiple choice questions at the end of each unit, in the style of the GOLA end of unit tests.

At the time of going to press both the author and publisher have used their best endeavours to ensure that any references to BS 7671 17th Edition Amendment 1 within these publications were correct. If, however, there are required amendments within the lifetime of these books, corrections will be made at reprint and corrected pages posted on Pearson's website.

Acknowledgements

JTL would like to express its appreciation to all those members of staff who contributed to the development of this book, ensuring that the professional standards expected were delivered and generally overseeing the high quality of the final product. Without their commitment this project would not have been seen through successfully.

Particular thanks to Dave Allan, who revised the content of the previous editions of this book and prepared extensive new material to cover the diploma specifications.

The Working life, Check your knowledge and Getting ready for assessment features were prepared by Bill Brady of EAL (EMTA Awards Limited). Bill also carried out a full review of the book and made many invaluable comments.

Pearson and JTL would like to thank Richard Swann of Loughborough College and Andy Jeffery of Oaklands College for their comprehensive and painstaking review work, which has again made an invaluable contribution towards ensuring accuracy in this book.

Pearson and JTL also wish to thank Paul, Rita and Jeff Hurt at P & R Hurt Education and Training for their patience, assistance, advice and support during the photo shoot.

Candidate handbook answers

The answers to all the questions in this book can be found on the Training Resource Disk. They can also be downloaded from Pearson's website at the following URL: **www.pearsonfe.co.uk/iestudentbookanswers**

If there are any problems downloading this PDF, please contact Pearson on 0845 6301111.

UNIT ELTK 04a

Understanding the principles of planning and selection for the installation of electrotechnical equipment and systems

An installation must be protected from overload or fault current and users of an installation must be protected from the risk of electric shock or fire. Selecting the correct cable types and sizes to meet the installation's requirements and environment is vital.

This unit will look at the different supply systems in use throughout the UK and the relevant protective measures used.

This unit covers four learning outcomes:

- Understand the characteristics and applications of consumer supply systems
- Understand the principles of internal and external earthing arrangements for electrical installations
- Understand the principles for selecting cables and circuit protection devices
- Understand the principles for selecting wiring systems, enclosures and equipment.

K1. Understand the characteristics and applications of consumer supply systems

In this unit, you will be introduced to the different supply systems in use throughout the UK and the relevant protective measures that we need to put in place. By connecting to earth all metalwork not intended to carry current, a path is provided for leakage current which can be detected and interrupted by fuses, circuit breakers and residual current devices.

In terms of protection, the installation must be protected from overload or fault current and users of an installation must be protected from the risk of electric shock or fire.

As regards the installation, we need to select the correct cable types and sizes to meet the installation requirements in line with the environment that they will be installed in. We then control and protect the installation as heavy currents that are not disconnected generate heat that can damage equipment, distort busbars, degrade insulation and possibly start a fire.

Characteristics and applications of earthing arrangements and supply systems

Chapter 31 of BS 7671, Regulation 312 deals with conductor arrangement and system earthing and goes on to classify the types of system earthing using a series of letters as shown in Table 4.01.

	First Letter (Relationship between supply and earth)
T	The supply is connected directly to earth ('T' coming from the Latin word **t**erra, meaning earth) at one or more points
I	The supply is **i**solated from earth, or one point is connected to earth through a high impedance
	Second Letter (Relationship between earth and the exposed-conductive parts of the installation)
N	Direct electrical connection of the exposed-conductive-parts to the earthed point of the supply system (Remember that for a.c. systems, the earthed point of the supply system is normally the **n**eutral point)
T	Direct electrical connection of exposed-conductive-parts to earth, independent of the earthing of any point of the supply system
	Subsequent letters (The arrangement of neutral and protective conductors)
C	The neutral and protective functions are **c**ombined in a single conductor (the PEN conductor)
S	The protective function provided by a conductor **s**eparate from the neutral conductor or from the earthed line (or, in a.c. systems, earthed phase) conductor.

Table 4.01 Classifications of system earthing arrangements

The most common variations of these letter codes are shown as follows:

TT system

For this system

- The first letter T means that the supply is connected directly to earth at the source (e.g. the generator or transformer at one or more points).
- The second letter T means that the exposed metalwork of the installation is connected to the earth by a separate earth electrode. The only connection between these two points is the general mass of earth (e.g. soil etc.) as shown in Figure 4.01.

This system is used where the customer installation has not been provided with an earth terminal by the electricity supply company. Although it can also be provided as an underground cable it is most commonly used in rural areas where there is often an overhead supply.

A typical consumer's intake position for a TT system is shown in Figure 4.02, where the RCD is external to the consumer unit. On more modern installations the RCD is included as part of the consumer unit.

Level 3 NVQ/SVQ Diploma Installing Electrotechnical Systems and Equipment Book B

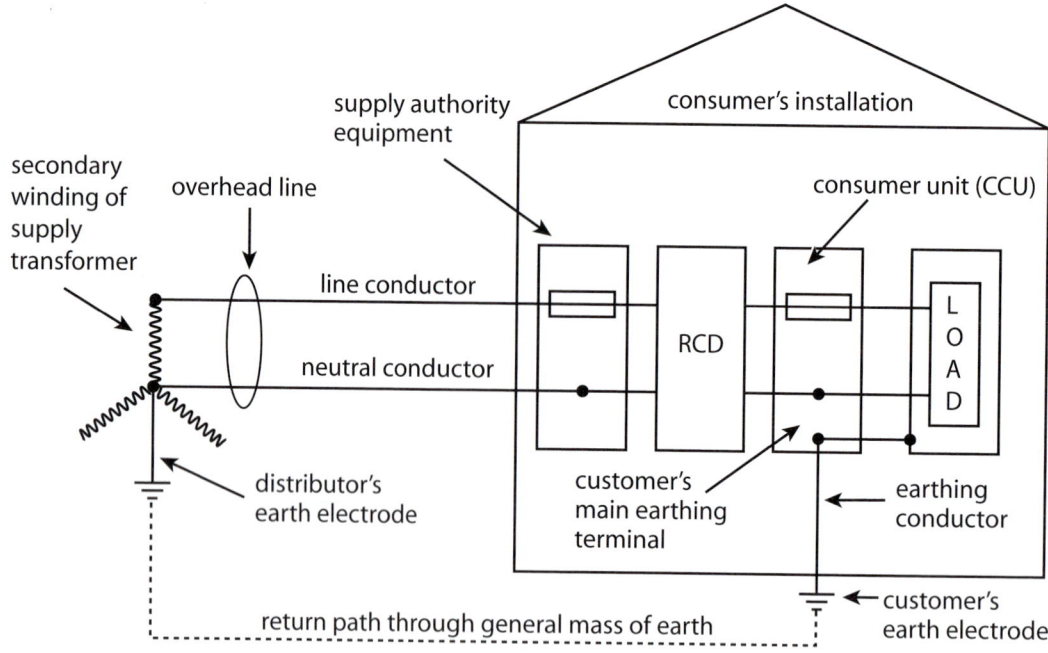

Figure 4.01 TT earthing system

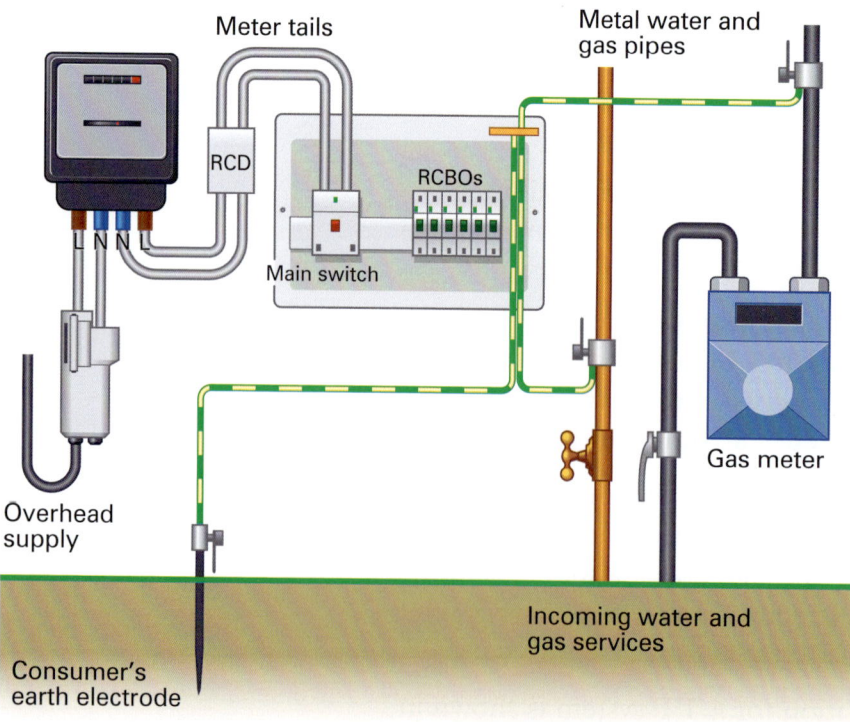

Figure 4.02 Typical consumer's intake position for a TT system

TN-S system

For this system:

- T means that the supply is connected directly to earth at one or more points.
- N means that the exposed metalwork of the installation is connected directly to the earthing point of the supply.
- S means that separate neutral and protective conductors are being used throughout the system from the supply transformer all the way to the final circuit, to provide the earth connection as shown in Figure 4.03.

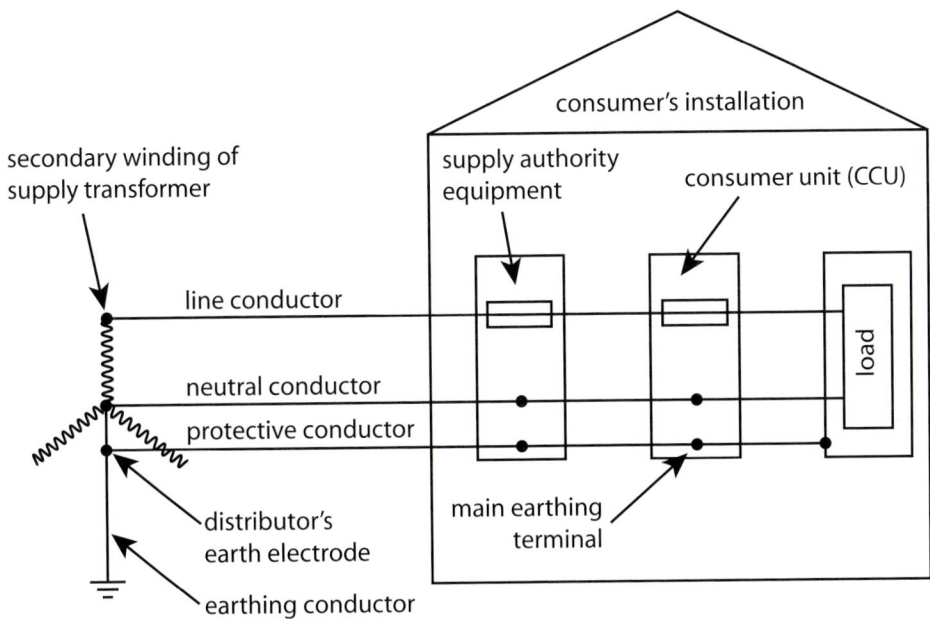

Figure 4.03 TN-S earthing system

This is probably the most common system in the UK. The earth connection is usually through the sheath or armouring of the supply cable and then by a separate conductor within the installation. As a conductor is used throughout the whole system to provide a return path for the earth-fault current, the return path should have a low value of impedance.

A typical consumer's intake position for a TN-S system is shown in Figure 4.04.

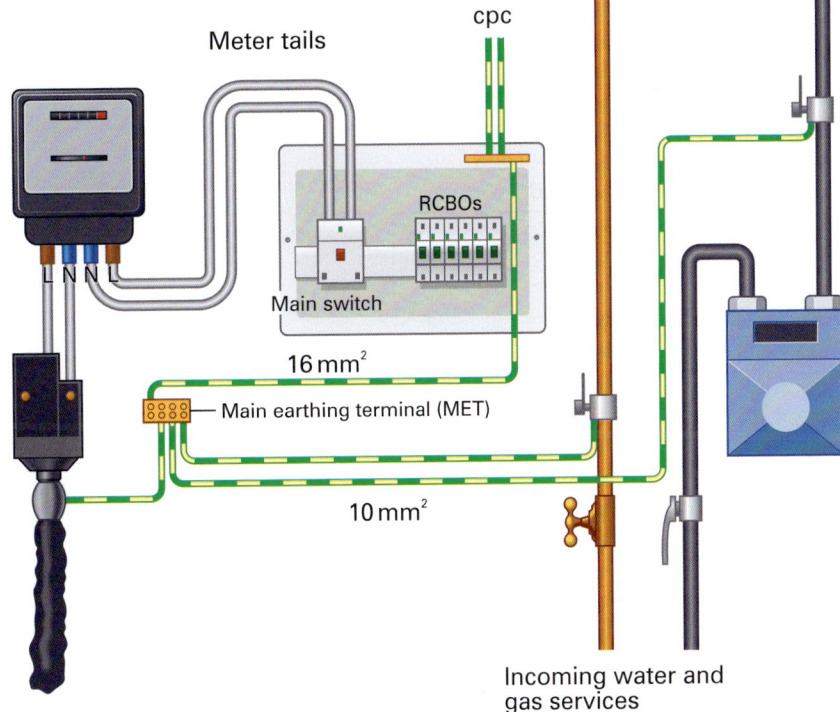

Figure 4.04 TN-S earthing system with metallic earth-return path

TN-C-S system (PME)

For this system:

- T means the supply is connected directly to earth at one or more points.
- N means the exposed metalwork of the installation is connected directly to the earthing point of the supply.
- C means that for some part of the system (generally in the supply section) the functions of neutral conductor and earth conductor are combined in a single common conductor.
- S means that for some part of the system (generally in the installation), the functions of neutral and earth are performed by separate conductors.

This system is shown in Figure 4.05.

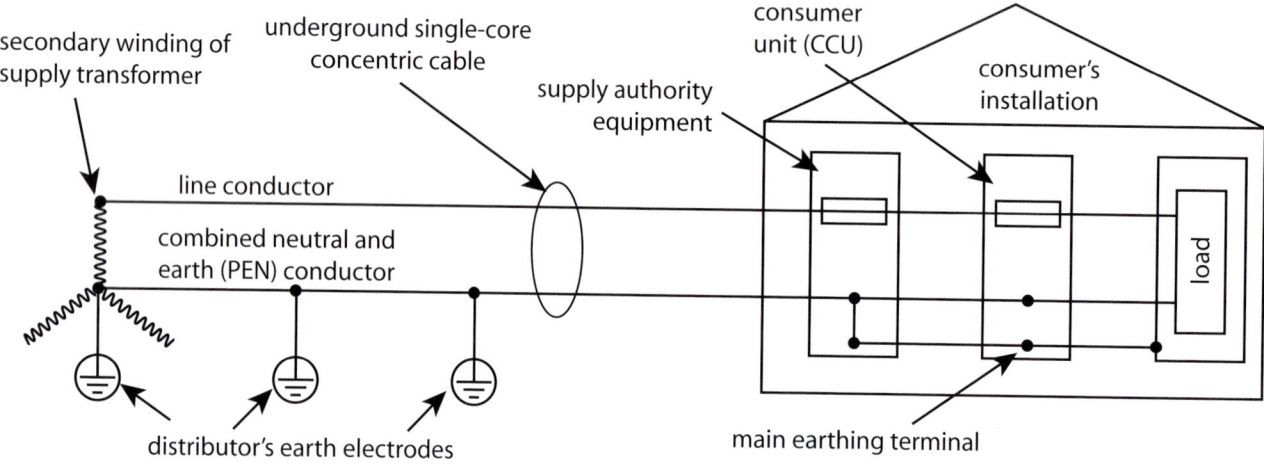

Figure 4.05 TN-C-S protective multiple earthing (PME) system

In a TN-C-S system, the supply uses a common conductor for both the neutral and the earth. This combined conductor is commonly known as the Protective Earthed Neutral (PEN) or also sometimes as the Combined Neutral and Earth (CNE) conductor.

The supply PEN is required to be earthed at several points, this type of system is also known as Protective Multiple Earthing (PME). As shown in Figure 4.05, this effectively means the distribution system is TN-C and that the consumer's installation is TN-S; this combination therefore giving us a TN-C-S system.

A typical consumer's intake position for a TN-C-S system is shown in Figure 4.06.

Protective Multiple Earthing (PME)

As discussed on page 6, this system of earthing is used on TN-C-S systems; it is an extremely reliable system and is consequently becoming more common as a distribution system in the UK.

With the PME system, the neutral of the incoming supply is used as the earth point and all cpcs connect all metal work, in the installation to be protected, to the consumer's earth terminal.

Consequently, all line-to-earth-faults are converted into line-to-neutral faults, which ensures that a heavier current will flow under fault conditions and protective devices operate rapidly.

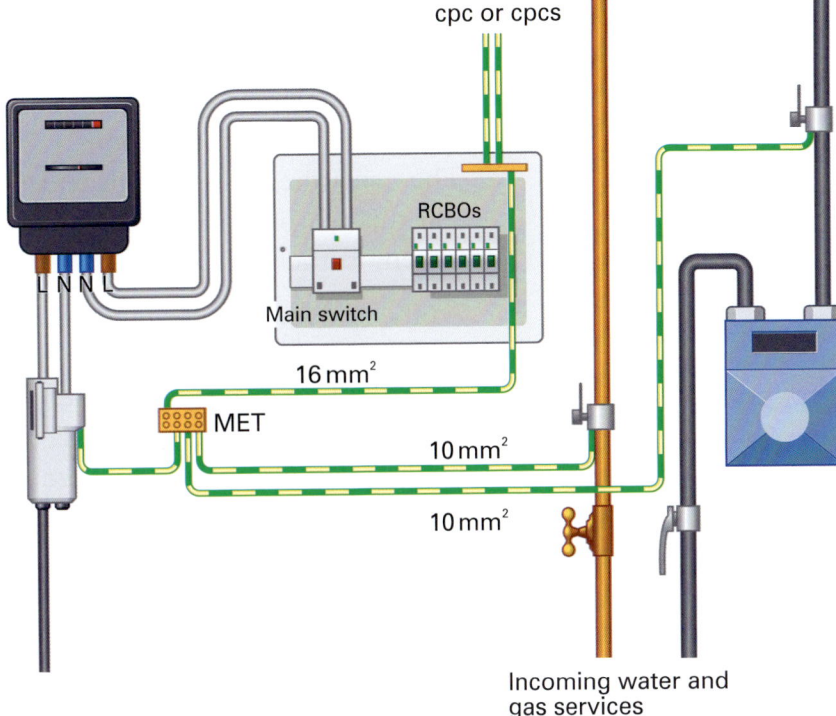

Figure 4.06 TN-C-S earthing-system

However this increase in fault current may produce two hazards:

1. The increased fault current results in an enhanced fire risk during the time the protective device takes to operate.

2. If the neutral conductor ever rose to a dangerous potential relative to earth, then the resultant shock risk would extend to all the protected metalwork on every installation that is connected to this particular supply distribution network.

Because of these possible hazards, certain conditions are laid down before a PME system is used. These include the following:

- PME can only be installed by the supply company if the supply system and the installations it will feed meet certain requirements.
- The neutral conductor must be earthed at a number of points along its length. It is this action that gives rise to the name 'multiple' earthing.
- The neutral conductor must have no fuse or other link that can break the neutral path.
- Where PME conditions apply, the main equipotential bonding conductor must be selected in accordance with the neutral conductor of the supply and Table 54.8 of BS 7671.

IT system

With the IT system, all exposed-conductive parts within the consumer's installation are connected to earth via an earth electrode. The supply is then either connected to earth via a deliberate impedance (about 1.5 kΩ) or is isolated from earth. The consumer therefore provides their own connection to earth. For this reason IT systems are not allowed in the public supply system in the UK.

As the system calls for fault tracking and elimination, insulation monitoring devices (IMD) are required. In many countries, the unearthed neutral is used whenever continuity of service is essential or when human life is at stake (e.g. hospitals). The method can also be used where a supply for special purposes is taken from a private generator.

> **Working life**
>
> While you are checking the specification with the consulting engineer you notice he has specified the supply as TN-S. From the site supply intake you know the supply earthing system is TN-C-S. What difference will this oversight make to:
>
> 1. The earthing arrangement?
> 2. The size of earthing and protective equipotential bonding conductors selected?

Supply systems

In 1988 the EU passed legislation to harmonise voltage levels between phase and neutral conductors throughout Europe to 230 V, and the UK supply regulations were amended to reflect this change. (Before this the European nominal voltage was 220 V and the UK voltage was 240 V.)

Effective from 1995, the 1988 agreement states a voltage tolerance of 230 V +10% – 6%, which gives us an effective range of 216 V – 253 V. This allows the European system to stay at 220 V and the UK to stay at 240 V, yet both are apparently harmonised!

Due to these tolerances, suppliers of electricity have been able to satisfy supply voltage regulations without altering their supply voltages and in many parts of the UK you will find the nominal supply voltage is still 240 V.

Most equipment is designed to operate satisfactorily between 230 V and 240 V. However, if an item is designed to operate at less than 240 V, and then used on a 240 V supply then this **could** cause it to operate at a reduced efficiency, use more power or operate at a higher temperature and subsequently reduce operational life.

> **Example**
>
> As an example to illustrate this concept, consider a 3 kW fan heater which had been specifically designed to operate at 230 V.
>
> Applying Ohm's Law we can say that if P = V ÷ I,
>
> Then I = P ÷ V giving a current of 13 A.
>
> However, if the fan heater is supplied with 240 V and we again apply Ohm's Law we can see that the power is increased:
>
> P = V × I = 240 × 13 = 3.1 kW

> **Remember**
>
> Electricity generation and distribution will be discussed more thoroughly in Unit ELTK 08 of this book.

Single-phase supply

In general, most domestic installations in the UK are provided with a single-phase supply; the most modern ones being rated at 80 A where there is no electric heating or 100 A where there is.

This capacity is normally more than adequate to meet the lighting and small power needs of a domestic or small commercial user. Single-phase means that the premises are supplied by a 2 core (1× line plus neutral) cable at 230 V.

Three-phase supply

Where commercial or industrial installations require a larger capacity, it is normal for a three-phase supply to be provided. Three-phase means 3 × line plus neutral and this would normally be supplied at 400 V for premises of this type.

With a three-phase installation, when we have a balanced three-phase load there is no current flowing in the neutral. This is why three-phase motors have only three conductors to their supply and all line conductors can be of the same cross-sectional area.

Three-phase and neutral

That said, most supplies to commercial or industrial properties will be three-phase and neutral. This allows for current flow in the neutral, should the load be unbalanced, and also allows the connection of single-phase loads (line and neutral).

Supply systems summary

With the arrangements of current-carrying conductors under normal operating conditions taken into account in Chapter 31 of BS 7671, we can summarise the most common systems as shown in Table 4.02.

Three-phase and neutral	400 V	4 wire	Commercial/industrial installations
Single-phase and neutral	230 V	2 wire	Domestic/light commercial installations
Three-phase	400 V	3 wire	Motors

Table 4.02 Common supply systems

Provision for isolation and switching

With all electrical supplies it is important to have a means of control. Chapter 53 of BS 7671 covers requirements for protection, isolation, switching and control with Table 53.4 giving guidance on the selection of protective, isolation and switching devices. Overcurrent and earth-fault protection will be covered in detail on pages 26–44. This section will look at isolation and switching.

BS 7671 makes the following definitions:

- **Isolation** – the ability to cut off, for safety reasons, all or part of an installation from every source of electrical energy
- **Isolator** – the device used to achieve isolation
- **Switches** – mechanical devices capable of carrying, making and breaking current under normal conditions.

Regulation 537.1.3 clearly states that every installation shall have provision for disconnection from the supply. Regulation 537.1.4 requires that a main linked switch or linked circuit-breaker shall be provided as near as practicable to the origin of every installation as a means of switching the supply on load and as a means of isolation.

In a domestic or similar type of single-phase installation, such a main switch intended for operation by ordinary persons must interrupt both live conductors of the single-phase supply. This is often done by the main switch in the consumer unit as shown in Figures 4.02 to 4.06 (pages 4–7).

Switching for isolation

BS 7671 requires that every circuit shall be capable of being isolated from both live conductors and that a suitable means shall be provided to prevent the circuit from being accidentally re-energised.

With a consumer unit this is done by using a non-interchangeable lock and key with the key held by the person carrying out the work. For a fused connection unit, for example supplying a wall-mounted heater, the fuses should be removed and kept by the person doing the work.

Devices that may be used for isolation include BS 88 and BS 1362 fuses, circuit breakers, RCDs, plugs and socket outlets, lighting support couplers and fused connection units.

> **Remember**
> There is a clear process that must be followed for safe isolation which is covered fully in Unit ELTK 01 in Book A.

Switching for mechanical maintenance

The main requirement for mechanical maintenance is to provide a means of switching off the supply where the work on electrically powered equipment could involve a risk of injury.

The requirements are similar to that of isolation, but Regulation 537.3.2.1 requires that any switch provided in the supply circuit for the purpose of switching off for mechanical maintenance must be capable of carrying the full load current.

Emergency switching

BS 7671 defines emergency switching as being a single action that removes any unexpected danger as quickly as possible from any part of an installation where it is deemed necessary.

Where there is a risk of electric shock, such a device must be an isolating device capable of breaking all live conductors. Devices for emergency switching must therefore be capable of breaking the full load current.

> **Safety tip**
> Devices should be hand operated. An emergency push button that disconnects supplies via a contactor is permitted, however, the release of an emergency switch must **not** re-energise the supply to that part of the installation.

Functional switching

This is defined as being the switching on or off of all or part of an installation as part of normal operating procedures. Consequently, (for example with light switches) such switches do not necessarily disconnect all live conductors, but they must be capable of handling the heaviest loads (Regulation 537.5.2.1).

Firefighter's switches

These are red switches for operation by firefighters and are required in the low voltage circuits supplying exterior installations and/or interior discharge lighting, where both operate in excess of low voltage.

For interior installations they must be mounted in the main entrance of the building and for exterior installations in a conspicuous location, accessible to firefighters and no higher than 2.75 m above the ground.

> **Progress check**
>
> 1. Explain the difference between the following supply earthing systems:
>   - TN-S
>   - TN-C-S
>   - TT
> 2. Briefly state where single- and three-phase supplies are utilised.
> 3. Identify the types of switching classified in BS 7671.

K2. Understand the principles of internal and external earthing arrangements for electrical installations

Key principles relating to earthing and bonding

In measuring electrical supply, we reference potential against the potential of the earth and the standard UK domestic supply has a potential of 230 V.

By this we mean the difference in potential between the supply (230 V) and the general planet earth, which is taken as being at zero potential (0 V). There must be a difference in potential for a current to flow.

The earth is therefore an important part of the UK supply system, as we make a connection between the two at the supply transformer, where the neutral conductor is connected to the star point of the transformer. This in turn is connected to earth using an earth electrode or the metal sheath and armouring of a buried cable. This effectively removes the chance of a **'floating' neutral** and locks it to zero potential, in other words 0 V.

Using the 230 V domestic situation as a basic reference point, we have a 3 wire supply system where:

- one conductor (Line) is at a 'high' potential and 'supplies' the property
- one conductor (Neutral) is maintained at zero potential because of its connection to the star point of the transformer where it therefore acts as a 'return' conductor and thus gives us our potential difference (230 V–0 V)
- one conductor is directly connected to earth and not intended to normally carry current.

If someone were to touch a live conductor then they would complete the electrical circuit and a current would flow through the person, the floor and through the actual earth back to the supply transformer via one or more earth connections of the transformer neutral. Should a fault occur, the current will always head for earth, taking the easiest route it can find to get there.

> **Remember**
>
> You will hear the phrase 'potential difference' used when discussing electricity. It may therefore be wise to read this section in conjunction with Unit ELTK 08, which covers the science and scientific concepts.

> **Key term**
>
> **'Floating' neutral** – where the star point of the transformer is not connected to the general mass of earth

> **Example**
>
> If someone came into contact with **both** an un-earthed exposed-conductive-part (e.g. steel conduit or steel trunking) which has become live under fault conditions **and** an earthed extraneous-conductive-part (e.g. a metal water pipe), then a serious shock could result because of the difference in potential between the two pieces of metalwork.

However, if we were to connect all such metalwork together and then connect it to earth using our third conductor, it would all be at 0 V.

This is the principle of bonding, and if the same fault situation now occurred, all the connected metalwork would rise to the same potential, and as there is no potential difference (because all are at equal potential) then there will be no dangerous 'shock' current.

By this connection of an extraneous-conductive part or an exposed-conductive part to earth using a protective conductor, we are providing a path for the fault current, which can then be detected and interrupted by protective devices such as fuses or circuit-breakers. We can therefore say in terms of protection that this path is called the earth-fault loop.

- The purpose of 'protective' earthing is to limit the duration of a fault (to allow a protective device to operate)
- The purpose of 'protective' bonding is to limit the size of the fault by equalising potential (this is also known as equipotential bonding).

The earth-fault loop

Figure 4.07 shows the path taken by the fault current. We call this the earth-fault loop, which, starting from the point of the fault, comprises:

- the circuit protective conductor (cpc)
- the consumer's main earth terminal (MET) and earth conductor
- the return path (this may be via electrodes or the cable armouring/sheath)
- the earthed neutral of the supply transformer and then the transformer winding (phase)
- the line conductor.

In the event of a fault, the protective device needs to operate as quickly as possible. Its ability to do this will be affected by the size of the fault current. This, in turn, is affected by the value of the combined resistances within the earth-fault loop. We therefore need this value to be as low as possible.

As the loop includes the resistance of the conductors and sheath in the loop plus the inductive effect of the transformer windings, we call this total system value the earth-fault loop impedance (Z_s).

We can calculate Z_s by either:

1. **Direct** measurement using an earth-fault loop impedance tester (see Unit ELTK 06 pages 114–118)

2. **Calculation**

We can calculate the value of Z_s with the following formula:

$$Z_S = Z_e + (R_1 + R_2)$$

Where:

- Z_e is that part of the loop impedance that is external to the consumer's installation (and therefore depends on the earthing system used e.g. TN-C-S)
- R_1 is the resistance of the line conductor
- R_2 is the resistance of the circuit protective conductor.

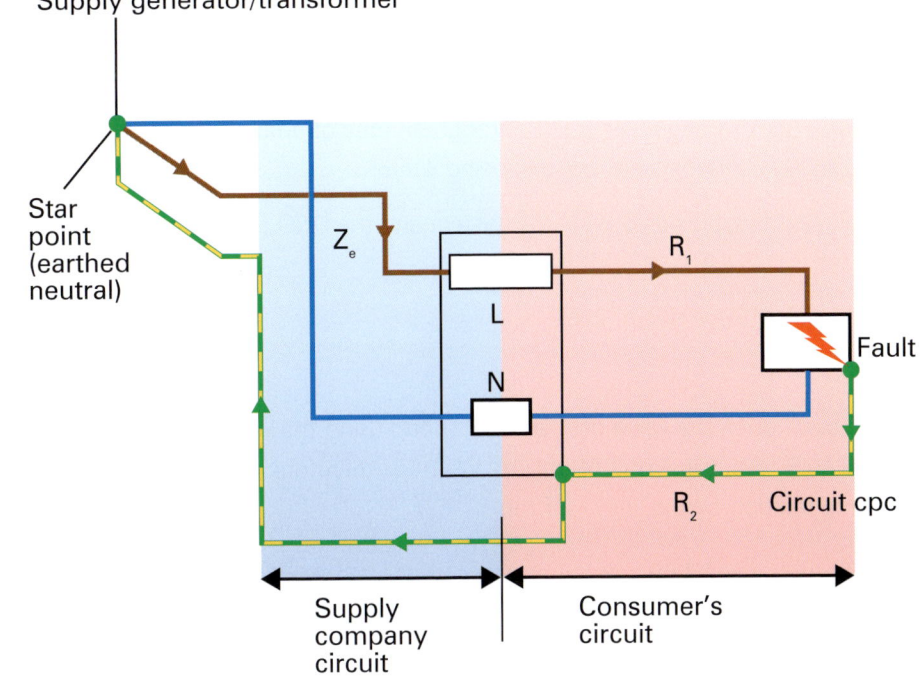

Figure 4.07 Earth-fault loop path

We can either obtain the values of R_1 and R_2 during the continuity testing stage (see Unit ELTK 06) or by making reference to Table 9A within the IET On Site Guide, which gives values of resistance/metre (at 20°C) for conductors up to 50mm².

Z_e can be obtained by measurement, calculation or asking the DNO (Distribution Network Operator), where typical maximum values for Z_e given by a DNO are shown in Table 4.03.

TN-C-S (PME)	0.35 Ω
TN-S	0.80 Ω
TT	21.00 Ω

Table 4.03 Typical maximum values for Z_e given by a DNO

Here is a typical Z_s calculation.

> **Example**
>
> An instantaneous water heater is fed, through an area of ambient temperature 20°C, by a 10 m length of 6 mm² PVC/PVC twin and earth cable, the cpc being 4 mm².
>
> The installation is supplied via a TN-C-S system and the DNO advise you that the value of Z_e is 0.35 Ω. Your circuit is protected by a 32 A Type B MCB.
>
> Will your value of Z_s be acceptable?
>
> $Z_s = Z_e + (R_1 + R_2)$
>
> We have been given the value of Z_e as 0.35 Ω and using the values in Table 9A of the OSG, a 6 mm² + 4 mm² cpc cable has an $R_1 + R_2$ value of 7.69 mΩ/m. We must therefore remember to convert mΩ to Ω by dividing by 1000 and also to multiply by the length of run.
>
> Therefore, $R_1 + R_2 = \left(\dfrac{7.69}{1000} \times 10\right) = 0.0769\ \Omega$
>
> We must now consider the multiplier values of Table 9C of the OSG that must be applied to give conductor resistance at maximum operating temperature. Our cable is 70°C thermoplastic with an incorporated cpc, therefore the multiplier will be 1.20.
>
> Therefore, our total value for $R_1 + R_2$ will be 0.0769 × 1.2 = 0.0923 Ω
>
> If $Z_s = Z_e + (R1 + R2)$
>
> Then Z_s = 0.35 + 0.0923 = **0.442 Ω**
>
> We now compare our value of Z_s with the appropriate tables in BS 7671 (Tables 41.2 and 41.3). For a circuit breaker the table will be 41.3 and for a 32 A Type B MCB the maximum permitted value of Z_s is 1.44 Ω.
>
> The calculated value of Z_s (0.442 Ω) is acceptable to meet required disconnection times.

Results of an unearthed appliance

We now know that by bonding together exposed-conductive parts (e.g. conduit, trunking, equipment housings) and extraneous-conductive parts (e.g. metallic gas and water pipes, radiators) and then connecting them to earth, we can begin to offer protection from shock, as the path formed will carry any leakage current which can then be detected and interrupted by protective devices such as fuses, circuit breakers and residual current devices.

This ability to detect and interrupt current is affected by the value of the impedance within the earth-fault loop. If the value of the earth-fault loop impedance is low enough then a fault current will operate the protective device.

The danger of having unearthed equipment is shown in Figure 4.08.

Current only flows when there is a difference in potential. Therefore, as shown in Figure 4.08, should a fault occur and the case of the appliance become live, then it has a potential of 230 V. If a person now touches the case and they are standing on the ground (the ground has a potential of 0 V) then they complete the circuit to earth. This creates a potential difference and the current will flow through the person to earth giving them an electric shock.

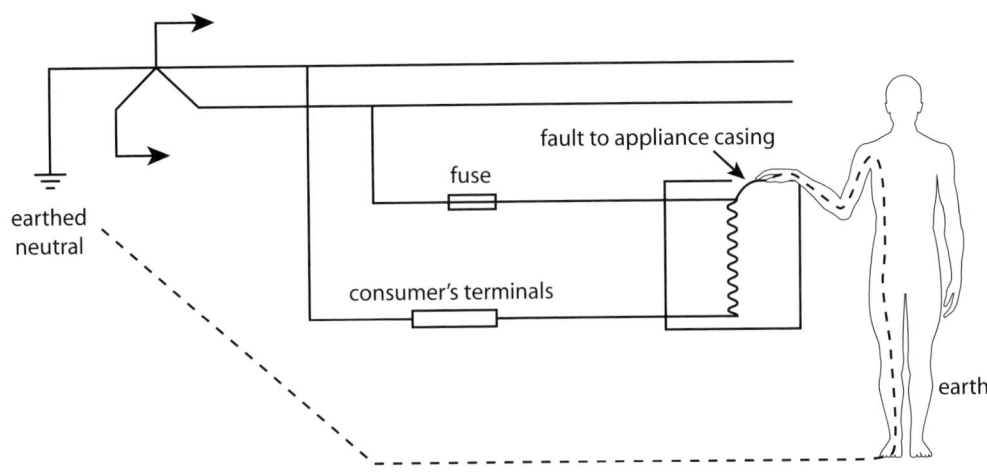

Figure 4.08 Electric shock

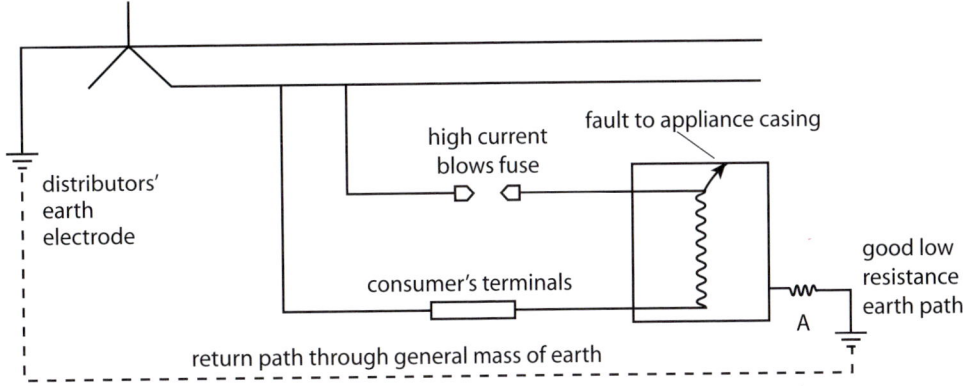

Figure 4.09 Good earth path

Figure 4.09 shows that, when a low resistance earth path is in place, it will allow a high current to flow, thus causing the protective device to operate quickly, which isolates the circuit and provides protection against electric shock.

Lightning protection

The protection of buildings and structures against lightning involves connecting various conductive parts to earth, hence its inclusion in this unit.

British Standard Code of Practice 326 (1965) and BS 6651 (1990) are the documents that cover the requirements for lightning protection and this type of installation is very specialised and is not something you would normally be involved with; however, an appreciation of the main requirements is useful.

Working life

Your company is working on a block of flats. The landlord asks you what all the green and yellow wires are in both the bathroom areas of the flats and the service intakes to electricity, water and gas. The landlord also asks you why they are all different sizes and wonders if this will be a problem.

1. How would you explain the purpose of the cables?
2. What illustrations could be used to make your explanation easier?
3. How can you explain the size differences in the cables?

What are lightning storms?

There are millions of lightning storms in the world every year, yet we are still unclear exactly how and when they form.

Generally it is believed that cooled water carried by the wind collides with ice crystals in the clouds, some of which form a slush. The heavier slush becomes negatively charged and the remaining ice crystals become slightly positively charged and rise to the top of the cloud. This separation of charges increases until there is sufficient potential difference to cause a lightning discharge.

Did you know?

Thunder is the shockwave radiating away from the strike path.

Lightning rapidly heats the air in its immediate vicinity to about 20 000°C (about three times the temperature of the surface of the Sun). The sudden heating effect and the expansion of heated air causes a supersonic shock wave in the surrounding clear air. It is this shock wave, once it decays to an acoustic wave, that we hear as thunder.

Effects of lightning

The discharge current created when lightning strikes a lightning conductor can be as high as 200 000 amps, which produces electrical, mechanical and thermal effects.

- The electrical effects happen when the lightning conductor's potential with respect to earth becomes very high and there is 'flash-over' to other metalwork in the structure itself.
- The mechanical effects are caused by air pressure waves when the lightning first strikes or by the high discharge currents flowing to earth, which can cause large mechanical forces to be exerted on to the conductor, its fixings and supports.
- The thermal effects are caused by the rise in temperature of the lightning conductor. When the lightning strike happens, a high value of current flows, but only for a short time and the effect on the system is normally negligible.

A lightning protective system typically consists of a network of conductors (copper or aluminium) positioned along the roof and/or walls of a structure at specified distances. These are bonded to other parts of the structure such as radio and TV masts, and ultimately connected to a common point discharging to earth. The maximum resistance of this network of conductors should not exceed 10 ohms.

Figure 4.10 Lightning storms can be both dramatic and dangerous

Key principles relating to the protection of electrical systems

To understand the key principles of protection in electrical systems, it is important that you know what you need protecting from. BS 7671 Part 4 answers this by providing guidance on the following:

- Chapter 41 – Protection against electric shock
- Chapter 42 – Protection against thermal effects
- Chapter 43 – Protection against overcurrent
- Chapter 44 – Protection against voltage disturbance and electromagnetic disturbance.

Protection against electric shock

BS EN 61140 is a basic safety standard that applies to the protection of persons and livestock. It states that the fundamental rule of protection against electric shock is that hazardous-live-parts shall not be accessible, and that accessible conductive parts shall not be hazardous-live, either in use without a fault, or in single fault conditions.

To comply with this and to ensure protection against electric shock, we could do something obvious such as insulating live parts. This is good, but in the event of a fault there must be some additional means of protection. BS 7671 Chapter 41 therefore states that we must provide the following two levels of protection for persons and livestock:

- **Basic protection** – protection when in normal use, i.e. without a fault
- **Fault protection** – protection under fault conditions.

To meet these levels of protection, BS 7671 Regulation 410.3.2 defines what must be in a protective measure, namely it shall consist of:

- an appropriate **combination** of provision for basic protection and an independent provision for fault protection **or**;
- an enhanced protective provision (e.g. reinforced insulation) that provides both basic and fault protection.

Before we consider these measures we should become familiar with the following definitions:

- **Exposed-conductive part** – these are the conductive parts of an installation that can be touched and, although not normally live, could become live under fault conditions. Examples of these are metal casings of appliances such as kettles and ovens or wiring enclosures such as steel conduit, trunking or tray plates.
- **Extraneous-conductive part** – these are the conductive parts within a building that don't form part of the electrical installation, but they could become live under fault conditions. Examples of these are copper water and gas pipes and metal air conditioning system ductwork.
- **Live part** – a conductor or conductive part that is meant to be live in normal use. This includes a neutral conductor.

BS 7671 Regulation 410.3.3 then requires that in each part of an installation, one or more protective measures shall be applied, bearing in mind the external influences. Generally, the following measures are permitted:

- Automatic Disconnection of Supply (ADS)
- Double or reinforced insulation
- Electrical separation from one item of current using equipment
- Extra Low Voltage (SELV and PELV).

In electrical installations Automatic Disconnection of Supply will probably be the most commonly used protective measure as it provides both basic and fault protection.

The requirements for basic and fault protection are given in the respective regulations for each of the protective measures. The following information is given as a guide, with the particular section or regulation within BS 7671 shown in brackets.

Automatic Disconnection of Supply (ADS) (411)

Automatic Disconnection of Supply (ADS) is a protective measure that provides both basic and fault protection as follows:

- **Basic protection** is given by basic insulation of live parts (416.1), barriers and enclosures (416.2) or, where appropriate, by obstacles (417.2) or placing out of reach (417.3).
- **Fault protection** is given by protective earthing (411.3.1.1), protective equipotential bonding (411.3.1.2) and by automatically disconnecting the supply in the event of a fault.

Be aware that protection by obstacles and placing out of reach is only providing basic protection and therefore only to be used in installations that are under the control or supervision of a skilled person (417.1 and 410.3.5).

In terms of fault protection, 'protective earthing' is where exposed-conductive parts are connected to the protective earthing terminal by means of the circuit protective conductors (cpc). Protective equipotential bonding is the interconnection of extraneous-conductive parts that are then connected to the protective earth terminal by means of protective bonding conductors. These conductors and their connection points are shown in Figure 4.11.

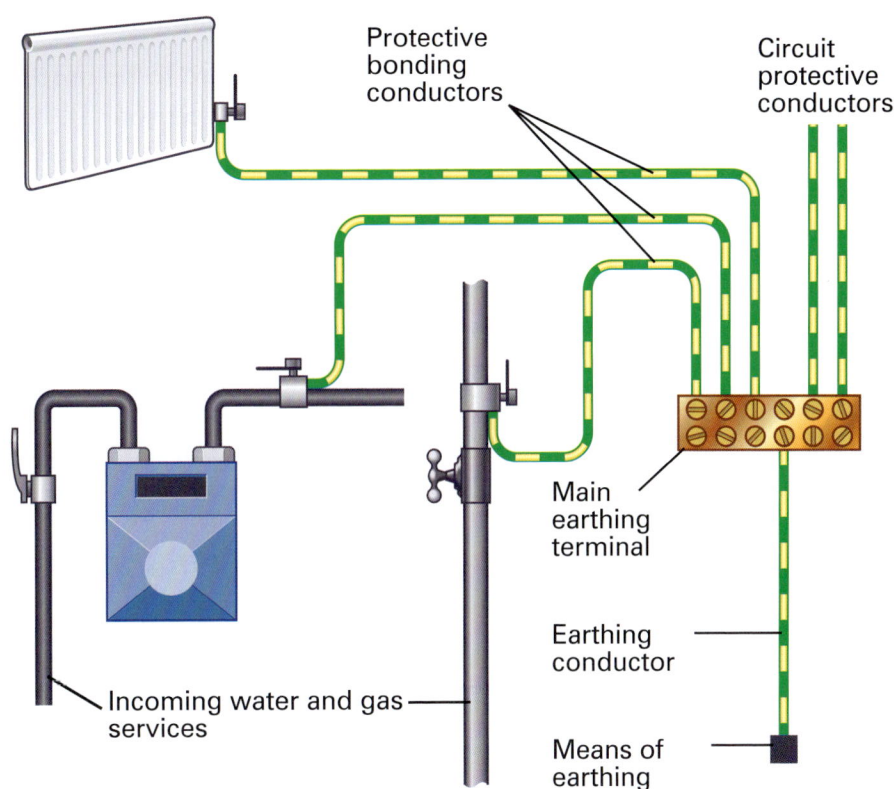

Figure 4.11 Circuit protective conductors (cpc) and their connection points

In order that ADS can automatically disconnect the supply, there needs to be a large enough earth fault current to operate the protective device (e.g. fuse, circuit breaker, RCD) within the times specified in Table 41.1 of BS 7671.

However, on a.c. systems for use by ordinary persons and for general use, we must provide 'additional protection' (411.3.3) for socket outlets with a rated current not exceeding 20 A. This additional protection is afforded by the use of a 30 mA RCD that must operate within 40 ms (415.1.1). The same applies to mobile equipment with a current rating not exceeding 32 A for use outdoors.

However, there are exceptions to the socket requirement:

- socket outlets that are to be used under the supervision of a skilled or instructed person
- socket outlets labelled or identified as being only for use with a particular piece of equipment, such as a fridge-freezer
- minor works associated with existing socket outlet circuits not provided with additional protection by means of an RCD where the designer is satisfied that there would be no increased risk from the installation of the addition or alteration. The decision being recorded under part 2 of the Minor Works Certificate or the comments section of the Electrical Installation Certificate.

Double (supplementary) or reinforced insulation (412)

These are protective measures where:

- basic protection is given by basic insulation and fault protection by supplementary insulation; **or**
- both basic and fault protection are given by reinforced insulation between live parts and accessible parts.

In other words the protection is achieved because the piece of equipment has no exposed metal parts that a dangerous voltage could appear on (Class II □). Although it is unlikely, where this protective measure is to be the only protective measure (that is the whole circuit consists entirely of equipment with double or reinforced insulation) then the circuit must be under effective supervision in normal use to make sure nothing can affect the protection.

This means that it cannot be used in a circuit where the user could change a piece of equipment without authorisation and therefore cannot be used where socket outlets, light support couplers or cable couplers would be used (412.1.3).

Electrical separation (413)

This is a protective measure where:

- basic protection is given by basic insulation of live parts or by barriers and enclosures (416); **and**
- fault protection is given by simple separation of the separated circuit from other circuits and earth.

To comply with fault protection requirements, the voltage of the separated circuit shall not exceed 500 V (413.3.2) and it is recommended that such circuits are in separate wiring systems (413.3.5) and no exposed-conductive part shall be connected to either the protective conductor or exposed-conductive parts of other circuits, or to earth (413.3.6).

Extra low voltage provided by SELV or PELV (414)

This is a protective measure consisting of either SELV or PELV, where both basic and fault protection are given when:

- the nominal voltage cannot exceed the upper limit of Band 1 voltage (414.1.1), i.e. 50 V a.c. or 120 V d.c.
- the supply is taken from one of the sources listed in Regulation 414.3 and the conditions of Regulation 414.4 are fulfilled.

Regulation 414.3 states the following sources of supply may be used:

- a safety isolating transformer
- a source of current providing a degree of safety equivalent to that of the safety isolating transformer (e.g. generator with windings providing equivalent isolation)
- an electro-chemical source (e.g. a battery)
- a source independent of a higher voltage circuit (e.g. a diesel-driven generator)
- certain electronic devices where it has been guaranteed that even in the case of a fault, the voltage on the outgoing terminals cannot exceed the values, or is immediately reduced to values, specified in Regulation 414.1.1.

Additional protection (415)

In addition to the four protective measures covered, additional protection may be specified. This is particularly so under certain external influences and in some special locations. Two methods are given:

- residual current devices (RCD) operating at 30 mA within 40 ms (415.1.1)
- supplementary equipotential bonding (415.2).

(Please note that an RCD is not recognised as the sole means of protection and therefore you would still need to apply one of the four protective measures (411 to 414). However, supplementary equipotential bonding is considered as an addition to fault protection.)

The various protective measures are summarised in Table 4.04.

Protective measures							
Automatic disconnection of supply (411)		Double or reinforced insulation (412)		Electrical separation (413)		Extra low voltage (414)	
Basic	Fault	Basic	Fault	Basic	Fault	Basic	Fault
Gives basic protection by basic insulation of live parts or by barriers/enclosures	Gives fault protection by protective earthing, protective equipotential bonding and ADS in case of fault	Gives basic protection by basic insulation	Gives fault protection by supplementary insulation	Provides basic protection by one or two methods in 416, i.e. basic insulation of live parts of barriers/enclosures	Provides fault protection by simple separation of the separated circuit from other circuits and earth		
There is an additional protection requirement by use of an RCD with socket outlets not exceeding 20 A		Gives basic and fault protection by reinforced insulation between live and accessible parts				Gives basic and fault protection by limiting voltage to the upper limit of Band 1 using a supply such as an isolating transformer with circuits in accordance with Regulation 411	

Table 4.04 Summary of protective measures

Basic protection

'Basic protection' is defined as providing protection against electric shock under normal operation, in other words in fault-free conditions. This can be done by preventing persons or livestock actually touching any live parts, for example by insulating the live parts or by preventing any access to them by using an enclosure or barrier.

At the start of this section it was stated that BS 7671 Regulation 410.3.2 requires:

- an appropriate combination of provision for basic protection and an independent provision for fault protection; **or**
- an enhanced protective measure (for example reinforced insulation) that provides both basic and fault protection.

This means that the protective measures previously described and summarised in Table 4.04 are applicable.

Protection against thermal effects

It is often reported in the news that the cause of a fire has been attributed to an electrical installation or piece of electrical equipment. Typically the causes of fire or heat could be:

- heat radiation, hot components or equipment
- failure of equipment such as protective devices, thermostats, seals and wiring systems
- overcurrent
- insulation faults
- arcs and sparks
- loose connections
- external influences such as a lightning surge.

BS 7671 Chapter 42 requires us to provide protection for persons and livestock against:

- the effects of heat or thermal radiation from electrical equipment
- the ignition, combustion or degradation of materials
- fire hazard spread from an electrical installation
- safety services being cut off by failure of electrical equipment.

Protection against fire caused by electrical equipment

Some fairly straightforward advice is given.

- Comply with manufacturer's instructions.
- Select and erect equipment so that its normal temperature cannot cause a fire.
- If the equipment's surface temperature could cause a fire then mount it on a surface that can withstand the heat or inside something that can withstand the temperature.
- Make sure any equipment that causes a focusing of heat (e.g. an electric fire), is far enough away from any object or building element (e.g. curtains) to avoid them reaching a dangerous temperature.
- Take precautions when the equipment location also contains flammable liquids.
- Make sure any termination, connection or joint of a live conductor is contained within an enclosure in accordance with Regulation 526.5.

Precautions where a particular risk of fire exists

The contents of this section are to be applied in addition to the previous section where any of the conditions of external influence covered in Regulations 422.2 to 422.6 apply:

- conditions for evacuation in an emergency
- locations, such as flour mills, where there is an increased risk from processed/stored materials
- combustible construction materials
- fire propagating structures (i.e. building shapes that could facilitate the spread of fire)
- installations of national, commercial or public significance (e.g. monuments or museums).

Protection against burns

Regulation 423.1 states that the accessible parts of fixed electrical equipment within arm's reach should not exceed the temperatures in Table 4.05.

Accessible part	Material of accessible surfaces	Maximum temperature (°C)
A hand-held part	Metallic	55
	Non-metallic	65
A part intended to be touched but not hand-held	Metallic	70
	Non-metallic	80
A part which need not be touched for normal operation	Metallic	80
	Non-metallic	90

Table 4.05 Surface temperatures of accessible parts of fixed electrical equipment

Safety tip

If the surface temperatures in Table 4.05 are exceeded, for even a short time, then guards must be fitted to avoid contact.

Operating principles, application and limitations of protective devices

BS 7671 Regulation 430.3 states that '... a protective device shall be provided to break any overcurrent in a circuit conductor before that current could cause danger due to thermal or mechanical effects ...'. An overcurrent is simply a current that is in excess of the rated value or of the current carrying capacity of a conductor, and is the result of either an overload or a fault current.

An **overload** is then defined as an overcurrent occurring within a circuit that is electrically sound. It is an otherwise healthy circuit carrying more current than it was designed for.

This could be as the result of faulty equipment such as a stalled motor or it may occur when too many pieces of equipment are added to the circuit, such as when several 13 A plugs are connected through an adaptor to a single 13 A socket outlet.

Fault current can be caused by either a short circuit or an earth-fault. A short circuit is defined as when there is a fault of negligible impedance between live conductors. An earth-fault current is defined as being a fault of negligible impedance between a line conductor and an exposed-conductive part or protective conductor.

BS 7671 Regulation 431 goes on to give the requirements for the protection of line conductors, stating that detection of overcurrent should be provided for all line conductors and should cause disconnection of the conductor in which the overcurrent is detected, and not necessarily in other line conductors, unless disconnecting one line conductor could cause danger.

> **Did you know?**
>
> An example of a time when disconnection could be dangerous is a electromagnetic crane in a foundry. It would not be wise to switch off the magnet while carrying heavy loads at height!

Regulation 431 also provides requirements for the protection of neutral conductors. It states that as the neutral conductor in standard single- and three-phase circuits will normally be the same cross-sectional area as the line conductors, it is not necessary to provide overcurrent detection and an associated disconnecting device. Where there is a reduced neutral, overcurrent detection will be required in line with the cable CSA.

Protective devices for overcurrent should then be installed at the point where there is a reduction in the current-carrying capacity of the circuit conductors caused by a change in CSA or perhaps the method of installation. The protective devices that we use to protect against overcurrent are designed to disconnect the supply automatically, such as fuses or MCBs. This section will also look at RCDs.

Residual current devices (RCD)

Residual current devices (RCDs) are a group of devices that provide extra protection to people and livestock by reducing the risk of electric shock. Although RCDs operate on small currents, there are circumstances where the combination of operating current and high earth-fault loop impedance could result in the earthed metalwork rising to a dangerously high potential.

The regulations draw attention to the fact that if the product of operating current (A) and earth-fault loop impedance (Z) exceeds 50 V, the potential of the earthed metalwork will be more than 50 V above earth potential and hence dangerous. This situation must not be allowed to occur.

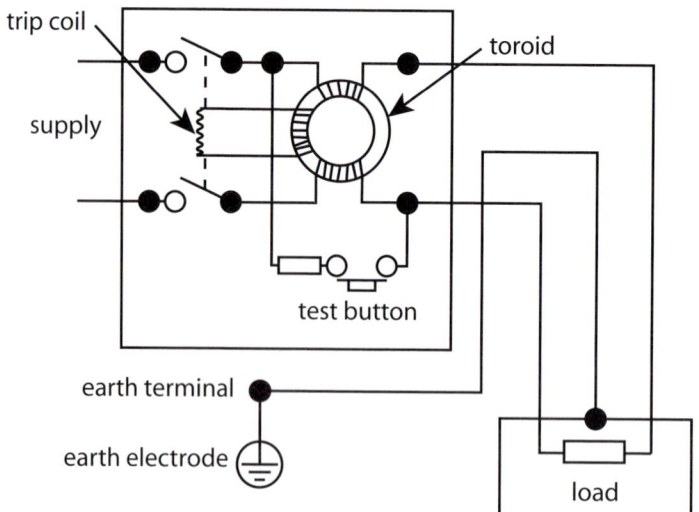

Figure 4.12 Fault detector coil

An RCD monitors the current flowing in a circuit by using a **toroid**, which is essentially a small current transformer specially designed to detect earth-fault currents with a winding taken to a trip mechanism.

As shown in Figure 4.12, all live conductors will pass through the toroid and the currents flowing in the live conductors of a healthy circuit will balance and therefore no current will be induced in the toroid.

> **Remember**
>
> The live conductors of a circuit include all line and neutral conductors.

Under normal safe working conditions, the current flowing in the line conductor into the load will be the same as that returning via the neutral conductor from the same load. When this happens, the line and neutral conductors produce equal and opposing magnetic fluxes in the toroid, meaning no voltage is induced in the trip coil.

If more current flows in the phase side than the neutral side (or less returns via the neutral because the current has 'leaked' to earth within the installation) an out-of-balance flux will be produced in the toroid which will induce a voltage in the fault detector coil (this earth-fault path could be through a person in contact with live parts as shown on page 17 or where basic insulation has failed at an exposed-conductive part).

The fault detector coil then energises the trip coil, which in turn opens the double pole (DP) switch, using the residual current produced by the induced voltage in the trip coil.

There are two types of residual current devices tripping mechanisms – electromagnetic and electronic.

Electromagnetic RCDs

Electromagnetic devices use a very sensitive toroid designed to operate the trip mechanism when it detects very small residual currents. These devices generally require no reference earth lead and are unaffected by temporary loss of supply, as the power to trip the device is derived directly from the fault current.

Electronic RCDs

Electronic devices do not need such a sensitive toroid, as electronic circuits within the device amplify the signal in order to operate the trip mechanism. However, these devices often require a safety earth reference lead to ensure that the device will continue to operate in the event of the supply neutral being lost. Without this the RCD would not work even though the phase supply is still connected.

The power to trip the device is taken from both the fault current and the mains supply, enabling the overall size of the devices to be smaller. These devices should also be disconnected while carrying out insulation resistance tests to prevent damage to the device and to avoid incorrect test results.

Functional testing of RCDs

Functional testing of an RCD's operation should be carried out by operating the test button at regular intervals. Operating the test button simulates a fault by creating an imbalance in the detection coil.

Initial and periodic inspection and testing procedures additionally require the RCD to be tested using the appropriate testing instrument (see Unit ELTK 06 on page 120). RCDs are available for single- and three-phase applications.

Practical application of RCDs (Regulation 531.2)

- An RCD shall disconnect all line conductors of the circuit at substantially the same time.
- The residual operating current of the RCD shall comply with Regulation 411 relative to the type of earthing system.

- The RCD shall be selected and circuits sub-divided so that any normally occurring earth leakage (such as with some computers) does not cause nuisance tripping of the RCD.
- The use of a 30 mA RCD with a circuit having a protective conductor is not deemed sufficient to afford fault protection.
- An RCD must be positioned so that its operation is not affected by the magnetic fields of other equipment.
- Where an RCD is used for fault protection with, but separately from, an overcurrent device, then it must be capable of withstanding (without damage) any stresses likely to be encountered in the event of a fault on the load side of the point at which it is installed.
- RCDs shall be selected and installed so that where required, if two or more RCDs are connected in series and discrimination is required to avoid danger, that discrimination is achieved.
- Where an RCD can be operated by someone other than a skilled or instructed person, it shall be selected and installed so that its settings cannot be altered without using a tool and the result of any alteration being clearly visible.

Sensitivity rated residual operating current	Level of protection	Applications
10 mA	Personnel	High-risk areas: schools, colleges, workshops, laboratories. Areas where liquid spillage may occur
30 mA	Personnel	Domestic, commercial and industrial
100 mA	Personnel, fire	Only limited personnel protection Excellent fire protection
300 mA	Fire	Commercial and industrial

Table 4.06 Overcurrent protection sensitivity

Fuses

BS 3036 Rewireable fuses

Early rewireable fuses had a very low short circuit capacity and were very dangerous when operating under fault conditions. The operating principle that a fuse element will melt, and therefore break, when an overcurrent flows means that the melted copper 'splashed' around and could cause fires.

Later rewireable fuses incorporated an asbestos pad to protect the fuse holder, thus reducing the risk of fire from scattering hot metal when rupturing.

The rewireable fuse consists of a fuse, holder, a fuse element and a fuse carrier. The holder and carrier are made of porcelain or Bakelite and the rating of the circuits for which this type of fuse is designed have a colour code (see Table 4.07) marked on the fuse holder.

5 A	White
15 A	Blue
20 A	Yellow
30 A	Red
45 A	Green

Figure 4.13 BS 3036 rewireable fuse

Table 4.07 Colour codes for fuse holders

This type of fuse was very popular in domestic installations. With the exception of older installations, it is now not normally used because it does have associated problems as shown in Table 4.08. They are, however, still available from electrical wholesalers.

Disadvantages of rewireable fuses	Advantages of rewireable fuses
• Easily abused when the wrong size of fuse wire is fitted • **Fusing factor** of around 1.8 – 2.0 means they cannot be guaranteed to operate up to twice the rated current is flowing. As a result cables protected by them must have a larger current-carrying capacity. • Precise conditions for operation cannot be easily predicted • Do not cope well with high short circuit currents • Fuse wire can deteriorate over time • Danger from hot metal scattering if the fuse carrier is inserted into the base where the circuit is faulty	• Low initial cost • Can easily see when the fuse has blown • Low element replacement cost • No mechanical moving parts • Easy storage of spare fuse wire

Table 4.08 Advantages and disadvantages of rewireable fuses

BS 1361/1362 Cartridge fuses

The cartridge fuse consists of a porcelain tube with metal end caps to which the element is attached and the tube then filled with granulated silica. The BS 1362 fuse is generally found in domestic appliance plugs used with 13 A BS 1363 domestic socket outlets.

Figure 4.14 BS 1361 and 1362 fuse

There are two common fuse ratings available, the 3 A, which is for use with appliances up to 720 watts (radios, table lamps, TVs) and the 13 A fuse which is used for appliances rated over 720 watts (irons, kettles, fan heaters, electric fires, washing machines).

There are other sizes of fuse, 1, 5, 7 and 10 A, but these are not so easily available. The physically larger BS 1361 fuse can be found in distribution boards and at main intake positions.

Disadvantages of cartridge fuses	Advantages of cartridge fuses
• They are more expensive to replace than rewireable fuses • They can be replaced with an incorrect size fuse (plug top type only) • The cartridge can be shorted out with wire • It is not possible to see if the fuse has blown • They require a stock of spare fuses to be kept	• They have no mechanical moving parts • The declared rating is accurate • The element doesn't weaken with age • They have a small physical size and no external arcing, which permits their use in plug tops and small fuse carriers • They have a low fusing factor – around 1.6–1.8 • They are easy to replace

Table 4.09 Advantages and disadvantages of cartridge fuses

BS 88 high breaking capacity (HBC) fuses

The HBC fuse (sometimes called an HRC fuse (high rupturing capacity) is a more sophisticated variation of the cartridge fuse and is normally found protecting motor circuits and industrial installations. It consists of a porcelain body filled with silica, a silver element and lug-type end caps. Another feature is the indicating bead, which shows when the fuse element has blown.

It is a reasonably fast-acting fuse and can discriminate between a starting surge and an overload.

These types of fuses would be used when there is a possibility of an abnormally high prospective short circuit current.

Figure 4.15 A sectional view of a typical BS 88 HBC fuse

Disadvantages of BS 88 fuses	Advantages of BS 88 fuses
• These are very expensive to replace • Stocks of these spares are costly and take up space • Care must be taken when replacing them to ensure that the replacement fuse has the same rating and also the same characteristics as the fuse being replaced	• They have no mechanical moving parts • The element doesn't weaken with age • Operation is very rapid under fault conditions • It is difficult to interchange the cartridge, since different ratings are made to different physical sizes

Table 4.10 Advantages and disadvantages of BS 88 fuses

Diazed and neozed fuses

Diazed (D type) fuse cartridges have a bottle-shaped ceramic body with metal end caps. Neozed (DO type) fuses are similar but, as shown in Figure 4.16, have a smaller more cylindrical body. Both are located into a fuse base by means of a screw cap. Fuse holders may then be secured by screws to a panel, attached to bus bars, or mounted on DIN rails. A diazed system is shown in Figure 4.18.

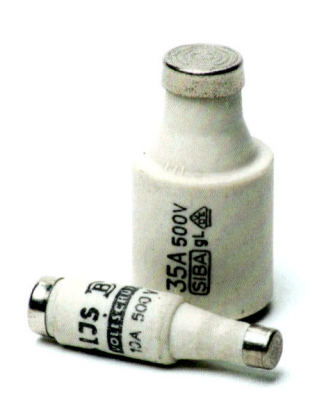

Figure 4.16 A neozed fuse   **Figure 4.17** A diazed fuse

The cap at the smaller end of the diazed fuse has a diameter that varies with the fuse rating: higher ratings have wider end caps. The fixed part of each fuse holder normally has a colour-coded ring that only allows a fuse of a certain diameter to be fitted and so only the correctly rated fuse can be fitted.

The cap at the larger end has an indicator to show whether a fuse has blown.

D- and DO-type fuses are used throughout Europe in domestic and light commercial installations and also for the protection of motors. In the course of your work you may come into contact with either type of fuse.

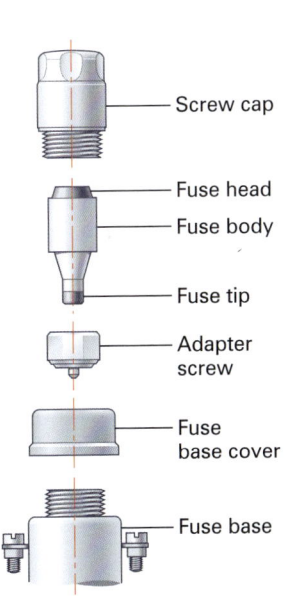

Figure 4.18 A diazed system

> **Remember**
>
> The types of fault in question are usually overloads and short circuits.

Figure 4.19 A miniature circuit-breaker

Miniature circuit-breakers (MCBs)

In essence, a circuit-breaker is a switch with contacts which automatically open when a fault occurs, in order to break an electrical circuit, this protects the attached equipment and wiring.

In many cases MCBs are a more flexible alternative to fuses; they can be easily reset and do not need replacement if an electrical overload or fault does occur.

Circuit-breaker contacts must carry the load current without excessive heating. Once a fault is detected the contacts inside the circuit-breaker must open to break the circuit and when a current is interrupted, an arc is generated.

This arc must be contained, cooled, and extinguished. Different circuit-breakers use a vacuum, air, insulating gas, or oil as the medium in which the arc forms, with MCBs traditionally using arc division and air to extinguish the arc. Finally, once the fault condition has been cleared, the contacts can again be closed to restore power to the interrupted circuit.

Due to improved design and performance, the modern MCB now forms an essential part of the majority of installations at the final distribution level and there are several different types to consider.

Thermal tripping

The most basic is a simple thermal trip breaker that incorporates a bi-metallic strip which carries the current of the circuit to be protected.

The current heats the strip, which bends by an amount commensurate with the current. Once overload conditions occur, the bi-metallic strip bends further than usual and activates the breaker's trip mechanism.

Thermal breakers are simple and inexpensive, but they provide only limited protection against short circuits. They should only be used where overload protection is the most important requirement and where short circuit currents will not exceed 1 000 A.

Thermal-magnetic (combined) tripping

Thermal-magnetic MCBs are more versatile and the most frequently used. These have a similar thermal trip mechanism to bi-metallic trips which provides effective protection against small and moderate overloads. However, in addition, they have an electromagnetic mechanism that provides almost instantaneous tripping for large overloads and short circuits.

The construction of a thermal-magnetic MCB is shown in Figure 4.20.

A	Actuator lever – used to manually trip and reset the circuit breaker. Also indicates the status of the circuit breaker (On or Off/tripped)
B	Actuator mechanism – forces the contacts together or apart
C	Contacts – Allow current when touching and break the current when moved apart
D	Supply and load terminals
E	Bi-metallic strip (thermal operation)
F	Calibration screw – allows the manufacturer to precisely adjust the trip current of the device after assembly
G	Solenoid (magnetic operation)
H	Arc divider / extinguisher

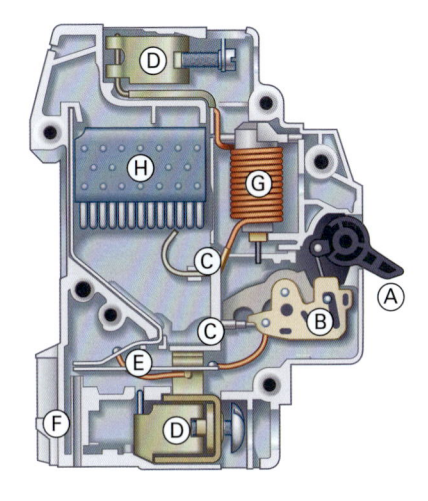

Figure 4.20 Structure of an MCB

This type of MCB can be designed to provide high breaking capacities, and to have various types of trip characteristic. They are ideal for applications where comprehensive protection is required at a moderate price, and where high short circuit levels may be encountered.

The current rating of an MCB is the maximum current that it will carry continuously without tripping and MCBs should always be chosen so that their current rating matches, as closely as possible, the maximum load current of the circuit they are protecting.

Trip characteristics

It would seem that an MCB that trips as quickly as possible under fault conditions would be ideal and this is what is required for short circuit faults.

For overload protection however, many items of equipment such as fluorescent lights, transformers and motors, draw a high peak current for a short period when they are switched on. An MCB that reacts instantaneously would therefore trip every time such a peak occurred, and would be unusable. Fortunately, the thermal element in MCBs does not react instantaneously because the bi-metallic strip takes time to heat up. It is therefore hardly affected by short-term current peaks. By changing the design of the bi-metallic elements, MCB manufacturers can determine what size of peak current a particular MCB will ignore, and for what length of time.

This relationship between current and tripping time is usually shown as a curve, known as the MCB trip characteristic. However, to avoid the need for consumers working with the curves, BS EN 60898 defines several types of standard characteristic, the most important of which are Types B, C and D.

There is no Type A to avoid confusion with A for Amperes, but you may comes across older MCBs classed as Types 1, 2, 3 and 4.

MCB type	Instantaneous trip current	Application
1 B	2.7 to 4 × I_n 3 to 5 × I_n	Domestic and commercial installations having little or not switching surge
2 C 3	4.0 to 7.0 × I_n 5 to 10 × I_n 7 × 10 I_n	General use in commercial/industrial installations, where the use of fluorescent lighting, small motors etc. can produce switching surges that would operate a Type 1 or Type B circuit-breaker. Type C or Type 3 may be necessary in highly inductive circuits such as banks of fluorescent lighting
4 D	10 to 50 × I_n 10 to 20 × I_n	Suitable for transformers, x-ray machines, industrial welding equipment etc. where high in-rush currents may occur

Table 4.11 MCB selection

RCBO and MCBs

Earlier in this unit we discussed the residual current device (RCD); a breaker that provides earth-fault protection. It should be noted that a new type of MCB is now available that incorporates earth-fault protection and they are called RCBOs (**R**esidual **C**urrent operated circuit-**B**reaker with integral **O**verload protection).

Generally a combination of a thermal-magnetic Type B MCB and integrated RCD, RCBOs enable both overcurrent protection and earth-fault current protection to be provided by a single unit.

The major advantage is that this allows earth-fault protection to be restricted to a single circuit and therefore only the circuit with the fault is interrupted. (In many circumstances a distribution board will be protected by an RCD, in which case all circuits would be disconnected in the event of a fault.)

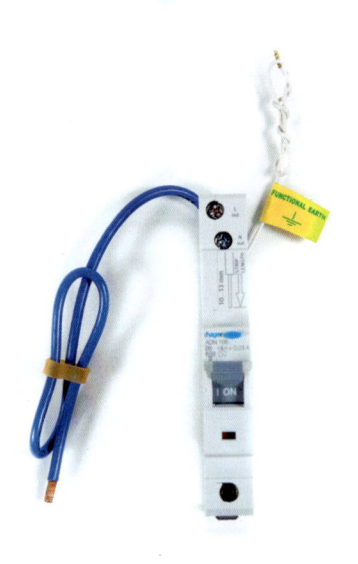

Figure 4.21 An RCBO

Disadvantages of MCBs	Advantages of MCBs
• These have mechanical moving parts • They are expensive • They must be regularly tested • Ambient temperature can change performance	• They have factory-set operating characteristics which cannot be altered • They will maintain transient overloads and trip on sustained overloads • Easily identified when they have tripped • The supply can be quickly restored

Table 4.12 Advantages and disadvantages of MCBs

Fuse characteristics

Fuse utilisation categories are used to define the particular application or characteristic of a fuse and each fuse is marked accordingly.

Table 4.13 shows the utilisation categories identified for fuses.

These replace the previous fuse classes of Class P, Q1 (similar to gG), Q2 and R (similar to aM).

These classes relate to the 'fusing factor'. This shows the operating speed of the fuse link, being the ratio between the conventional fusing current and the rated current. The conventional fusing current (I_f) is the lowest value of current that will cause the device to operate within the 'conventional time' (1–4 hours dependent on the device rating). The rated current is the number on the side of the fuse (e.g. 3 A or 13 A) and represents the maximum value of current that the fuse may carry continuously without deterioration under specified conditions.

As the protective device must carry this rated current, we can see that the fusing factor will be a number greater than 1. The higher this number, the less accurate and reliable the selected device will be.

gG	Full range breaking capability, general applications
gM	Full range breaking capability, motor circuit protection (dual rating)
aM	Partial range breaking capability, motor applications
aR/gR	Semiconductor protection, fast acting
gS	Semiconductor protection including cable overload protection

Table 4.13 Utilisation categories for fuses

Protective device	Fusing factor
BS 3036	1.8 – 2.0
BS 88	1.25 – 1.7
BS 1361	1.6 – 1.9
MCB	Up to 1.5

Table 4.14 Fusing factors for protective devices

It is worth noting that category gG fuses do have the ability to protect motor circuits and when selected correctly can withstand

Figure 4.22 A category gG fuse

motor starting surges and full load currents without deterioration as well as offering short circuit protection to associated motor starter components.

Category gM type fuse links have a dual rating which is characterised by two current values. The first is the maximum continuous current of the fuse and associated fuse holder; the second indicates the equivalent electrical characteristic to which the fuse conforms.

These two ratings are normally separated by the letter M which defines the application. For example a 20M32 fuse link is intended for use in the protection of motor circuits and has a maximum continuous rating of 20 A but the electrical characteristics of a 32 A rating. This means that the associated equipment need only be rated at 20 A thereby providing significant economies against 32 A equipment.

Breaking capacity

When looking at fuses, you will see a number followed by the letters kA stamped onto the end cap of an HBC fuse or printed onto the body of a BS 1361 fuse. This is known as the breaking capacity of fuses and circuit-breakers.

When a short circuit occurs, the current may, for a fraction of a second, reach hundreds or even thousands of amperes. The effects of a short circuit current are thermal, which can cause melting of conductors and insulation, fire or the alteration of the properties of materials and mechanical parts where large magnetic fields can build up, resulting in distortion of conductors and breaking of supports and insulators.

Each protective device must therefore be able to safely break, or make, such a current without damage to its surroundings by arcing, overheating or the scattering of hot particles.

Regulation 434.5 of BS 7671 requires that the breaking capacity of every fault current protective device shall be no less than the maximum prospective fault current (this includes both short circuit and earth fault conditions) at the point at which it is installed; this value of prospective fault current having been established by measurement, calculation or enquiry.

Most manufacturers now offer ranges of MCBs with breaking capacities of 10 kA – 15 kA and can be extended to 25 kA for certain products. This higher breaking capacity can reduce costs substantially in some applications, by allowing MCBs to be used

Did you know?

When making an enquiry to the DNO, a design value of 16 kA could be expected for single-phase 230 V supplies up to 100 A.

where previously more costly moulded case breakers would have been required.

The breaking capacities of MCBs are indicated by an 'M' number, e.g. M6. This means that the breaking capacity is 6 kA or 6 000 A. The breaking capacity will be related to the prospective short circuit current.

Time/current characteristics of overcurrent protective devices

The time/current characteristic is essentially a curve showing the operating time of a fuse in relation to the prospective fault current under set conditions.

Appendix 3 of BS 7671 provides tables giving such time/current characteristics for overcurrent protective devices and RCDs. These tables take the form of graphs that have a horizontal axis representing the prospective current in amps and a vertical axis representing the time in seconds. Both scales are logarithmic by which we mean that the scale increases by multiples of 10.

Using the table, we can illustrate this concept, as shown in Figure 4.23. In Figure 4.23 we can see that for a 10 A rated BS 88 fuse, a fault current of 60 A will cause the device to operate in 0.1 seconds.

Time/current characteristics for fuses in BS 88.2.2 and BS 88.6					
Fuse rating	Current for time				
	0.1 sec	0.2 sec	0.4 sec	1 sec	5 sec
10 A	60 A	51 A	45 A	39 A	31 A
16 A	120 A	95 A	85 A	72 A	55 A
25 A	220 A	180 A	160 A	130 A	100 A
40 A	400 A	340 A	280 A	240 A	170 A
63 A	710 A	590 A	500 A	400 A	280 A
100 A	1400 A	1150 A	980 A	790 A	550 A
160 A	2400 A	2000 A	1700 A	1400 A	900 A

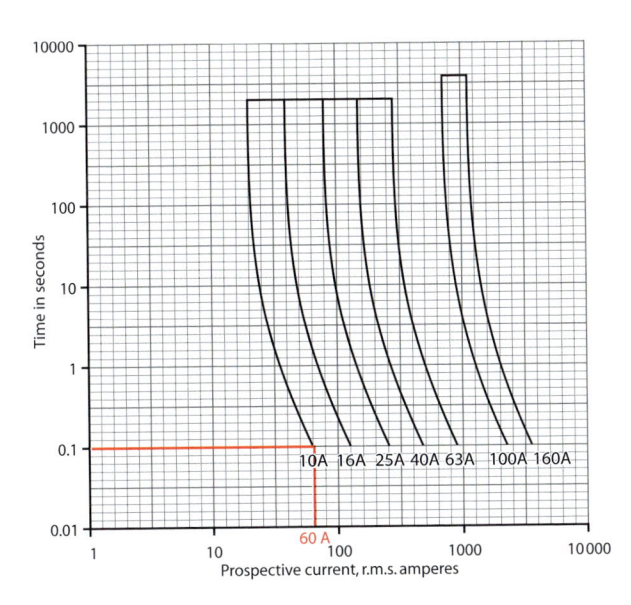

Figure 4.23 Time/current characteristics of overcurrent protection devices

Protection against overload current

Coordination between conductor and overload device

Regulation 433.1 of BS 7671 requires every circuit to be designed so that a small overload of a long duration is unlikely to happen. Therefore, should an overload occur, the idea is that the protective device operates and disconnects the circuit, protecting the cable insulation, termination or equipment from any great rise in conductor temperature.

To do this, the circuit has to be designed to coordinate the current-carrying capacity, the load current and the characteristics of the protective device. Regulation 433.1.1 therefore states that the operating characteristics of a device protecting a conductor against overload shall satisfy all of the following conditions.

(a) The rated current of the protective device (I_n) is not less than the design current (I_b) of the circuit.

(b) The rated current of the protective device (I_n) does not exceed the lowest of the current-carrying capacities (I_z) of any of the conductors of the circuit.

(c) The current (I_2) causing effective operation of the protective device does not exceed 1.45 times the lowest current-carrying capacity (I_z) of any of the conductors in the circuit.

Where:

I_b	The design current for that circuit
I_z	The current-carrying capacity of the conductor
I_n	The rated current or current setting of the protective device
I_2	The current ensuring effective disconnection by the protective device in the conventional time, i.e. 0.4 or 5 seconds.

We generally summarise this with the formula: $I_b \leq I_n \leq I_Z$

However, for point (b), coordination will be met with the following formula: $I_2 \leq 1.45 \times I_Z$

Specific issues for types of fuses

Where the protective device is either a general purpose type (gG) fuse to BS 88, a fuse to BS 1361, a circuit-breaker to BS EN

60898, a circuit-breaker to BS EN 60947-2 or an RCBO to BS EN 61009-1, compliance with conditions (a) and (b) above will also result in compliance with condition (c).

Where the protective device is a semi-enclosed fuse to BS 3036, this type of device can have a fusing factor as high as 2. Consequently, compliance with condition (c) above of Regulation 433.1.1 above is met if the rated current (I_n) doesn't exceed $0.725 \times I_z$.

We can show this term is established as follows:

$I_2 = 2 \times I_n$ which in turn must be $\leq 1.45 \times I_z$. In other words $I_2 = 2 I_n \leq 1.45 I_z$.

If $I_2 = 2 I_n$ then we can rewrite this as $2 I_n \leq 1.45 I_z$.

Therefore by transposition $I_n \leq 1.45 I_z$ or in other words $I_n \leq 0.725 \times I_z$.

Protection against fault current

The operating characteristics for every fault current protective device must satisfy the following conditions.

Regulation 434.5.1 of BS 7671 requires that the breaking capacity of every fault current protective device shall be no less than the maximum prospective fault current (pfc), and this includes both short circuit and earth fault conditions at the point at which it is installed; this value of prospective fault current having been established by measurement, calculation or enquiry (434.1).

Remember that the further away from the intake position you are, the more resistance will be experienced and therefore the value of pfc will decrease. Some designers therefore simply check that the breaking capacity of the lowest rated fuse in the installation exceeds the pfc at the intake position.

Remember also that the cable must be able to carry the full amount of current that can pass through the protective device before it operates, not just the design current, and a fault occurring at any point in a circuit must be interrupted before the fault current causes the limiting temperature of any conductor to be exceeded.

Therefore, effectively giving the maximum disconnection time, Regulation 434.5.2 states that the maximum permitted time in which a fault current can raise live conductors from the highest operating temperature to the limiting temperature is calculated using the formula on page 42.

$$t = \frac{k^2 S^2}{I^2}$$

Where:

t	The duration in seconds
S	The conductor CSA in mm²
I	The fault current in amperes
k	A value used to denote conductor insulation, resistivity and heat capacity

The values of k are given in Table 43.1 of BS 7671 and, as an example, we traditionally take k as being 115 for a copper conductor with 70°C thermoplastic (PVC) insulation.

Discrimination

In most installations there is usually a series of fuses and/or circuit-breakers between the incoming supply and the electrical outlets, and the relative rating of the protective devices used will decrease the nearer they are located to the current-using equipment.

Discrimination is the ability of protective devices to operate selectively to ensure that, in the event of a fault, only the faulty circuit is isolated from the system, therefore allowing healthy circuits to remain in operation. So for example in a house, a fault on an appliance should cause the fuse in the plug top connected to the appliance to operate before the protective device in the consumer unit.

Similarly, a fault on a lighting circuit should only result in the MCB in the consumer unit for that circuit to trip; all other MCBs in the consumer unit and the main DNO fuse should remain intact and energised.

However, just because we have a 3 A fuse in the plug top and a higher rated device in the consumer unit does not mean all is well. We also need to consider the following points.

Discrimination between fuses

It is standard practice to find HRC fuse links in series with one another to provide protection at different levels in an electrical installation. Discrimination between fuse links can be checked by ensuring that the time/current characteristics do not overlap at any point.

Discrimination between fuses and other protective devices

Protective devices other than fuses, for example circuit-breakers, are generally electro-mechanical devices that have operating times longer than those of similar rated fuses, except at low values of overcurrent.

Therefore to achieve discrimination, ensure that the tripping time characteristic curves of the electro-mechanical device do not intersect with the time/current characteristics of the fuse.

Remember that when using BS 7671 Appendix 3, you will see a difference in appearance between the characteristics for a fuse and a circuit-breaker, in that the circuit-breaker curves (see Figure 4.24 below) have a vertical section to them. You may have therefore noticed that fuses have different values of current to achieve the different disconnection times from 0.1 to 5s, whereas circuit-breakers only have one value of current as represented by the vertical section.

This is because circuit-breakers have both magnetic and thermal components. The vertical part of the graph shows the operation of the magnetic element dealing with short circuits (faults), whereas the actual curved part is showing the thermal component that deals with overloads. Figure 4.23 (page 39) showed the disconnection time for fuses. Figure 4.24 shows the disconnection time for circuit-breakers, using the information from the table.

> **Working life**
>
> Your site supervisor asks you to explain how a protective device may be used for protection against both overcurrent and fault current.
>
> 1. What is the difference between these two currents?
> 2. How would the same device offer protection against both situations?

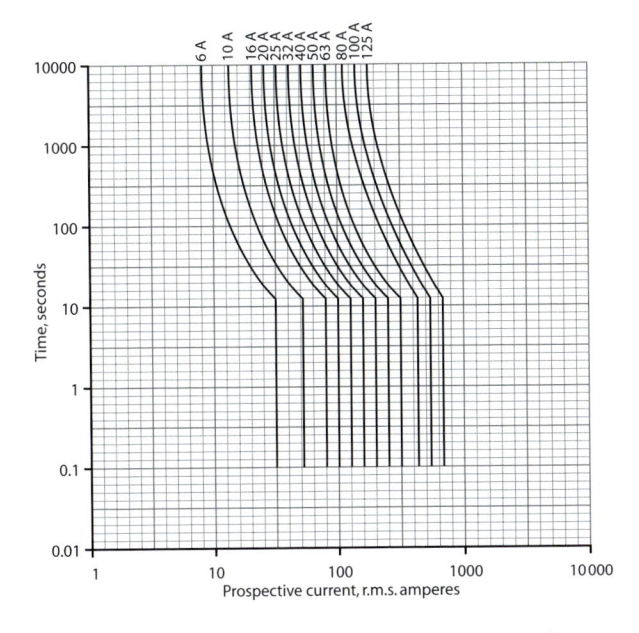

Figure 4.24 Disconnection time for circuit breakers

Time/current characteristics for Type B circuit-breakers to BS EN 60898 and the overcurrent characteristics of RCBOs to BS EN 61009-1	
Current for time, 0.1 sec to 5 secs	
Rating	Current
6 A	30 A
10 A	50 A
16 A	80 A
20 A	100 A
25 A	125 A
32 A	160 A
40 A	200 A
50 A	250 A
63 A	315 A
80 A	400 A
100 A	500 A
125 A	625 A

Disconnection times

Regulation 411.3.2.1 requires that when a fault of negligible impedance occurs between the line conductor and an exposed-conductive part or a protective conductor in the circuit or equipment, the protective device should operate in the required time.

Table 41.1 of Regulation 411.3.2.2 applies to final circuits not exceeding 32 A and requires a disconnection time of 0.4 seconds on a final circuit supplied by a TN system at a nominal voltage of 230 V and 0.2 seconds for a TT system.

Should the final circuit be in excess of 32 A, Regulation 411.3.2.3 permits a disconnection time of 5 seconds where supplied by a TN system at a nominal voltage of 230 V and Regulation 411.3.2.4 permits 1 second for a TT system.

> **Progress check**
>
> 1 Describe the component parts of the earth-fault loop.
>
> 2 Explain the difference between an exposed-conductive part and a extraneous-conductive part.
>
> 3 What are the requirements for additional protection for a socket outlet rated at less than 20 A for use by ordinary persons.
>
> 4 In relation to protective devices explain the term discrimination.

K3. Understand the principles for selecting cables and circuit protection devices

Once the supply is ready and protective measures have been put in place, with the correct isolation and switching requirements, then an electrician is ready to select and install cables. There are several important rules that must be followed when determining the correct size of cable. This section will look at these rules.

To select cables there are some important symbols and definitions that you need to be aware of. These are shown in Table 4.15.

I_z	The maximum current carrying capacity of the cable in the situation where it is installed	
I_t	The value of current given in BS 7671 Appendix 4 for a single circuit of the relevant cable at an ambient temperature of 30°C	
I_b	The design current of the circuit, i.e. the current intended to be carried by the circuit in normal use	
I_n	The rated current or current setting of the protective device	
I_2	The operating current (i.e. the fusing current or tripping current for the conventional operating time) of the device protecting the circuit against overload	
$C_?$	A rating factor to be applied where the installation conditions differ from those shown in the tables of Appendix 4. Shown as the letter C followed by a subscript letter, the various rating factors are identified as follows:	
	C_a	Ambient temperature
	C_g	Grouping
	C_i	Thermal insulation
	C_t	Operating temperature of conductor
	C_f	Semi-enclosed fuse to BS 3036
	C_s	Thermal resistivity of soil
	C_c	Buried circuits
	C_d	Depth of buried circuit

Table 4.15 Symbols and definitions for cable selection

How external influences affect the choice of wiring system and enclosure

Appendix 5 of BS 7671 lists the 'Classification of External Influences', where each condition of external influence is designated with a code that comprises a group of two capital letters and a number.

The first letter relates to the general category of the external influence (for example the environment), the second letter then relates to the circumstances or nature of that external influence (for example the presence of water) and finally, the number relates to the specific degree or level of that condition within each external influence (for example splashes).

To explain this, let us look at the influences mentioned in the above paragraph and use BS 7671 Table 5 to convert them into a code.

> **Example**
>
> | General category of influence | Environment | A |
> |---|---|---|
> | Nature of the influence | Water | D |
> | Level of that influence | Splashes | 3 |
>
> The designated code for these conditions would be **AD3**.

Working in the opposite direction and covering each of the three general categories of external influence, here are three further codes and their explanations.

> **Example**
>
> **AA4**
> A   Environment
> AA   Environment – Ambient Temperature
> AA4   Environment – Ambient Temperature – in the range –5°C to +40°C
>
> **BC1**
> B   Utilisation
> BC   Utilisation – Contact with earth
> BC1   Utilisation – Contact with earth – None
>
> **CA2**
> C   Construction of buildings
> CA   Construction of buildings – Materials
> CA1   Construction of buildings – Materials – Combustible

Once you have decided on the type of cable suitable for the environmental conditions, the following process is used to establish the size of conductor to be used. The first stage to consider is the design current.

Procedure for selecting a suitably sized cable

Design current (I_b)

You will have to calculate this value. It is the normal resistive load current designed to be carried by the circuit and the following formulae apply to single- and three-phase supplies:

Single-phase supplies (U_o = 230 V)

$$I_b = \frac{Power}{U_0}$$

where U_o is the line voltage to earth (supply voltage).

Three-phase supplies (U_o = 400 V)

$$I_b = \frac{Power}{\sqrt{3} \times U_0}$$

In a.c. circuits, the effects of either highly inductive or highly capacitive loads can produce a poor power factor (PF) and you will have to allow for this. To find the design current in such cases you may need to use the following equations where PF is the power factor of the circuit concerned:

Single-phase circuits:

$$I_b = \frac{Power}{U_0 \times PF}$$

Three-phase circuits:

$$I_b = \frac{Power}{\sqrt{3} \times U_0 \times PF}$$

Rating of the protective device (I_n)

Once you have calculated the design current of the circuit (I_b), the next stage is to work out the required current rating or setting (I_n) of the protective device.

Regulation 433.1.1 of BS 7671 states that current rating (I_n) of the protective device must be no less than the design current (I_b) of the circuit. The reason for this is that the protective device must be able to pass enough current for the circuit to operate at full load, but without the protective device operating and disconnecting the circuit.

Protective devices are supplied with standard values, such as 13 A, 20 A etc. and these values can be either obtained from manufacturers or found in BS 7671, Tables 41.2, 41.3 and 41.4 or in Appendix 3.

For example, if we had a load producing a design current of 39 A and it is to be protected by a Type B circuit-breaker, using Table 41.3 we would choose a 40 A as our nominal rated device as this would be the next largest size available (nominal in this context means unadjusted).

Installation and reference methods

Within Appendix 4 of BS 7671 there are two tables that should be referred to: 4A1 and 4A2, shown here as Tables 4.16 and 4.17.

We need to use these tables because the method of installation can affect the chosen cable's ability to get rid of any heat generated in normal operation and this in turn affects its current carrying capacity.

It is not practical to calculate the current ratings for every single installation method and many of these would have the same rating. Therefore a range of current ratings have been calculated covering all of the installation methods within Table 4A2 and these are known as reference methods.

However, we should first look at Table 4A1 opposite to find out whether the proposed installation method will be allowable for the type of cable that needs to be installed.

As an example, we want to install a PVC/PVC (thermoplastic/multicore) twin and earth cable, of a yet to be determined size, by clipping it directly to a wall. Looking at Table 4.16 we can see that this installation method is acceptable (shown by the red circle).

> **Remember**
>
> The table you use to select the value of your protective device will depend on the type of equipment or circuit to be supplied and the requirements for disconnection times.

Conductors and cables		Installation methods							
		Without fixings	Clipped direct	Conduit systems	Cable trunking systems*	Cable ducting systems	Cable ladder, cable tray, cable brackets	On insulators	Support wire
Bare conductors		np	np	np	np	np	np	P	np
Non-sheathed cable		np	np	P[1]	P[1 2]	P[1]	np[1]	P	np
Sheathed cables (including armoured and mineral insulated)	Multicore	P	(P)	P	P	P	P	n/a	P
	Single-core	n/a	P	P	P	P	P	n/a	P

* including skirting, trunking and flush floor trunking
[1] Non-sheathed cables which are used as protective conductors or protective bonding confuctors need not be laid in conduits or ducts
[2] Non-sheathed cables are acceptable if the trunking system provides at least the degree of protection IP4X or IPXXD and if the cover can only be removed by means of a tool of deliberate action

Table 4.16 BS 7671 Table 4A1 – Schedule of installation methods in relation to conductors and cables

Now that we know this, we have to use Table 4A2 (shown in Table 4.17 on page 50) to establish the specific installation method and resultant reference method that will govern the current-carrying capacity of our cable.

As can be seen, the specific installation method for our example is 20 and this is classed as falling within reference method C (again shown here with a red circle).

Rating factors (C)

Our next stage in the process is to understand the rating factors, when they are required and how to apply them to the nominal rating of our selected protective device (I_n). The available factors from BS 7671 are shown in Table 4.18.

The rated current or current setting of the protective device (I_n) must not be less than the design current (I_b) of the circuit, and the rated current or current setting of the protective device (I_n) must not exceed the lowest of the current-carrying capacities (I_z) of any of the conductors of the circuit.

As stated earlier, we generally summarise this with the formula:

$I_b \leq I_n \leq I_z$

Installation method			Reference method to be used to determine the current carrying capacity
Number	Examples	Description	
15		Non-sheathed cables in conduit or single-core or multicore cable in architrave	A
16		Non-sheathed cables in conduitor or single-core or multicore cable in window frames	A
(20)		Single core or multicore cables fixed on (chipped direct), or spaced less than 0.3 × cable diameter from a wooden or masonry wall	(C)
21		Single core or multicore cables fixed directly under a wooden or masonry ceiling	C Higher than standard ambient temperatures may occur with this installation method.
22		Single-core or multicore cables spaced from a ceiling	E, F or G Higher than standard ambient temperatures may occur with this installation method.

Table 4.17 BS 7671 Table 4A2

C_a	Ambient temperature	Tables 4B1 and 4B2
C_g	Grouping	Tables 4C1 to 4C5
C_i	Thermal insulation	Regulation 523.9
C_t	Operating temperature of conductor	
C_f	Semi-enclosed fuse to BS 3036	0.725 – Regulation 433.1.101
C_s	Thermal resistivity of soil	Table 4B3
C_c	Buried circuits	0.9 – where the cable installation method is 'in a duct in the ground' or 'buried direct'. For cables installed above ground $C_c = 1$
C_d	Depth of buried circuit	
n/a	Mineral insulated cable	0.9 – Table 4G1A

Table 4.18 Rating Factors (C)

Where the overcurrent device is intended to afford protection against overload, I_2 must not exceed $1.45 \times I_z$ and I_n must not exceed I_z. However, where the overcurrent device is intended to afford fault current protection only, I_n can be greater than I_z and I_2 can be greater than $1.45 \times I_z$.

The protective device must be selected for compliance with Regulation 434.5.2.

Ambient temperature

Ambient temperature is the temperature of the environment surrounding the cable, such as the temperature of the room or building or ground in which the cable is installed. When a cable is carrying a current it gives off heat. The hotter its surroundings then the harder it is to get rid of that heat. If the temperature of the surroundings is low then the cable can get rid of its heat more easily and could carry more current. (This is also why we move into the shade when we get hot on a summer's day and why we fit bigger fans on overclocked computers.)

The current-carrying capacities that are given in Appendix 4 of BS 7671 for cables direct in ground or in ducts in the ground are based on ambient temperature of 20°C. However, the above-mentioned factor of 1.45 that is applied in Regulation 433.1.1, when considering overload protection, assumes that the tabulated current-carrying capacities are based on an ambient temperature of 30°C.

To get the same degree of overload protection when cables are 'in a duct in the ground' or 'buried direct', as compared with other installation methods, a rating factor of 0.9 is now applied as a multiplier to the tabled current-carrying capacity.

The ambient temperature cannot be ignored; as an ambient temperature below 30°C (e.g. 25°C) will give a correction factor that will allow your choice of cable to be improved. A lower ambient temperature will allow the cable to **dissipate** heat more easily and therefore you may be able to reduce the size of cable. Should a cable pass through areas of different ambient temperature then only apply a correction factor based on the highest temperature.

> **Key term**
>
> **Dissipate** – get rid of through dispersion or scattering

Grouping

For grouping, the concept is that when cables are installed so that they are touching each other it is more difficult for them to dissipate any heat generated in normal use.

The rating factors within Appendix 4 of BS 7671 are based upon groups of 'similar and equally loaded cables'. However, where we have groups of different sizes of sheathed/non-sheathed cable installed in conduit, trunking or ducting systems, we can use the following formula to calculate the group rating factor:

$$F = \frac{1}{\sqrt{n}}$$

Where F is the group rating factor and n is the number of circuits in that group. Although not ideal, this formula can also be used when these cables are installed on a tray.

Thermal insulation

We know that thermal insulation is used throughout the construction industry. Like putting on an extra thick woolly coat on a hot summer's day, thermal insulation reduces the ability of the cable to dissipate any heat being generated.

Consequently Regulation 523.9 of BS 7671 states that:

- For a cable installed in a thermally insulated wall or above a thermally insulated ceiling, where the cable has one side in contact with the thermally conductive surface, current-carrying capacities are given in Appendix 4 of BS 7671.
- For a single cable that is totally covered by thermal insulation for more than 0.5 m and where no more precise information is available, the current-carrying capacity shall be taken as half the value of that same cable when clipped to a surface and open (Reference Method C).
- For a single cable that is totally covered by thermal insulation for less than 0.5 m, the current-carrying capacity shall be reduced as appropriate depending on the size of the cable, its length in the insulation and the thermal properties of that insulation. De-rating factors are therefore given in BS 7671 Table 52.2 (reproduced as Table 4.19) for conductor sizes up to 10 mm^2.

Length in insulation (mm)	De-rating factor
50	0.88
100	0.78
200	0.63
400	0.51

Table 4.19 Cable rating (BS 7671 Table 52.2)

Protective device

Regulation 433.1.100 states that where overload protection is given by either a fuse to BS 88 or 1361, an MCB to BS EN 60898 or an RCBO to BS EN 61009, then the requirements for coordination of conductor and protective device in Regulation 433.1.1 have been complied with and the rating factor (C_f) is therefore given as 1.0.

However, if the protective device is a rewireable fuse to BS 3036, due to potentially poor fuse performance we always use a rating factor for C_f of 0.725.

As mentioned earlier, should we have a cable that is installed either in a duct in the ground or is buried directly in the ground, then we apply a rating factor for C_f of 0.9.

Therefore, if we had a situation where a cable buried directly in the ground was protected by a BS 3036 fuse, both rating factors would apply and our new rating would be:

$C_f = 0.725 \times 0.9$ and therefore $C_f = 0.653$

Application of rating factors

The purpose of applying these factors is to make sure that a cable is large enough to carry the current without too much heat being generated. If the cable's ability to give off heat is reduced by external conditions, then the only solution is to increase the size of the cable.

Appendix 4 of BS 7671 requires us to ensure that the current-carrying capacity of the cable (I_z) is no less than the circuit's design current (I_b). Equally, the rating of the fuse or circuit-breaker (I_n) must be at least as big as the circuit design current (I_b). The reason for doing this is to make sure that the cable is not overloaded and that the protective device does not operate immediately the load is turned on.

Therefore, to find I_z we start by dividing the rated current of the protective device (I_n) by the applicable factors that can de-rate the cable. The result of this calculation is the true rated current of the cable required, which we use to select our cable from the appropriate table in Appendix 4 of BS 7671.

This value of current obtained from the tables in Appendix 4 of BS 7671 is given the symbol I_t and it must be greater than the value resulting from the above calculation (I_z).

We therefore use the following formula to establish I_z:

$$I_z = \frac{I_n}{C_a \; C_i \; C_s \; C_d \; C_f \; C_c}$$

The best way to see this all in operation is with some worked examples. Checks must also be made for voltage drop, shock protection and thermal constraints and all these will be considered as we proceed through the examples.

Example 1

A single-phase circuit has a design current of 30 A. It is to be wired in flat two core 70°C PVC insulated and sheathed cable with cpc to BS 6004. It will have copper conductors and the cable is enclosed in trunking with four other similar cables. If the ambient temperature is 30°C and the circuit is to be protected by a BS 3036 fuse, what should be the nominal current rating of the fuse and the minimum CSA of the cable conductor?

In answering these types of questions, it is a good idea to construct a table of information. This will help you understand and retain the process of cable selection. It also makes cable selection easier to understand.

Installation Method number will be 6/7 (installed in trunking), therefore Reference Method = B

As ambient temperature equals 30°C, then from Table 4B1 for 70° Celsius thermoplastic, C_a = 1.0.

As grouped, then from Table 4C1, C_g = 0.6 (there are five circuits and Reference Methods A to F apply).

Design current (I_b) was given as 30 A.

As the protective device rating (I_n) must be ≥ the rating of the design current (I_b) we can say that:

I_n = 30 A.

But as the protective device is a BS 3036 fuse, a factor C_f equal to 0.725 must be applied

$$I_z = \frac{I_n}{C_a \times C_g \times C_f} = \frac{30}{1.0 \times 0.6 \times 0.725} = 69 \text{ A}$$

We said that I_t must be ≥ I_z, therefore using Table 4D2A (column 4), we see that a 16 mm² conductor size has a tabulated current rating of 69 A.

Therefore the minimum conductor csa that can be used is **16 mm²**.

Checking voltage drop

Regulation 525.101 of BS 7671 states that 'the voltage drop between the origin of the installation (usually the supply terminals) and a socket outlet (or the terminals of fixed current using equipment) doesn't exceed that stated in Appendix 12'.

Appendix 12 goes on to state that for low voltage installations supplied from a public distribution system, then voltage drop should be no greater than 3% of the supply voltage for lighting circuits and no greater than 5% for other circuits. This means that where the supply voltage is 230 V, a maximum voltage drop of 6.9 V for lighting or 11.5 V for other circuits is allowed.

In unusual circumstances where the final circuit has a length in excess of 100 m, BS 7671 now allows an increase in voltage drop of 0.005% per metre of that circuit above 100 m, up to a maximum value of 0.5%.

The voltage drops either because the resistance of the conductor becomes greater as the length of the cable increases or the cross-sectional area of the cable is reduced. This means that on long cable runs, the cable cross-sectional area may have to be increased, reducing the resistance allowing current to 'flow' more easily and reducing the voltage drop across the circuit.

We can calculate the maximum allowed voltage drop like this:

1. For single-phase 230 volt lighting circuits:

$$3\% = 230 \times \frac{3}{100} = 6.9 \text{ volts}$$

2. For three-phase 400 volt systems:

$$5\% = 400 \times \frac{5}{100} = 20 \text{ volts}$$

Therefore, when carrying out circuit calculations it is worth remembering we are allowed to have a minimum system voltage as low as 223.1 V on lighting circuits and 218.5 V on other circuits. It is better to keep the voltage drops as low as possible, because low voltages can reduce the efficiency of the equipment being supplied, thus lamps will not be as bright and heaters may not give off full heat. The values for cable voltage drop are given in the accompanying tables of current-carrying capacity in Appendix 4 of the IET Wiring Regulations. The values are given in millivolts per ampere per metre (mV/A/m).

Use the formula below to calculate the actual voltage drop.

$$\text{Voltage drop (VD)} = \text{mV/A/m} \times \frac{I_b \times L}{1000}$$

Where   mV/A/m   is the value given in the Regulation Tables.
I_b is the circuit's design current.
L   is the length of cable in the circuit measured in metres.

NB: As mV/A/m uses the prefix milli (thousandths) then you are going to have to convert I_b and L into the common unit and you do this by dividing them by 1000.

Note: In some areas of the country people prefer the following equation for working out voltage drop.

Voltage drop (VD) = mV/A/m × I_b × L × 10^{-3}

An example of this is shown opposite.

Shock protection

On pages 20–22 we looked at the protection measure Automatic Disconnection of Supply (ADS). With this measure, all metalwork including extraneous-conductive parts and exposed-conductive parts are connected to earth. Therefore, should an earth-fault occur, we need the protective device to work quickly. The current must then be large enough to operate the device within given times and the earth-fault loop impedance must be low enough to allow this to happen.

BS 7671 Regulation 411.3.2.1 states that a protective device shall automatically interrupt the supply to the line conductor (of a circuit or of equipment) in the event of a fault of negligible impedance between the line conductor and an exposed-conductive part or protective conductor in the circuit or equipment, within the times given in Table 41.1 of BS 7671 (see Table 4.20).

From this table we can see that for a 230 V final circuit (e.g. lighting, socket outlets, cooker circuits etc.) on a TN system the disconnection time is 0.4 seconds, but on a TT system is 0.2 seconds. Only for distribution circuits (i.e. a circuit supplying a DB or switchgear) is a time on TN of 5.0 seconds acceptable.

To achieve these times, the value of earth-fault loop impedance (Z_s) should not be larger than the values given in Tables 41.2, 41.3 and 41.4 of BS 7671.

Example 2

To reinforce understanding, this is the same question as in Example 1, but this time you are also required to work out the voltage drop for the cable and consider its effect.

A single-phase circuit has a design current of 30 A. It is to be wired in flat two core 70°C PVC insulated and sheathed cables with cpc to BS 6004. It will have copper conductors and the cable is enclosed in trunking with four other similar cables. If the ambient temperature is 30°C, the circuit is to be protected by a BS 3036 fuse and the circuit length is 27 m, what should be the nominal current rating of the fuse (I_n) and the minimum CSA of the cable conductor?

Installation Method number will be 6/7 (installed in trunking) therefore Reference Method = B

As ambient temperature = 30°C then, from Table 4B1, for 70°C thermoplastic: C_a = 1.0

As grouped, then from Table 4C1, C_g = 0.6 (there are five circuits and Reference Methods A to F applies)

Design current (I_b) was given as 30 A.

As the protective device rating (I_n) must be ≥ the rating of the design current (I_b) we can say that I_n = 30 A.

But as the protective device is a BS 3036 fuse, the factor C_f = 0.725 must be applied

$$I_z = \frac{I_n}{C_a \times C_g \times C_f} = \frac{30}{1.0 \times 0.6 \times 0.725} = 69 \text{ A}$$

We said that I_t must be ≥ I_z, therefore using Table 4D2A (column 4), we see that a 16 mm² conductor size has a tabulated current rating of 69 A and is therefore acceptable in this respect. We must now check the voltage drop for this cable.

From Table 4D2B (column 3) we see that the voltage drop is given as 2.8 mV/A/m. We need to work in common units, so we must divide I_b and L by 1000.

$$\text{Voltage drop} = mV/A/m \times \frac{I_b \times L}{1000} = 2.8 \times \frac{(30 \times 27)}{1000} = \textbf{2.27 V}$$

BS 7671 states that for other than a lighting circuit, voltage drop must not exceed 5% of the supply voltage, which in this case would be 5% of 230 V or 11.5 V. As the actual voltage drop of our proposed cable was calculated to be 2.27 V, this means that the minimum CSA of the cable conductor will be **16 mm²**.

System	50 V < U_o < 120 V seconds		120 V < U_o < 230 V seconds		230 V < U_o < 400 V seconds		U_o < 400 V seconds	
	a.c.	d.c.	a.c.	d.c.	a.c.	d.c.	a.c.	d.c.
TN	0.8	NOTE 1	0.4	5	0.2	0.4	0.1	0.1
TT	0.3	NOTE 1	0.2	0.4	0.07	0.2	0.04	0.1

Table 4.20 Maximum disconnection times (Table 41.1 BS 7671)

You need to do a calculation to check that the circuit protective device will operate within the required time.

You do this by checking that the actual Z_s is lower than the maximum Z_s given in the relevant table, for the protective device you have chosen.

Maximum Z_s can also be found from IET Wiring Regulations or manufacturers' data. Use the following formula to calculate the actual Z_s:

$$\text{Actual } Z_s = Z_e + \left(\{R_1 + R_2\} \times (\text{mf}) \times \frac{\text{length}}{1000} \right)$$

Where:

- Z_e = the external impedance on the supply authority's side of the earth-fault loop. You can get this value from the supply authority. Typical maximum values are: TN-C-S (PME) system 0.35 ohms, TN-S (cable sheath) 0.8 ohms, TT system 21 ohms.
- $R_1 + R_2$ = the resistance of the phase conductor plus the cpc resistance. You can find values of resistance/metre for $R_1 + R_2$ for various combinations of phase and cpc conductors up to and including 50 mm² in Table 11 of the IET On Site Guide.
- Table 4.21 gives you multipliers (mf) from Table 11 of the On Site Guide to calculate the resistance under fault conditions. If the conductor temperature rises, resistance in the conductors will increase. This table is based on the type of insulation used and whether the cpc isn't incorporated in a cable or bunched with cables (54.2), or is incorporated as a core in a cable or bunched with cables (54.3).

Conductor installed as	Insulation material		
	70° thermoplastic (pvc)	85° thermosetting (rubber)	90° thermosetting
54.2 (not incorporated)	1.04	1.04	1.04
54.3 (incorporated)	1.20	1.26	1.28

Table 4.21 Multipliers to be applied to Table 11 of the On Site Guide

Example 3

Again this is the same question that was used in Examples 1 and 2, but this time you are also required to work out the requirement for shock protection and consider its effect.

A single-phase circuit supplying a domestic cooker has a design current of 30 A and is to be wired in flat two core 70°C PVC insulated and sheathed cables with cpc to BS 6004. It will have copper conductors and the cable is enclosed in trunking with four other similar cables. If the ambient temperature is 30°C, the circuit is to be protected by a BS 3036 fuse and the circuit length is 27 m, what should be the nominal current rating of the fuse (I_n) and the minimum CSA of the cable conductor? You are informed that Z_e is 0.8 Ω.

Installation Method number will be 6/7 (installed in trunking) therefore Reference Method = B.

As ambient temperature = 30°C then, from Table 4B1, for 70°C thermoplastic: C_a = 1.0

As grouped, then from Table 4C1, C_g = 0.6 (there are five circuits and Reference Methods A to F applies)

Design current (I_b) was given as 30 A.

As the protective device rating (I_n) must be ≥ the rating of the design current (I_b) we can say that I_n = 30 A.

But as the protective device is a BS 3036 fuse, the factor C_f = 0.725 must be applied.

$$I_z = \frac{I_n}{C_a \times C_g \times C_f} = \frac{30}{1.0 \times 0.6 \times 0.725} = 69 \text{ A}$$

We said that I_t must be ≥ I_z, therefore using Table 4D2A (column 4), we see that a 16 mm² conductor size has a tabulated current rating of 69 A and is therefore acceptable in this respect. We must now check the voltage drop for this cable.

From Table 4D2B (column 3) we see that the voltage drop is given as 2.8 mV/A/m. We need to work in common units, so we must divide I_b and L by 1000.

$$\text{Voltage drop} = mV/A/m \times \frac{I_b \times L}{1000} = 2.8 \times \frac{(30 \times 27)}{1000} = \mathbf{2.27 \text{ V}}$$

BS 7671 states that for other than a lighting circuit, voltage drop must not exceed 5% of the supply voltage, which in this case would be 5% of 230 V or 11.5 V. As the actual voltage drop of our proposed cable was calculated to be 2.27 V, this is acceptable. We must now check for shock protection.

From Table 41.2, the maximum permitted Z_s for our 30 A BS 3036 fuse is 1.09 Ω and from Table 11 of the On Site Guide, for a 16 mm² with 6 mm² cpc, the Resistance/m ($R_1 + R_2$/m) is given as 4.23 mΩ/m.

Additionally we need the multiplier for a grouped cable, which from Table 13 of the On Site Guide (Table 4.21) on page 58 is 1.2. We also need to convert length from m to mm by dividing by 1000. Therefore:

$$Z_s = Z_e + \left([R_1 + R_2] \times mf \times \frac{\text{length}}{1000}\right) = 0.8\left([4.23] \times 1.2 \times \frac{27}{1000}\right) = \mathbf{0.937 \text{ ΩΩ}}$$

As actual Z_s (0.937 Ω) is less than the permitted value (1.09 Ω), this means that 16 mm² cable with 6 mm² cpc is the minimum CSA of the cable conductor.

This information, together with the length of the circuit, can now be applied to the formula. Then you can check to see that the actual Z_s is less than the maximum Z_s given in the appropriate tables.

When you have done this, you can prove that the protective device will disconnect the circuit in the time that is specified in the appropriate table. In other words, if the actual Z_s is less than the maximum Z_s then we have compliance for shock protection.

Thermal constraints

Now that you have chosen the type and size of cable to suit the conditions of the installation, we must look at 'thermal constraints'. This is a check to make sure that the size of the cpc, 'the earth conductor', complies with the IET Wiring Regulations.

If there is a fault on the circuit, which could be a short circuit, or earth-fault, a fault current of hundreds or thousands of amperes could flow. Imagine that this is a 1 mm² or 2.5 mm² cable; if this large amount of current was allowed to flow for a short period of time, i.e. a few seconds, the cable would melt and a fire could start.

We need to check that the cpc will be large enough to be able to carry this fault current without causing any heat/fire damage. The formula that is used to check this situation is the **adiabatic equation**. The cpc will only need to carry the fault current for a short period of time, until the protective device operates.

Regulation 543.1.1 states that 'The cross-sectional area of every protective conductor shall be calculated in accordance with Regulation 543.1.3 (adiabatic equation) or selected in accordance with Regulation 543.1.4 (Table 54.7)'. You will see that Regulation 543.1.4 asks that reference should be made to Table 54.7. This table shows that, for cables 16 mm² and below with the cpc made from the same material as the line conductor, the cpc should be the same size as the line conductor. A line conductor between 16 mm² and 35 mm² requires a cpc to be 16 mm². A line conductor above 35 mm² requires a cpc to be at least half the cross-sectional area.

Multicore cables have cpcs smaller than their respective line conductors, except for 1 mm², which has the same-sized cpc. Regulation 543.1.2 of BS 7671, gives two options, calculation or selection. If we were to apply option (ii) of this Regulation then selection would make these cables contravene the Regulations. Clearly, it is not intended that composite cables should have their cpcs increased in accordance with the table and therefore calculation by the adiabatic equation required in option (i) of the same Regulation should be applied.

> **Working life**
>
> You have been asked by your employer to investigate an 'earthing' problem that has been reported by a customer. The customer concerned has just moved into a large 80-year-old house and plans to work from home as a computer software designer. The house has large gardens, a separate garage, three outside security lights, a gas-fired central-heating system and the customer will be working from one of the bedrooms on his computers.
>
> The customer has reported that he is getting intermittent RCD trips on the main consumer unit that seem to occur only at night, and only when it has been raining. Rain during the day does not cause the same problems. There is also a circuit out to the garage that has its own RCD and this does not trip.
>
> What do you think could be causing the problem?

The adiabatic equation referred to in the introduction enables a designer to check the suitability of the cpc in a composite cable. If the cable does not incorporate a cpc, a cpc installed as a separate conductor may also be checked.

The equation is as follows:

$$S = \frac{\sqrt{I_f^2 \times t}}{k}$$

Where:

- S = the cross-sectional area of the cpc in mm²
- I_f = the value of the fault current in amperes
- t = the operating time of the disconnecting device in seconds
- k = a factor that takes into account resistivity, temperature coefficient and heat capacity.

In order to apply the adiabatic equation, we first need to calculate the value of I (fault current) from the following equation:

$$I_f = \frac{U_0}{Z_s}$$

Where:

- U_0 is the nominal supply line voltage to earth
- Z_s is the earth-fault loop impedance.

If you are using method (i) from Reg 543.1.2 and applying the adiabatic equation, you must find out the time/current characteristics of the protective device. A selection of time/current characteristics for standard overcurrent protective devices is given in Appendix 3 of IET Wiring Regulations. You can get the time (t) for disconnection to the corresponding earth-fault current from these graphs.

If you look at the time/current curve you will find that the scales on both the time (seconds) scale and the prospective current (amperes) scale are logarithmic and the value of each subdivision depends on the major division boundaries into which it falls.

For example, on the current scale, all the subdivisions between 10 and 100 are in quantities of 10, while the subdivisions between 100 and 1000 are in quantities of 100 and so on. This also occurs with the time scale, subdivisions between 0.01 and 0.1 being in hundredths and the subdivisions between 0.1 and 1 being in tenths, etc.

As an example, if you look at the graph shown in Figure 4.25 you will see that for a BS 88 fuse with a rating of 32 A, a fault current of 200 A will cause the fuse to clear the fault in 0.6 seconds.

In addition to this graph, the IET has produced a small table showing some of the more common sizes of protective devices and the fault currents for a given disconnection time.

Next, you need to select the k factor using Tables 54.2 to 54.6. The values in these tables are based on the initial and final temperatures shown in each table. This is where you may need to refer to the cable's operating temperature shown in the cable tables.

Now you must substitute the values for I, t and k into the adiabatic equation. This will give you the minimum cross-sectional area for the cpc. If your calculation produces a non-standard size, you must use the next largest standard size.

From a designer's point of view, it is advantageous to use the calculation method as this may lead to savings in the size of cpc.

Now try the example on page 64 to give you a complete understanding of cable selection. The example is similar to those used in Cable Selection Examples 1, 2 and 3, but this time you also need to complete calculations for thermal constraints.

Figure 4.25 Time/current characteristics graph and table

Fuse rating	Current for time				
	0.1 sec	0.2 sec	0.4 sec	1 sec	5 sec
6 A	36 A	31 A	27 A	23 A	17 A
20 A	175 A	150 A	130 A	110 A	79 A
32 A	320 A	260 A	220 A	170 A	125 A
50 A	540 A	450 A	380 A	310 A	220 A
80 A	1100 A	890 A	740 A	580 A	400 A
125 A	1800 A	1500 A	1300 A	1050 A	690 A
200 A	3000 A	2500 A	2200 A	1700 A	1200 A

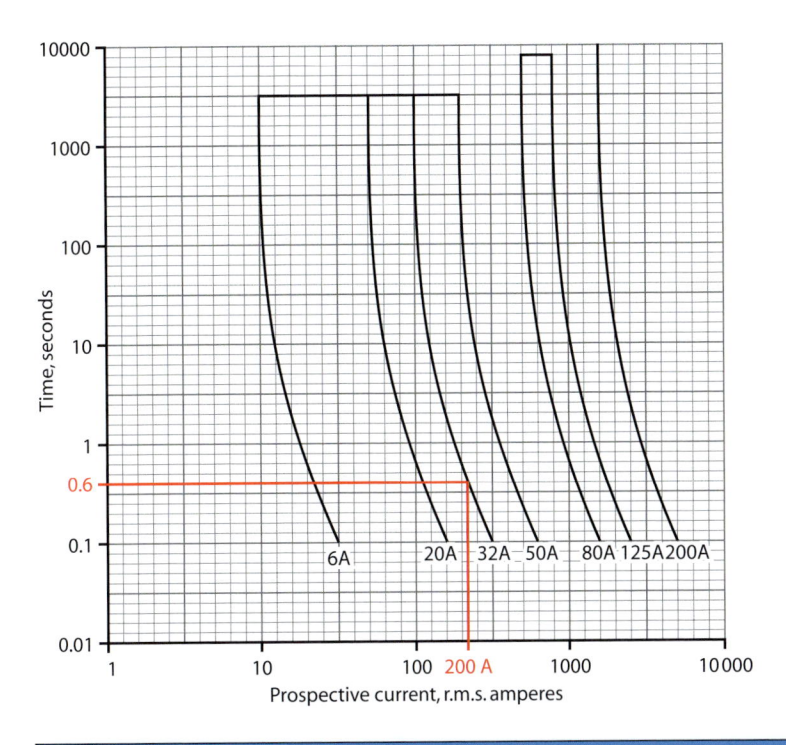

Time/current characteristics for fuses to BS 88.2.2 and BS 88.6

Diversity

In this section we will look at Maximum Demand and Diversity as considered in BS 7671 Chapter 31. The current demand for a final circuit is determined by adding up the current demands of all points of utilisation and equipment in the circuit and, where appropriate, making an allowance for diversity. Diversity makes allowances on the basis that not all of the load or connected items will be in use at the same time.

Example 4

To reinforce understanding, this is the same question that was used in Examples 1, 2 and 3, but this time we will only concern ourselves with the calculations relevant to thermal constraints. You are also required to work out the requirement for shock protection and consider its effect.

A single-phase circuit supplying a domestic cooker has a design current of 30 A and is to be wired in flat two core 70°C PVC insulated and sheathed cables with cpc to BS 6004. It will have copper conductors and the cable is enclosed in trunking with four other similar cables. If the ambient temperature is 30°C, the circuit is to be protected by a BS 3036 fuse and the circuit length is 27 m, what should be the nominal current rating of the fuse (I_n) and the minimum CSA of the cable conductor?

You are informed that Z_e is 0.8 Ω.

From our previous calculations in Example 3, we know that I_n = **30 A** and that **16 mm²** cable with 6 mm² cpc **is the minimum CSA of the cable conductor.**

For thermal constraints, we need to know the fault current before we can apply the adiabatic equation. We find this by:

$$I_f = \frac{U_0}{Z_s} = \frac{230}{0.937} = 245.5 \text{ A}$$

We also need to know the value of t and the value of k. Therefore from Table 3.2A within Appendix 3 of BS 7671 we can see that our fault current of 245.5 A will operate the device in a time of 0.3 second. We therefore state t as being 0.3 second.

From Table 54.3 we can see that the value of k will be 115, as the cable is under 300 mm².

Now using the adiabatic equation:

$$S = \frac{\sqrt{I_f^2 \times t}}{K} = \frac{\sqrt{245.5^2 \times 0.3}}{115} = \frac{\sqrt{18081.08}}{115} = \frac{134.5}{115} = \textbf{1.17 mm}^2$$

As 1.17 mm² is less than our 6 mm² cpc, our choice of cable remains as a cable with 16 mm² conductors and a 6 mm² cpc to ensure compliance with thermal constraints.

> **Remember**
>
> The three main advantages of applying diversity are:
> 1. Reduced size of sub-main cables
> 2. Reduced sizes of isolation controls and protective devices
> 3. Overall cost

In most cases main or sub-main cables will supply a number and/or variety of final circuits. Use of the various loads must now be considered, otherwise if all the loads are totalled, a larger cable than necessary will be selected at considerable extra cost. Therefore a method of assessing the load must be used.

Section 3 of the Unite Guide to Good Electrical Practice contains such a method, which allows diversity to be applied depending upon the type of load and installation premises. The individual circuit/load figures are added together to determine the total 'Assumed Current Demand' for the installation. This value can then be used as the starting point to determine the rating of a suitable protective device and the size of cable, considering any influencing factors in a similar manner to that applied to final circuits.

Please also remember that the calculation of maximum demand is not an exact science and a suitably qualified electrical engineer may use other methods of calculating maximum demand.

Example 5

A 230 V domestic installation consists of the following loads:
15 × filament lighting points
6 × fluorescent lighting points, each rated at 40 watts
4 × fluorescent lighting points each rated at 85 watts
3 × ring final circuits supplying 13 A socket outlets
1 × radial circuit protected by a 20 A device supplying 13 A sockets for the adjoining garage
1 × 3 kW immersion heater with thermostatic control
1 × 13.6 kW cooker with a 13 A socket outlet incorporated in the control unit.

Determine the maximum current demand for determining the size of the sub-main cable required to feed this domestic installation? The circuit protection is by the use of BS 1361 fuses.

Answer
Lighting
Tungsten light points (See page 40 of the Unite Guide to Good Electrical Practice)
   15 × 100 W minimum   1500

Fluorescent light points (See page 40 and note of the Unite Guide to Good Electrical Practice)
   6 × 40 W with multiplier of 1.8 (40 × 1.8 = 72)   432
   4 × 85 W with multiplier of 1.8 (85 × 1.8 = 153)   612
   Total   2544 W

Using Item 1 of the table on page 41 of the Unite Guide to Good Electrical Practice, we can apply diversity as being 66% of the total current demand. Therefore:

   66% of 2544 = 1679 W and since $I = \dfrac{P}{V}$ this gives us $\dfrac{1679}{230} = 7.3$ A

Power (Item 9 on page 41 of the Unite Guide to Good Electrical Practice)

3 ring final circuits
   1 × ring at 100% rating (30 A)   30 A
   2 × ring at 40% rating (40% of 30 A)   24 A

20 A radial circuit (see item 9 on page 41 of the Unite Guide to Good Electrical Practice)
   1 × radial at 40% rating (40% of 20 A)   8 A

Example 5 continued

3 kW immersion heater (See Item 6 on page 41 of the Unite Guide to Good Electrical Practice)

3 kW heater with no diversity $I = \dfrac{P}{V} = \dfrac{3000}{230}$ which gives us 13 A

13.6 kW Cooker with socket outlet (See item 3 on page 41 of the Unite Guide to Good Electrical Practice)
The first 10 A, plus 30% of the remainder of the overall rated current, plus 5 A for the socket.

$I = \dfrac{P}{V}$ giving $\dfrac{136000}{230}$ which gives us a total rated current of 59 A

59 A − 10 A = 49 A and therefore 30% of 49 A = 14.7 A

The allowable total cooker rating is therefore 10 A + 14.7 A + 5 A = 29.7 A

Our total assumed current demand is therefore:

7.3 + 30 + 24 + 8 + 13 + 29.7 giving a total of **112 A**

Working life

An additional circuit is to be added to the installation your company is completing. The cable leaves the distribution board in a bunch. Along the run the cable passes through thermal insulation for 1.5 m before termination in an ambient temperature higher than anticipated. Draft a brief explanation to the site supervisor covering:

1. How these conditions affect cables in service.
2. How the conditions are accounted for when selecting suitable cables.
3. How could these effects on the cable to be selected be avoided.

Determining the size of conduit and trunking as appropriate to the size and number of cables to be installed

The number of cables that can be drawn into or laid in any enclosure of a wiring system must be such that no damage can occur to the cables or the enclosure during installation.

Therefore, the size of conduit used to enclose the cables needs to be calculated and this is done by using a 'factor' system to compare the number of cables against the overall CSA of the conduit.

The minimum conduit size will be a one that has a higher factor than that of the cables.

The tables shown are applicable to both plastic and metal conduit and use information found in Section 5 of the Unite Guide to Good Electrical Practice. These tables only give guidance to the maximum number of cables that should be drawn in and the electrical effects of grouping are not taken into account.

Therefore, as the number of circuits increases, the current carrying capacity of the cables will decrease and therefore cable sizes would have to be increased, with a consequent increase in cost of cable and conduit. It may therefore be more economical to divide the circuits concerned between two or more enclosures.

Tables 4.22 and 4.23 are for use with short, straight runs of up to 3 m.

Type of conductor	Conductor CSA mm²	Cable factor
Solid	1	22
Solid	1.5	27
Solid	2.5	39
Stranded	1.5	31
Stranded	2.5	43
Stranded	4	58
Stranded	6	88
Stranded	10	146
Stranded	16	202
Stranded	25	385

Table 4.22 Cable factors for use in short (up to 3 m) straight runs

16	290
20	460
25	800
32	1400
38	1900
50	3500
63	5600

Table 4.23 Conduit factors for use in short (up to 3 m) straight runs

Here is an example of how to use these tables.

> **Example 6**
>
> If 10 × 2.5 mm² cables with stranded copper conductors were to be installed in a 2 m straight length of conduit, their cable factor will be 10 × 43 = 430. You would need to select a conduit with a factor greater than 430.
>
> From the conduit factor table, you can see that 16 mm conduit is too small (290) but that 20 mm conduit has a factor of 460, so this is what you would install.

This method is acceptable when using short, straight runs. However, when longer runs are involved, or the run has bends in it, you will need to use other tables.

> **Remember**
>
> One double set equates to one 90 degree bend.

Cable and conduit factors for runs over 3 m or with bends

Tables 4.24 and 4.25 apply to cable and conduit runs over 3 m or with bends. However, conduit runs in excess of 10 m or with more than 4 bends should be divided into separate sections and then have the table values applied to each of those sections.

Tables with factors for conduit sizes larger than 32 mm aren't available. Therefore these are calculated by multiplying the 32 mm factor as follows:

- for 38 mm conduit (1.4 × 32 mm term)
- for 50 mm conduit (2.6 × 32 mm term)
- for 63 mm conduit (4.2 × 32 mm term).

	1	16
	1.5	22
	2.5	30
Solid or stranded cables	4	43
	6	58
	10	105
	16	145
	25	217

Table 4.24 Cable factors for use in long (over 3 m) straight runs, or runs of any length with bends

Unit ELTK 04a

Length in mm	Conduit diameter (mm)																			
	Straight				One bend				Two bends				Three bends				Four bends			
	16	20	25	32	16	20	25	32	16	20	25	32	16	20	25	32	16	20	25	32
1	Covered by short straight run tables				188	303	543	947	177	286	514	900	158	256	463	818	130	213	388	692
1.5					182	294	528	923	167	270	487	857	143	233	422	750	111	182	333	600
2					177	286	514	900	158	256	463	818	130	213	388	692	97	159	292	529
2.5					171	278	500	878	150	244	442	783	120	196	358	643	86	141	260	474
3					167	270	487	857	143	233	422	750	111	182	333	600	Divide into 2 or more parts with draw in boxes			
3.5	179	290	521	911	162	263	475	837	136	222	404	720	103	169	311	563				
4	177	286	514	900	158	256	463	818	130	213	388	692	97	159	292	529				
4.5	174	282	507	889	154	250	452	800	125	204	373	667	91	149	275	500				
5	171	278	500	878	150	244	442	783	120	196	358	643	86	141	260	474				
6	167	270	487	857	143	233	422	750	111	182	333	600	Divide into 2 or more parts with draw in boxes							
7	162	263	475	837	136	222	404	720	103	169	311	563								
8	158	256	463	818	130	213	388	692	97	159	292	529								
9	154	250	452	800	125	204	373	667	91	149	275	500								
10	150	244	442	783	120	196	358	643	86	141	260	474								

Table 4.25 Conduit factors for use in long (over 3 m) straight runs, or runs of any length with bends

> **Example 7**
>
> A lighting circuit in a school requires 12 × 1.5 mm² cables inside a conduit of 8 m with two right angled bends.
>
> From the cable factor table, you can see the cable has a factor of 264 (12 × 22).
>
> From the conduit factor table, you can see that 20 mm is too small (159) to handle this, but a 25 mm conduit will be acceptable, with a factor of 292.

The number of cables that can be drawn into or laid in any enclosure of a wiring system must be such that no damage can occur to the cables or the enclosure during installation.

Therefore, the size of trunking used to enclose the cables also needs to be calculated and this is done by using a 'factor' system in similar way to conduit.

Tables 4.26 and 4.27 are applicable to both plastic and metal trunking and use information found in Section 5 of the Unite Guide to Good Electrical Practice. These tables only give guidance to the maximum number of cables that should be drawn in.

Type of conductor	Conductor CSA mm²	Thermoplastic cable factor	Thermosetting cable factor
SOLID	1.5	8.0	8.6
	2.5	11.9	11.9
STRANDED	1.5	8.6	9.6
	2.5	12.6	13.9
	4	16.6	18.1
	6	21.2	22.9
	10	35.3	36.3
	16	47.8	50.3
	25	73.9	75.4

Table 4.26 Cable factors for trunking

Dimensions of trunking mm × mm	Term	Dimensions of trunking mm × mm	Term
50 × 38	767	200 × 100	8 572
50 × 50	1 037	200 × 150	13 001
75 × 25	738	200 × 200	17 429
75 × 38	1 146	225 × 38	3 474
75 × 50	1 555	225 × 50	4 671
75 × 75	2 371	225 × 75	7 167
100 × 25	993	225 × 100	9 662
100 × 38	1 542	225 × 150	14 652
100 × 50	2 091	225 × 200	19 643
100 × 75	3 189	225 × 225	22 138
100 × 100	4 252	300 × 38	4 648
150 × 38	2 999	300 × 50	6 251
150 × 50	3 091	300 × 75	9 590
150 × 75	4 743	300 × 100	12 929
150 × 100	6 394	300 × 150	19 607
150 × 150	9 697	300 × 200	26 285
200 × 38	3 082	300 × 225	29 624
200 × 50	4 145	300 × 300	39 428
200 × 75	6 359		

Table 4.27 Trunking factors

Example 8

The following XLPE insulated, stranded copper conductor cables are to be installed in trunking: 25 × 1.5 mm², 20 × 2.5 mm², 6 × 4.0 mm², 2 × 10 mm² and 2 × 16 mm².

This gives you factor values of 240 (25 × 9.6), 278 (20 × 13.9), 108.6 (6 × 18.1), 73.8 (2 × 36.9) and 100.6 (2 × 50.3), which gives you a total 'term' value of 801.

If you now look at the trunking factors table, you can see that the best option is a **50 mm × 50 mm trunking**, which has a term capacity of 1037.

Progress check

1. What is the purpose of applying rating factors to cables during the design stage?
2. What is meant by the term voltage drop and how does this occur in electrical installations? What is the maximum voltage drop allowed for a single-phase lighting circuit?
3. What is meant by the term diversity and why is this not allowed for heating circuits?

K4. Understand the principles for selecting wiring systems, enclosures and equipment

The construction, application, advantages and disadvantages of the most common cable types are covered in Unit ELTK 04 in Book A of this series.

Getting ready for assessment

All installations need to be protected from any overload or fault current. Learning how to do this is a vital part of being able to carry out installation safely. In order to know the correct cable types and sizes to meet installation requirements, you will also need to revise your knowledge from Units ELTK 04 and 05. This unit will help you to apply that knowledge with the correct protective measures to use for the different supply systems you will encounter in your professional career.

For this unit you will need to be familiar with:

- characteristics and applications of consumer supply systems
- principles of internal and external earthing arrangements for electrical installations for buildings, structures and the environment
- principles for selecting cables and circuit protection devices
- principles and procedures for selecting wiring systems, equipment and enclosures

For each learning outcome, there are several skills you will need to acquire, so you must make sure you are familiar with the assessment criteria for each outcome. For example, for Learning Outcome 3 you will need to be able to explain how external influences can affect the choice of wiring systems and enclosures and state the current ratings for different circuit protection devices. You will also need to specify the procedure for selecting appropriate overcurrent protection devices and state what is meant by diversity factors, explaining how a circuit's maximum demand is established after diversity factors are applied. You will need to be able to specify the procedure for selecting a suitably sized cable following a range of factors and determine the size of conduit or trunking needed for the size and number of cables to be installed.

It is important to read each question carefully and take your time. Try to complete both progress checks and multiple choice questions, without assistance, to see how much you have understood. Refer to the relevant pages in the book for subsequent checks. Always use correct terminology as used in BS 7671. There are some simple tips to follow when writing answers to exam questions:

- **Explain briefly** – usually a sentence or two to cover the topic. The word to note is 'briefly' meaning do not ramble on. Keep to the point.
- **Identify** – refer to reference material, showing which the correct answers are.
- **List** – a simple bullet list is all that is required. An example could include listing the installation tests required in the correct order.
- **Describe** – a reasonably detailed explanation to cover the subject in the question.

This unit has a large number of practical skills in it, and you will need to be sure that you have carried out sufficient practice and that you feel you are capable of passing any practical assessment. It is best to have a plan of action and a method statement to help you work. It is also wise to check your work at regular intervals. This will help to ensure that you are working correctly and help you to avoid any problems developing as you work. Remember, don't rush the job as speed will come with practice and it is important that you get the quality of workmanship right.

Good luck!

CHECK YOUR KNOWLEDGE

1. The supply earthing system which has a combined neutral and earth in part of the system is designated as what?
   a) TT
   b) IT
   c) TN-S
   d) TN-C-S

2. Why are three-phase supplies usually provided?
   a) Client requests it
   b) A large capacity is required
   c) Site engineer thinks it is a good idea
   d) Wiring is easier on three-phase systems

3. Which of the following would **not** be regarded as an extraneous-conductive part?
   a) Gas service pipe
   b) Water service pipe
   c) Metallic trunking
   d) Metal radiator

4. What is the formula used to calculate earth-loop impedance?
   a) $Z_S = Z_e + (R_1 + R_2)$
   b) $Z_e = Z_S + (R_1 + R_2)$
   c) $Z_S = Z_e - (R_1 + R_2)$
   d) $Z_S = Z_e + (R_1 - R_2)$

5. What is an RCD used to provide?
   a) Overload protection
   b) Short circuit protection
   c) Thermal protection
   d) Additional protection

6. What gives the rated current of a protective device?
   a) I_b
   b) I_z
   c) I_n
   d) I_2

7. The maximum permitted time a fault current can be allowed, before permanent damage is done to cable insulation is given in the formula

   $$t = \frac{k^2 S^2}{I^2}$$

   What does **S** represent in this formula?
   a) Time in seconds
   b) Fault current
   c) Conductor CSA
   d) Factor for insulation, resistivity and heat capacity

8. The coordination of protective device rating, design current of a circuit and the current-carrying capacity of conductors is given by which formula?
   a) $I_n \leq I_b \leq I_z$
   b) $I_b \leq I_n \leq I_z$
   c) $I_z \leq I_b \leq I_n$
   d) $I_n \leq I_z \leq I_b$

9. Diversity would **not** be applied to which of the following circuits?
   a) Cooker
   b) Immersion heater
   c) Ring
   d) Lighting

10. What is the calculated full load current of an 8.5 kW single-phase shower unit?
    a) 35.4 A
    b) 36.9 A
    c) 40 A
    d) 45 A

UNIT ELTK 06

Understanding the principles, practices and legislation for the inspection, testing, commissioning and certification of electrotechnical systems and equipment

Electrical installations should be designed into separate circuits, e.g. power, lighting, cooker circuits, to avoid danger in the event of faults and to facilitate the safe operation of the inspection and testing process. It is a requirement of the Electricity at Work Regulations 1989 that this information will be available as an on-site record of the design.

This unit examines the requirements for inspecting and testing an installation, making reference to BS 7671, the IET On Site Guide (OSG) and IET Guidance Note 3 (GN3) and it covers the following learning outcomes:

- Understand the principles, regulatory requirements and procedures for completing safe isolation of an electrical circuit
- Understand the principles and regulatory requirements for inspecting, testing and commissioning electrical systems, equipment and components
- Understand the regulatory requirements and procedures for completing the inspection of electrical installations
- Understand the regulatory requirements and procedures for the safe testing and commissioning of electrical installations
- Understand the procedures and requirements for the completion of electrical installation certificates and related documentation.

K1. Understand the principles, regulatory requirements and procedures for completing safe isolation of an electrical circuit

The methods used for safe isolation have been covered in Unit ELTK 01 in Book A, which covers the correct procedure for safe isolation and the implications of safe isolation to other personnel, customers, the public and building systems. The implications of not carrying out safe isolation are also fully explored.

Requirements of the Electricity at Work Regulations (EaWR) 1989

The EaWR ensure that duties rest with employers, the self-employed and employees; where the provisions relate to matters which are within their control, each becomes a '**duty holder**'. Any duty holder is then required to assess work activities which utilise electricity, or which may be affected by it. Duty holders are required to have regard to all foreseeable of electrical systems for specific tasks; not merely the prevention of electric shock.

Regulation 16 of the EaWR states that no person shall engage in work that requires technical knowledge or experience to prevent danger or injury, unless he or she has that knowledge or experience, or is under appropriate supervision.

Part 6 of BS 7671 states that every electrical installation shall, during erection and on completion before being put into service, be inspected and tested to verify, so far as is reasonably practicable, that the requirements of the Regulations have been met. Equally, precautions shall be taken to avoid danger to any person or livestock and to avoid damage to both property and installed equipment during the inspection and test process.

Health and safety requirements

Bearing in mind the above points it is the responsibility of the person carrying out the test to ensure the safety of themselves and others. So for example, where testing does not require the installation or part of it to be made live, then it should be isolated safely and securely (Unit ELTK 01 in Book A has more information on health and safety issues during safe isolation procedures).

We can also aid safety when using any test instrument by ensuring that the person carrying out the tests:

> **Key term**
>
> **Duty holders** – people who have duties under a particular piece of legislation, in this case EaWR

- has a thorough understanding of the equipment being used and its rating
- ensures that all safety procedures are being followed, e.g. erection of warning notices and barriers where appropriate
- ensures that all instruments being used conform to the appropriate British Standard safety specifications, namely BS EN 61010 (or for older instruments, BS 5458) and are in good condition and have been recently calibrated
- checks that test leads including probes and clips are in good condition, are clean and have no cracked or broken insulation. Where appropriate the requirements of GS38 should be observed.

Particular care should be taken when using instruments capable of generating a test voltage in excess of 50 volts, e.g. insulation resistance testers. Touching the live terminals of such an instrument may result in a shock. Although this may not be harmful in itself, it could cause a loss of concentration that could be dangerous, especially if working at height.

Care must also be taken when working with instruments that use the supply voltage for the purpose of the test, such as earth loop impedance testing or when testing a residual current device (RCD). Either of these tests can impose a voltage on associated earthed metalwork, and precautions must be taken to avoid the risk of electric shock.

Competence and responsibility

A further consideration is the competence of the inspector. Any person called upon to carry out an inspection and test of an installation must be skilled and experienced and have sufficient knowledge of the type of installation to be inspected and tested so that they can ensure there is no risk of injury or damage as outlined in our opening paragraph.

It is therefore the responsibility of the person carrying out the inspection and testing to:

- ensure no danger occurs to persons, property or livestock
- compare the installation design against the results of the inspection and testing
- take a view on the installation's condition and advise remedial work if necessary
- recommend immediate isolation of any defective part and inform the client.

K2. Understand the principles and regulatory requirements for inspecting, testing and commissioning electrical systems, equipment and components

Purpose and requirements of initial verification and periodic inspection

In addition to any work necessary as part of the fault finding process, the process of inspection and testing will occur under the following situations.

Information required for initial verification

Initial verification is intended to confirm that the installation complies with the designer's requirements and has been constructed, inspected and tested in accordance with BS 7671.

Inspection is therefore a very important part of this procedure and should be carried out prior to electrical tests being applied and normally with that part of the installation under inspection disconnected from the supply.

The results of all such inspections and tests must be recorded and compared with the relevant design criteria and the responsibility for doing so lies with the person responsible for inspecting and testing the installation.

The relevant criteria will for the most part be the requirements of BS 7671, although there may be some instances where the designer has specified requirements that are particular to the installation concerned. In these cases the person carrying out the inspection and test should be provided with the necessary data for comparison purposes, but in the absence of such data should apply the requirements set out in BS 7671.

Periodic inspection and testing

BS 7671 Regulation 610.1 requires that every electrical installation during its construction and upon its completion, shall, before being put into service, be inspected and tested to verify, so far as is reasonably practicable, that the requirements of the Regulations have been met.

> **Did you know?**
>
> The final act of the commissioning process is to ensure the safe and correct operation of all circuits and equipment which have been installed, and that the customer's requirements have been met. All of these points will be discussed in detail later.

The Electricity at Work Regulations 1989 state that 'as may be necessary to prevent danger, all systems shall be maintained so as to prevent, as far as is reasonably practicable, such danger'.

BS 7671 Regulation 621.1 then requires every electrical installation to undergo regular inspection and testing to make sure that the installation remains in a satisfactory and safe condition.

Equally, The Landlords and Tenant Act 1985 requires landlords to 'keep in proper working order the installations in dwelling houses' and this affects rented domestic and residential accommodation such as student housing.

Initial inspection

BS 7671 Regulations 610.1, 611.1 and 611.2 require that inspection must precede testing and shall normally be done with that part of the installation disconnected from the supply. Additionally it must be carried out progressively as the installation is installed, and must be done before it is energised.

An initial inspection should be carried out to verify that:

- all equipment and material is of the correct type and complies with applicable British Standards or acceptable equivalents
- all parts of the fixed installation are correctly selected and erected
- no part of the fixed installation is visibly damaged or otherwise defective
- the equipment and material used are suitable for the installation relative to the environmental conditions.

The most important considerations prior to carrying out any inspection and test procedure are that all the required information is available, the person carrying out the procedure is competent to do so and that all safety requirements have been met.

Forward planning is also a major consideration and it is essential that suitable inspection checklists have been prepared and that appropriate certification document is ultimately available for completion.

It is also important to realise that a large proportion of any new installation will be hidden from view once the building fabric has been completed and therefore it is good practice to carry out a certain amount of visual inspection throughout the installation process, e.g. conduit, cable tray or trunking is often installed either above the ceiling or below the floor and once the ceiling or floor tiles have been fitted it is difficult and often expensive to gain access for inspection purposes.

The same principle applies to testing and it may be advisable to carry out tests such as earth continuity during construction rather than after the building has been completed. It must be remembered, however, that when visual inspection and/or tests are carried out during the construction phase, the results must be recorded on the appropriate checklists or test certificates.

> **Remember**
> One of the tools used as part of an inspection is yourself! Remember to use your senses. Can you see something wrong, smell something burning or hear sparking from an enclosure?

Requirements of relevant documents for inspection, testing and commissioning

Statutory and non-statutory requirements

For domestic electrical installations, compliance with BS 7671, a non-statutory document, is the only requirement. For commercial or industrial installations the requirements of the Electricity Supply Regulations 1988 and the Electricity at Work Regulations 1989, both of which are statutory instruments, should also be taken into account.

Compliance with BS 7671 will in most cases satisfy the requirements of statutory Regulations such as the EaWR but this cannot be guaranteed. It is essential to establish which statutory and other Regulations apply, and to carry out the design, the construction and the inspection and testing accordingly.

> **Did you know?**
> For certain installations where there are increased risks or occupation by members of the public, such as cinemas, public houses, restaurants and hotels etc. the local licensing authority may impose additional requirements, especially in the area of regular inspection and testing.

Guidance Note 3 requirements for inspection and testing

Guidance Note 3 3.8.1 General Procedure

In old installations where information such as drawings, distribution board schedules, charts etc. are not available, some exploratory work may be necessary to enable inspection and testing to be carried out safely and without damage to existing equipment. A survey should be carried out to identify all items of switchgear and control gear and their associated circuits.

During the survey a note should be made of any known changes in either the structure of the building, the environmental conditions or of any alterations or additions to the installation which may affect the suitability of the wiring or the method of installation.

Guidance Note 3 3.8.2

The requirements of BS 7671 for periodic inspection and testing are for a thorough **Inspection** of the installation supplemented by **Testing** where necessary. For safety reasons a visual inspection must be carried out before opening any enclosures, removing covers or carrying out any tests.

To comply with the Electricity at Work Regulations 1989 the inspection should be carried out with the supply de-energised wherever possible. A thorough visual inspection should be carried out of all electrical equipment that is not concealed and, where damage or deterioration has occurred, this must be recorded on the inspection schedule. The inspection should include a check on the condition of all equipment and materials used in the installation with regard to the following:

- safety
- damage
- external influences
- wear and tear
- overload
- suitability
- corrosion
- age
- correct operation.

Information required to correctly conduct initial verification

Before carrying out the inspection and test of an installation it is essential that the person carrying out the work be provided with the following information:

- the maximum demand of the installation expressed in amperes per phase together with details of the number and type of live conductors both for the source of energy and for each circuit to be used within the installation (e.g. single-phase, two-wire a.c. or three-phase, four-wire a.c. etc.)
- the general characteristics of the supply such as:
  - the nominal voltage (U_o)
  - the nature of the current (I) and its frequency (Hz)
  - the prospective short circuit current at the origin of the installation (kA)
  - the earth-fault loop impedance (Z_e) of that part of the system external to the installation
  - the type and rating of the overcurrent device acting at the origin of the installation.

(If this information is not known it must be established either by calculation, measurement, inquiry or inspection.)

- the type of earthing arrangement used for the installation, e.g. TN-S, TN-C-S, TT etc.
- the type and composition of each circuit (i.e. details of each sub-circuit, what it is feeding, the number and size of conductors and the type of wiring used)
- the location and description of all devices installed for the purposes of protection, isolation and switching (e.g. fuses/circuit-breakers etc.)
- details of the method selected to prevent danger from shock in the event of an earth-fault. (This will invariably be protection by earthed equipotential bonding and automatic disconnection of the supply.)
- the presence of any sensitive electronic device which may be susceptible to damage by the application of 500 V d.c. when carrying out insulation resistance tests.

It is a requirement of BS 7671 that this information shall be available as an on-site record of the design. This information may be found from a variety of sources such as the project specification, contract drawings, as fitted drawings or distribution board schedules. If such documents are not available, then the person ordering the work should be approached.

Did you know?

Such details would normally be included in the project Health and Safety File required by the Construction (Design and Management) Regulations.

Working life

You are tasked with mentoring a new colleague who has joined the company to help with the inspection and testing.

1. What information should you check to ascertain they are competent?
2. How would you check to ensure they are safe?
3. What information should you ensure is available to them before commencing the inspection and test process?

Progress check

1. A site supervisor has requested an apprentice to carry out the inspection and test alone on the area he has been working, as he had to leave site and pick up more equipment. Should the apprentice carry out the inspection?
2. Explain the purpose of the initial inspection on an electrical installation.

K3. Understand the regulatory requirements and procedures for completing the inspection of electrical installations

Items to be checked during the inspection process

The following text provides a detailed description of the procedures required to carry out an inspection of an electrical installation. Substantial reference has been made to the IET Wiring Regulations (BS 7671), the IET On-Site Guide and IET Guidance Note 3 and it is recommended that these documents should be referred to whenever clarification is needed.

BS 7671 Regulation 611.3 states that, as a minimum, the inspection shall include the checking where applicable and including as appropriate all particular requirements for special installations or locations (BS 7671 Part 7) of the items below.

> **Remember**
> You are one of the tools used as part of an inspection so learn to use your senses. Can you see something obviously wrong, smell something burning or hear sparking inside an enclosure?

Regulation 611.3 inspection requirements

1. Connection of conductors
2. Identification of conductors
3. Routing of cables in safe zones or protection against mechanical damage
4. Selection of conductors for current carrying capacity and voltage drop
5. Connection of single pole devices in line conductors only
6. Correct connection of accessories and equipment
7. Presence of fire barriers and protection against thermal effects
8. Methods of protection against electric shock:
   - **(i)** both basic and fault protection
     - SELV
     - PELV
     - double insulation
     - reinforced insulation

- (ii) basic protection (including measurement of distances)

    protection by insulation of live parts

    protection by a barrier or enclosure

    protection by obstacles

    protection by placing out of reach

- (iii) fault protection

    - (a) automatic disconnection of supply

        Presence of adequately sized:

        - earthing conductor
        - cpcs
        - protective bonding conductors
        - supplementary bonding conductors
        - earthing arrangements for combined protective and functional purposes

        presence of arrangements for alternative sources

        FELV

        choice and setting of protective devices

    - (b) non-conducting location

        absence of protective conductors

    - (c) earth-free local equipotential bonding

        presence of earth-free protective bonding conductors

    - (d) electrical separation

- (iv) additional protection

9. Prevention of mutual detrimental influence
10. Presence of appropriate devices for isolation and switching
11. Presence of undervoltage protective devices
12. Labelling of protective devices
13. Selection of equipment/protective devices appropriate to external influences
14. Adequacy of access to switchgear and equipment
15. Presence of danger notices and other signs
16. Presence of diagrams, instructions and similar information
17. Erection methods

The following section seeks to give additional guidance on each of the previous points.

1. Connection of conductors

Every connection between conductors or between conductors and equipment must be electrically continuous and mechanically sound. We must also make sure that all connections are adequately enclosed but accessible where required by the Regulations.

2. Identification of conductors

A check should be made that each conductor is identified in accordance with the requirements of BS 7671 Table 51. Although numbered sleeves may be used in special circumstances, the most common form of identification is by means of coloured insulation or sleeving (but not green). Remember that only protective conductors can be identified by a combination of the colours green and yellow.

3. Routing of cables in safe zones or protection against mechanical damage

Cables should be routed out of harm's way and protected against mechanical damage where necessary. Permitted cable routes are clearly defined (note the RCD situation) or alternatively cables should be installed in earthed metal conduit or trunking.

4. Selection of conductors for current carrying capacity and voltage drop

Where practicable the size of cable used should be checked for current-carrying capacity and voltage drop based upon information provided by the installation designer.

5. Connection of single pole devices in line conductors only

A check must be made that all single pole devices are connected in the phase conductor only.

6. Correct connection of accessories and equipment

Accessories and equipment should be checked to ensure they have been connected correctly including correct polarity.

7. Presence of fire barriers and protection against thermal effects

A check must be made (preferably during construction) that fire barriers, suitable seals and/or other means of protection against thermal effects have been provided as necessary to meet the requirements of the Regulations.

8. Methods of protection against electric shock

A check must be made that the requirements of BS 7671 Chapter 41 have been met for the protection method being used.

9. Prevention of mutual detrimental influence

Account must be taken of the proximity of other electrical services in a different voltage band and of non-electrical services and influences. For example, fire alarm and emergency lighting circuits must be separated from other cables and from each other, and Band 1 and Band 2 circuits must not be present in the same enclosure or wiring system unless they are either segregated or wired with cables suitable for the highest voltage present.

Mixed categories of circuits may be contained in multicore cables, subject to certain requirements.

Band 1 circuits are circuits that are nominally extra-low voltage, i.e. not exceeding 50 volts a.c. or 120 volts d.c., e.g. telecommunications or data and signalling. Band 2 circuits are circuits that are nominally low voltage, for example exceeding extra-low voltage but not exceeding 1000 volts a.c. between conductors or 600 volts a.c. between conductors and earth.

10. Presence of appropriate devices for isolation and switching

BS 7671 requires that effective means suitably positioned and ready to operate shall be provided so that all voltage may be cut off from every installation, every circuit within the installation and from all equipment, as may be necessary to prevent or remove danger.

This means that switches and/or isolating devices of the correct rating must be installed as appropriate to meet the above requirements. It may be advisable where practicable to carry out an isolation exercise to check that effective isolation can be achieved. This should include switching off, locking off and testing to verify that the circuit is dead and no other source of supply is present.

11. Presence of undervoltage protective devices

Sometimes referred to in starters as no-volt protection, suitable precautions must be taken where a loss or lowering of voltage or a subsequent restoration of voltage could cause danger. The most common situation would be where a motor-driven machine stops due to a loss of voltage and unexpectedly restarts when the voltage is restored unless precautions, such as the installation of a motor starter containing a contactor, are employed. Regulations require that where unexpected restarting of a motor may cause danger, the provision of a motor starter designed to prevent automatic restarting must be provided.

12. Labelling of protective devices

A check should be carried out to ensure that labels and warning notices as required by BS 7671 have been fitted, e.g. labelling of circuits, MCBs, RCDs, fuses and isolating devices, periodic inspection notices advising of the recommended date of the next inspection, and warning notices referring to earthing and bonding connections.

13. Selection of equipment protective devices appropriate to external influences

All equipment must be selected as suitable for the environment in which it is likely to operate. Items to be considered are ambient temperature, presence of external heat sources, presence of water, likelihood of corrosion, ingress of foreign bodies, impact, vibration, flora, fauna, radiation, building use and structure.

14. Adequacy of access to switchgear and equipment

BS 7671 requires that every piece of equipment that requires operation or attention must be installed so that adequate and safe means of access and working space are provided.

15. Presence of danger notices and other signs

Suitable notices are required to be suitably located and give warnings relative to voltage, isolation, periodic inspection and testing, RCDs and earthing and bonding connections.

16. Presence of diagrams, instructions and other similar information

All distribution boards should be provided with a distribution board schedule that provides information regarding types of circuits, number and size of conductors, type of wiring etc. These should be attached within or adjacent to each distribution board.

17. Erection methods

Correct methods of installation should be checked. In particular fixings of switchgear, cables, conduit etc., must be adequate and suitable for the environment.

Inspection schedule and inspection checklist

All the previous items should be inspected and the results noted on an inspection schedule. Then to ensure that all the requirements of the Regulations have been met, an inspection checklist should be drawn up and used appropriate to the type of installation being inspected.

Appendix 6 of BS 7671 gives examples of model certificates and schedules. Figures 6.01 and 6.02 show examples of an inspection schedule and an inspection checklist.

INSPECTION SCHEDULE

Methods of protection against electric shock

Both basic and fault protection:
- [ ] (i) SELV
- [ ] (ii) PELV
- [ ] (iii) Double insulation
- [ ] (iv) Reinforced insulation

Basic protection:
- [ ] (i) Insulation of live parts
- [ ] (ii) Barriers or enclosures
- [ ] (iii) Obstacles
- [ ] (iv) Placing out of reach

Fault protection:

(i) Automatic disconnection of supply:
- [ ] Presence of earthing conductor
- [ ] Presence of circuit protective conductors
- [ ] Presence of protective bonding conductors
- [ ] Presence of supplementary bonding conductors
- [ ] Presence of earthing arrangements for combined protective and functional purposes
- [ ] Presence of adequate arrangements for alternative source(s), where applicable
- [ ] FELV
- [ ] Choice and setting of protective and monitoring devices (for fault and/or overcurrent protection)

(ii) Non-conducting location:
- [ ] Absence of protective conductors

(iii) Earth-free local equipotential bonding:
- [ ] Presence of earth-free local equipotential bonding

(iv) Electrical Separation:
- [ ] Provided for **one item** of current-using equipment
- [ ] Provided for **more than one item** of current-using equipment

Additional protection:
- [ ] Presence of residual current devices(s)
- [ ] Presence of supplementary bonding conductors

Prevention of mutual detrimental influence
- [ ] (a) Proximity of non-electrical services and other influences
- [ ] (b) Segregation of Band I and Band II circuits or use of Band II insulation
- [ ] (c) Segregation of safety circuits

Identification
- [ ] (a) Presence of diagrams, instructions, circuit charts and similar information
- [ ] (b) Presence of danger notices and other warning notices
- [ ] (c) Labelling of protective devices, switches and terminals
- [ ] (d) Identification of conductors

Cables and conductors
- [ ] Selection of conductors for current-carrying capacity and voltage drop
- [ ] Erection methods
- [ ] Routing of cables in prescribed zones
- [ ] Cables incorporating earthed armour or sheath, or run within an earthed wiring system, or otherwise adequately protected against nails, screws and the like
- [ ] Additional protection provided by 30 mA RCD for cables in concealed walls (where required in premises not under the supervision of a skilled or instructed person)
- [ ] Connection of conductors
- [ ] Presence of fire barriers, suitable seals and protection against thermal effects

General
- [ ] Presence and correct location of appropriate devices for isolation and switching
- [ ] Adequacy of access to switchgear and other equipment
- [ ] Particular protective measures for special installations and locations
- [ ] Connection of single-pole devices for protection or switching in line conductors only
- [ ] Correct connection of accessories and equipment
- [ ] Presence of undervoltage protective devices
- [ ] Selection of equipment and protective measures appropriate to external influences
- [ ] Selection of appropriate functional switching devices

Inspected by .... Date ....

Notes:
- ✓ to indicate an inspection has been carried out and the result is satisfactory
- ✗ to indicate an inspection has been carried out and the result is not satisfactory (applicable to a periodic inspection only)
- N/A to indicate the inspection is not applicable to a particular item
- LIM to indicate that, exceptionally, a limitation agreed with the person ordering the work prevented the inspection being carried out (applicable for a periodic inspection only).

Figure 6.01 Inspection schedule

INSPECTION CHECKLIST	
General	
1 Complies with requirements 1–3 in Section 2.1 (133.1, 134.1)	
2 Accessible for operation, inspection and maintenance (513.1)	
3 Suitable for local atmosphere and ambient temperature. (Installations in potentially explosive atmospheres are outside the scope of BS 7671)	
4 Circuits to be separate (no borrowed neutrals) (314.4)	
5 Circuits to be identified (neutral and protective conductors in same sequence as line conductors) (514.1.2, 514.8.1)	
6 Protective devices adequate for intended purpose (BS 7671 – Ch. 53)	
7 Disconnection times likely to be met by installed protective devices (Ch. 41)	
8 Sufficient numbers of conveniently accessible socket outlets are provided in accordance with the design (553.1.7). Note that in Scotland section 4.6.4 (socket outlets) in the domestic technical handbook published by the Scottish Building Standards Agency (SBSA) gives specific recommendations for the number of socket outlets for various locations within an installation.	
9 All circuits suitably identified (514.1, 514.8, 514.9)	
10 Suitable main switch provided (Ch. 53)	
11 Supplies to any safety services suitably installed, e.g. Fire Alarms to BS 5839 and emergency lighting to BS 5266	
12 Environmental IP requirements accounted for (B5 EN 60529)	
13 Means of isolation suitably labelled (514.1, 537.2.2.6)	
14 Provision for disconnecting the neutral (537.2.1.7)	
15 Main switches to single-phase installations, intended for use by an ordinary person, e.g. domestic, shop, office premises, to be double-pole (537.1.4)	
16 RCDs provided where required (411.1, 411.3, 411.4, 411.5, 522.6.7, 522.6.8, 532.1, 701.411.3.3, 701.415.2, 702.55.4, 705.411.1, 705.422.7, 708.553.1.13, 709.531.2, 711.410.3.4, 711.411.3.3, 740.410.3, 753.415.1)	
17 Discrimination between RCDs considered (314, 531.2.9)	
18 Main earthing terminal provided (542.4.1) readily accessible and identified (514.13.1)	
19 Provision for disconnecting earthing conductor (542.4.2)	
20 Correct cable glands and gland plates used (BS 6121)	
21 Cables used comply with British or Harmonised Standards (Appendix 4 of the Regulations, 521.1)	

Figure 6.02 Inspection checklist

General (continued)	
22 Conductors correctly identified (Section 514)	
23 Earth tail pots installed where required on mineral insulated cables (134.1.4)	
24 Non-conductive finishes on enclosures removed to ensure good electrical connection and if necessary made good after connecting (526.1)	
25 Adequately rated distribution boards (BS EN 60439 may require de-rating)	
26 Correct fuses or circuit-breakers installed (Sections 531 and 533)	
27 All connections secure (134.1.1)	
28 Consideration paid to electromagnetic effects and electromechanical stresses (Ch. 52)	
29 Overcurrent protection provided where applicable (Ch. 43)	
30 Suitable segregation of circuits (Section 528)	
31 Retest notice provided (514.12.1)	
32 Sealing of the wiring system including fire barriers (527.2).	

Switchgear	
1 Suitable for the purpose intended (Ch. 53)	
2 Meets requirements of BS EN 61008, BS EN 61009, BS EN 60947-2, BS EN 60898 or BS EN 60439 where applicable, or equivalent standards (511)	
3 Securely fixed (134.1.1) and suitably labelled (514.1)	
4 Non-conductive finishes on switchgear removed at protective conductor connections and if necessary made good after connecting (526.1)	
5 Suitable cable glands and gland plates used (526.1)	
6 Correctly earthed (Ch. 54)	
7 Conditions likely to be encountered taken account of, i.e. suitable for the foreseen environment (522)	
8 Where relevant correct IP rating applied (BS EN 60529)	
9 Suitable as means of isolation, where applicable (537.2.2)	
10 Complies with the requirements for locations containing a bath or shower (Section 701)	
11 Need for isolation, mechanical maintenance, emergency and functional switching met (Section 537)	

Figure 6.02 Inspection checklist (cont.)

Switchgear (continued)	
12 Firefighter's switch provided where required (537.6.1)	
13 Switchgear suitably coloured where necessary (537.6.4)	
14 All connections secure (Section 526)	
15 Cables correctly terminated and identified (Sections 514 and 526)	
16 No sharp edges on cable entries, screw heads, etc. which could cause damage to cables (522.8)	
17 All covers and equipment in place and secure (Section 522.6.3)	
18 Adequate access and working space (132.12 and Section 513).	

General (applicable to each type of accessory)	
1 Complies with BS 5733, BS 6220 or other appropriate standard (Section 511)	
2 Box or other enclosure securely fixed (134. 1.1)	
3 Metal box or other enclosure earthed (Ch. 54)	
4 Edge of flush boxes not projecting above wall surface (134.1.1)	
5 No sharp edges on cable entries, screw heads, etc. which could cause damage to cables (522.8)	
6 Non-sheathed cables, and cores from which sheath removed, not exposed outside the enclosure (526.9)	
7 Conductors correctly identified (514.6)	
8 Bare protective conductors having a cross-sectional area of 6mm^2 or less to be sleeved green and yellow (514.4.2, 543.3.2)	
9 Terminals tight and containing all strands of the conductors (Section 526)	
10 Cord grip correctly used or clips fitted to cables to prevent strain on the terminals (522.8.5, 526.6)	
11 Adequate current rating (133.2.2)	
12 Suitable for the conditions likely to be encountered (Section 522).	
Lighting controls	
1 Light switches comply with BS 3676 (Section 511)	
2 Suitably located (Section 512.2)	
3 Single-pole switches connected in line conductors only (132.14.1)	

Figure 6.02 Inspection checklist (cont.)

Lighting controls *(continued)*	
4 Correct colour coding or marking of conductors (514.6)	
5 Earthing of exposed metalwork, e.g. metal switch plate (Ch. 54)	
6 Complies with the requirements for locations containing a bath or shower (Section 701)	
7 Adequate current rating (133.2.2)	
8 Suitable for inductive circuits or de-rated where necessary (512.1.2)	
9 Switch labelled to indicate purpose, where this is not obvious (514.1.1)	
10 Appropriate controls suitable for the luminaires (559.6.1.9).	
Lighting points	
1 Correctly terminated in a suitable accessory or fitting (559.6.1.1)	
2 Ceiling rose complies with BS 67 (559.6.1.1)	
3 Not more than one flex unless designed for multiple pendants (559.6.1.3)	
4 Flex support devices used (559.6.1.5)	
5 Switch-lines identified (514.3.2)	
6 Holes in ceiling above rose made good to prevent spread of fire (527.2.1)	
7 Not connected to a supply exceeding 250 V (559.6.1.2)	
8 Suitable for the mass suspended (559.6.1.5)	
9 Lamp-holders to BS EN 60598 (559.6.1.1)	
10 Luminaire couplers comply with BS 6972 or BS 7001 (559.6.1.1)	
11 Track systems comply with BS EN 60570 (559.4.4).	
Socket outlets	
1 Complies with BS 196, BS 546, BS 1363, BS EN 60309 2 (553.1.3) and shuttered for household and similar installations (553.1.4)	
2 Mounting height above the floor or working surface suitable (553.1.6)	
3 Correct polarity (612.6)	
4 If installed in a location containing a bath or shower, installed beyond 3 m horizontally of the bath or shower unless shaver supply unit or SEW (701.512.3)	

Figure 6.02 Inspection checklist (cont.)

Socket outlets *(continued)*	
5 Protected where mounted in a floor (Section 522)	
6 Not used to supply a water heater having un-insulated elements (554.3.3)	
7 Circuit protective conductor connected directly to the earthing terminal of the socket outlet, on a sheathed wiring installation (543.2.7)	
8 Earthing tall from the earthed metal box, on a conduit installation to the earthing terminal of the socket outlet (543.2.7).	
Joint box	
1 Joints accessible for inspection (526.3)	
2 Joints protected against mechanical damage (526.7)	
3 All conductors correctly connected (526.1).	
Fused connection unit	
1 Correct rating and fuse (533.1)	
2 Complies with BS 1363-4 (559.6.1.1 vii).	
Cooker control unit	
1 Sited to one side and low enough for accessibility and to prevent flexes trailing across radiant plates (522.2.1)	
2 Cable to cooker fixed to prevent strain on connections (522.8.5).	

Conduits	
General	
1 Securely fixed, box lids In place and adequately protected against mechanical damage (522.8)	
2 Inspection fittings accessible (522.8.6)	
3 Number of cables for easy draw not exceeded (522.8.1 and see On Site Guide Appendix 5)	
4 Solid elbows and tees used only as permitted (522.8.1 and 522.8.3)	
5 Ends of conduit reamed and bushed (522.8)	
6 Adequate boxes suitably spaced (522.8 and see On Site Guide Appendix 5)	
7 Unused entries blanked off where necessary (412.2.2)	
8 Conduit system components comply with a relevant British Standard (Section 511)	

Figure 6.02 Inspection checklist (cont.)

General *(continued)*	
9 Provided with drainage holes and gaskets as necessary (522.3)	
10 Radius of bends such that cables are not damaged (522.8.3)	
11 Joints, scratches, etc. In metal conduit protected by painting (134.1.1, 522.5).	
Rigid metal conduit	
1 Complies with BS EN 50086 or BS EN 61386 (Section 511)	
2 Connected to the main earth terminal (411.4.2)	
3 Line and neutral cables contained in the same conduit (.521.5.2)	
4 Conduit suitable for damp and corrosive situations (522.3 and 522.5)	
5 Maximum span between buildings without intermediate support (522.8 and see Guidance Note 1 and On Site Guide Appx 5).	
Rigid non-metallic conduit	
1 Complies with BS 4607, BS EN 60423, BS EN 50086-2-1 or the BS EN 61386 series (521.6)	
2 Ambient and working temperatures within permitted limits (522.1 and 522.2)	
3 Provision for expansion and contraction (522.8)	
4 Boxes and fixings suitable for mass of luminaire suspended at expected temperature (522.8, 559.6.1.5).	
Flexible metal conduit	
1 Complies with BS EN 60423 and BS EN 50086-1 or the BS EN 61386 series (521.6)	
2 Separate protective conductor provided (543.2.1)	
3 Adequately supported and terminated (522.8).	

Trunking	
General	
1 Complies with BS 4678 or BS EN 50085-1 (521.6)	
2 Securely fixed and adequately protected against mechanical damage (522.8)	
3 Selected, erected and routed so that no damage is caused by ingress of water (522.3)	
4 Proximity to non-electrical services (528.2)	

Figure 6.02 Inspection checklist (cont.)

General *(continued)*	
5 Internal sealing provided where necessary (527.2.4)	
6 Holes surrounding trunking made good (527.2.1)	
7 Band 1 circuits partitioned from Band 2 circuits or insulated for the highest voltage present (528.1)	
8 Circuits partitioned from Band 1 circuits or wired in mineral-insulated metal sheathed cables (528.1)	
9 Common outlets for Band 1 and Band 2 provided with screens, barriers or partitions	
10 Cables supported for vertical runs (522.8).	

Metal trunking	
1 Line and neutral cables contained in the same metal trunking (521.5.2)	
2 Protected against damp or corrosion (522.3 and 522.5)	
3 Earthed (411.4.2)	
4 Joints mechanically sound and of adequate continuity (543.2.4).	

Busbar trunking and powertrack systems	
1 Busbar trunking to comply with BS EN 60439-2 or other appropriate standard and powertrack system to comply with BS EN 61534 series or other appropriate standard (521.4)	
2 Securely fixed and adequately protected against mechanical damage (522.8)	
3 Joints mechanically sound and of adequate continuity (543.2.4).	

Insulated cables	
Non-flexible cables	
1 Correct type (521)	
2 Correct current rating (523)	
3 Protected against mechanical damage and abrasion (522.8)	
4 Cables suitable for high or low ambient temperature as necessary (522.1)	
5 Non-sheathed cables protected by enclosure in conduit, duct or trunking (521.10)	
6 Sheathed cables: ○ routed in allowed zones or mechanical protection provided (522.6.6) ○ in the case of domestic or similar installations not under the supervision of skilled or instructed persons, additional protection is provided by RCD having $I_{\Delta n}$ not exceeding 30 mA (522.6.7)	

Figure 6.02 Inspection checklist (cont.)

Non-flexible cables (contined)	
7 Cables in partitions containing metallic structural parts in domestic or similar installations not under the supervision of skilled or instructed persons should be: ○ provided with adequate mechanical protection to suit both the installation of the cable and its normal use ○ provided with additional protection by RCD having $I_{\Delta n}$ not exceeding 30 mA (522.6.8)	
8 Where exposed to direct sunlight, of a suitable type (522.11)	
9 Not run in lift shaft unless part of the lift installation and of the permitted type (BS 5655 and BS EN 81-1) (528.3.5)	
10 Buried cable correctly selected and installed for use (522.6.4)	
11 Correctly selected and installed for use overhead (521)	
12 Internal radii of bends not sufficiently tight as to cause damage to cables or to place undue stress on terminations to which they are connected (relevant BS, BS EN and 522.8.3)	
13 Correctly supported (522.8.4 and 522.8.5)	
14 Not exposed to water, etc. unless suitable for such exposure (522.3)	
15 Metal sheaths and armour earthed (411.3.1.1)	
16 Identified at terminations (514.3)	
17 Joints and connections electrically and mechanically sound and adequately insulated (526.1 and 526.2)	
18 All wires securely contained in terminals, etc. without strain (522.8.5 and Section 526)	
19 Enclosure of terminals (Section 526)	
20 Glands correctly selected and fitted with shrouds and supplementary earth tags as necessary (526.1)	
21 Joints and connections mechanically sound and accessible for inspection, except as permitted otherwise (526.1 and 526.3).	
Flexible cables and cords (521.9)	
1 Correct type (521)	
2 Correct current rating (Section 523)	
3 Protected where exposed to mechanical damage (522.6 and 522.8)	
4 Suitably sheathed where exposed to contact with water (522.3) and corrosive substances (522.5)	
5 Protected where used for final connections to fixed apparatus, etc. (526.9)	
6 Selected for resistance to damage by heat (522.1)	

Figure 6.02 Inspection checklist (cont.)

Flexible cables and cords (521.9) *(continued)*	
7 Segregation of Band 1 and Band 2 circuits (BS 6701 and Section 528)	
8 Fire alarm and emergency lighting circuits segregated (BS 5839, BS 5266 and Section 528)	
9 Cores correctly identified (514.3.2)	
10 Joints to be made using appropriate means (526.2)	
11 Where used as fixed wiring, relevant requirements met (521.9.3)	
12 Final connections to portable equipment, a convenient length and connected as stated (553.1.7)	
13 Final connections to other current-using equipment properly secured or arranged to prevent strain on connections (Section 526)	
14 Mass supported by cable to not exceed values stated (559.11.6).	
Protective conductors	
1 Cables incorporating protective conductors comply with the relevant BS (Section 511)	
2 Joints in metal conduit, duct or trunking comply with Regulations (543.3)	
3 Flexible or pliable conduit to be supplemented by a protective conductor (543.2.1)	
4 Minimum cross-sectional area of copper conductors (543.1)	
5 Copper conductors, other than strip, of 6 mm^2 or less protected by insulation (543.3.2)	
6 Circuit protective conductor at termination of sheathed cables insulated with sleeving (543.3.2)	
7 Bare circuit protective conductor protected against mechanical damage and corrosion (542.3 and 543.3.1)	
8 Insulation, sleeving and terminations identified by colour combination green and yellow (514.3.1, 514.4.2)	
9 Joints electrically and mechanically sound (526.1)	
10 Separate circuit protective conductors not less than 4 mm^2 if not protected against mechanical damage (543.1.1)	
11 Main and supplementary bonding conductors of correct size (Section 544).	
Enclosures	
General	
1 Suitable degree of protection (IP Code in BS EN 60529) appropriate to external influences (416.2, Section 522 and Part 7).	

Figure 6.02 Inspection checklist (cont.)

Working life

A client is visiting site while your company is completing inspection and testing. He remarks 'I thought you would have finished by now' and states that it surely must be a quick look round to complete the work.

1. What documentation can you show him to demonstrate the extent of work to be done?
2. What else could be referred to, highlighting why the process is so involved and time-consuming?

Progress check

1. During the inspection process, checks are needed on the routing of cables. When would that normally be done and what are the checks to ascertain?
2. Effective means of isolation and switching suitably positioned and ready to operate are required. What inspection checks should be made?
3. On completion of an installation labelling and information should all be present, what would you check during inspection?

K4. Understand the regulatory requirements and procedures for the safe testing and commissioning of electrical installations

This section will look at the requirements for tests to be carried out on an electrical installation, covering each of these in the order you would need to carry them out on a new installation. It will then look at the actions to take in the event of unsatisfactory results being recorded, before covering the test instruments that will be used.

Tests to be carried out on an electrical installation in accordance with the IET Wiring Regulations and IET Guidance Note 3

Initial testing

BS 7671 Regulation 610.1 states that 'Every installation shall, during erection and/or on completion before being put into service, be inspected and tested to verify, so far as is reasonably practicable, that the requirements of the Regulations have been met. Precautions shall be taken to avoid danger to persons, livestock, and to avoid damage to property and installed equipment during inspection and testing.'

BS 7671 Regulation 612 lists the sequence in which tests should be carried out. If any test indicates a failure to comply, that test and any preceding test, the results of which may have been influenced by the fault indicated, must be repeated after the fault has being rectified.

This is because installation testing can be dangerous and the danger level can increase if tests are not carried out in the correct sequence. To clarify, after inspecting the installation we would start the testing process by establishing the continuity of any protective conductors. For safety, protective conductors should be in place before injecting current or carrying out any live tests. Protective conductors should certainly be in place and insulation resistance must be satisfactory before carrying out an earth loop impedance test.

For practical reasons some tests will be carried out using links between known conductors established from earlier tests in the sequence.

The sequence of tests

Initial tests should be carried out in the following sequence, **where applicable**, before the supply is connected or with the supply disconnected as appropriate:

- Continuity of protective conductors including main and supplementary bonding (612.2.1)
- Continuity of ring final circuit conductors (612.2.2)
- Insulation resistance (612.3)
- Protection by SELV, PELV or electrical separation (612.4)
- Protection by barriers/enclosures provided during erection (612.4.5)
- Insulation resistance/impedance of non-conducting floors and walls (612.5)
- Polarity (612.6)
- Earth electrode resistance (612.7)
- Protection by automatic disconnection of supply (ADS) (612.8)
- Earth-fault loop impedance (612.9)
- Additional protection (612.10)
- Prospective fault current (612.11)
- Phase sequence (612.12)
- Functional testing (612.13)
- Verification of voltage drop (612.14).

The test results should be recorded on an installation schedule (see page 83) and compared to the design criteria.

Continuity of protective conductors including main and supplementary bonding

BS 7671 states that every protective conductor, including each bonding conductor, shall be tested to verify that it is electrically sound and correctly connected.

This test is carried out with a low resistance ohmmeter. As well as checking the continuity of the protective conductor, the meter will also measure $R_1 + R_2$ which, when corrected for temperature and added to the value of Z_e will allow the designer to verify the calculated earth-fault loop impedance Z_s.

Test method 1

Before carrying out this test, as shown in Figure 6.03, the leads should be 'nulled out'. If the test instrument does not have this facility, the resistance of the leads should be measured and deducted from the readings.

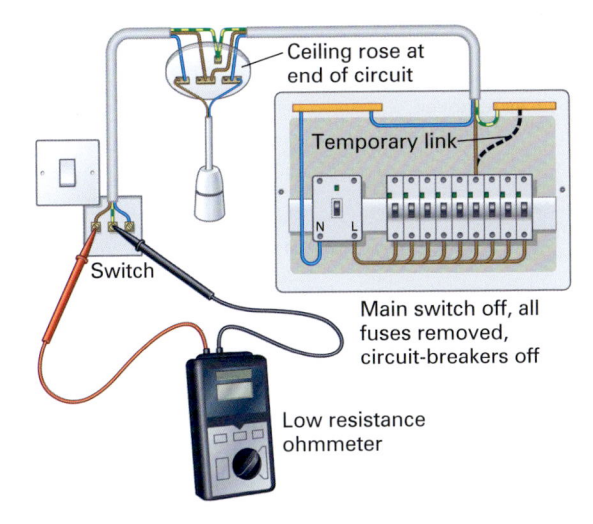

Figure 6.03 Test method 1

The line conductor and the protective conductor are then linked together at the consumer unit or distribution board.

The ohmmeter is used to test between the line and earth terminals at each outlet in the circuit. The measurement at the circuit's extremity should be recorded and this is the value of $R_1 + R_2$ for the circuit under test.

On a lighting circuit the value of R_1 should include the switch wire at the luminaires. This method should be carried out before any supplementary bonds are made.

Test method 2

One lead of the continuity tester is connected to the consumer's main earth terminals and then the other lead is connected to a trailing lead, which we then use to make contact with the protective conductor at light fittings, switches, spur outlets etc.

As we can see from Figure 6.04, the resistance of the test leads and wandering lead will be included in the result; therefore their resistance must be measured and subtracted from the reading obtained if the instrument does not have a nulling facility.

As in this method we are only testing the protective conductor, only R_2 is recorded on the installation schedule.

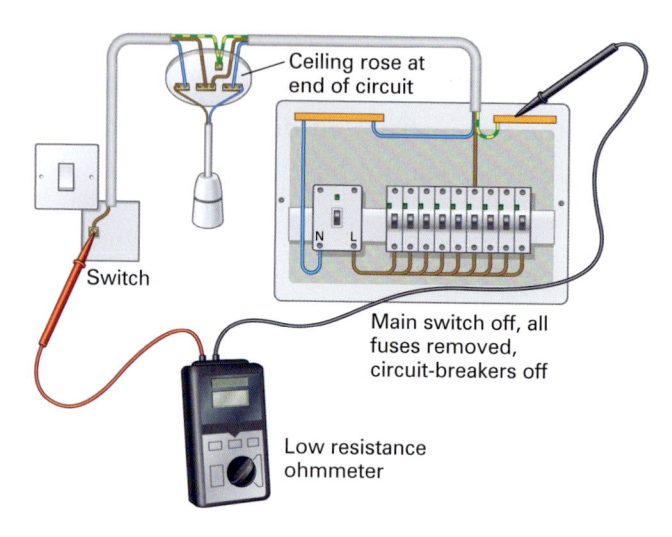

Figure 6.04 Test method 2

Test of the continuity of supplementary bonding conductors

We use test method 2 for this purpose, where the ohmmeter leads are connected between the points being tested, between simultaneously accessible extraneous-conductive parts, e.g. pipework, sinks etc. or between simultaneously accessible extraneous-conductive parts and exposed-conductive parts (the metal parts of the installation).

This test will verify that the conductor is sound. To check this, move the probe to the metalwork to be protected as shown in Figure 6.05. This method is also used to test the main equipotential bonding conductors.

Where ferrous enclosures have been used as the protective conductors, e.g. conduit, trunking, steel-wire armouring etc., the following special precautions should be followed:

- Inspect the enclosure along its length to verify its integrity.
- Perform the standard ohmmeter test using the appropriate test method described above.

Figure 6.05 Test of the continuity of supplementary bonding conductors

If there is any doubt as to the soundness of this conductor a further test should be performed using a phase-earth loop impedance tester after the connection of the supply.

If there is still doubt, a further test may be carried out using a high-current, low-impedance ohmmeter, which has a test voltage not exceeding 50 volts and can provide a current approaching 1.5 times the design current of the circuit, but the current need not exceed 25 A.

Continuity of ring final circuit conductors

A test is required to verify the continuity of each conductor including the circuit protective conductor (cpc) of every ring final circuit. The test results should establish that the ring is complete, has no interconnections and that the ring is not broken.

Such faults are shown in Figure 6.06.

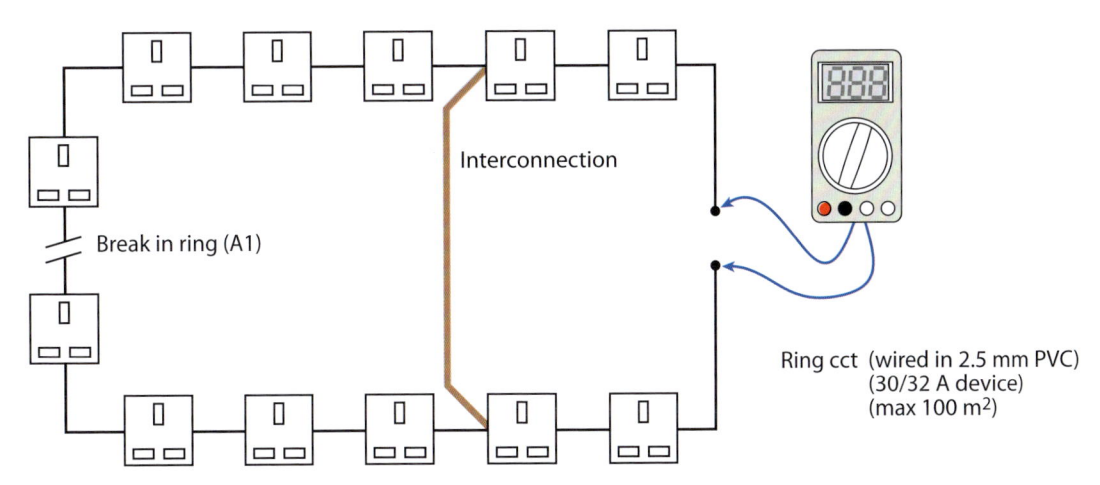

Figure 6.06 Test of continuity of ring final circuits

The inspector may be able to check visually each conductor throughout its entire length. This is an alternative and establishes that no interconnected multiple loops have been made in the ring circuit. In most circumstances however, this will not be practicable and the following test method for checking ring circuit continuity is recommended.

Step 1

The line, neutral and protective conductors are identified and the end to end resistance of each is measured separately (Figure 6.07).

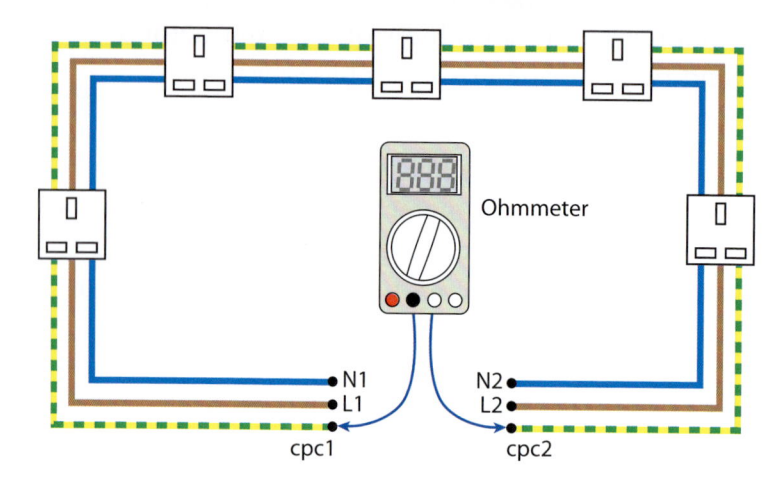

Figure 6.07 Step 1 Measurement of line, neutral and protective conductors

A finite reading confirms that there is no open circuit on the ring conductors under test.

The resistance values obtained should be the same (within 0.05 Ω) if the conductors are of the same size. If the protective conductor has a reduced CSA, then its resistance will be proportionally higher than that of the line or neutral loop. If these relationships are not achieved then either the conductors are incorrectly identified or there is a problem at one of the accessories.

Step 2

The line and neutral conductors are then connected together so that the outgoing line conductor is connected to the returning neutral conductor and vice versa as shown in Figure 6.08.

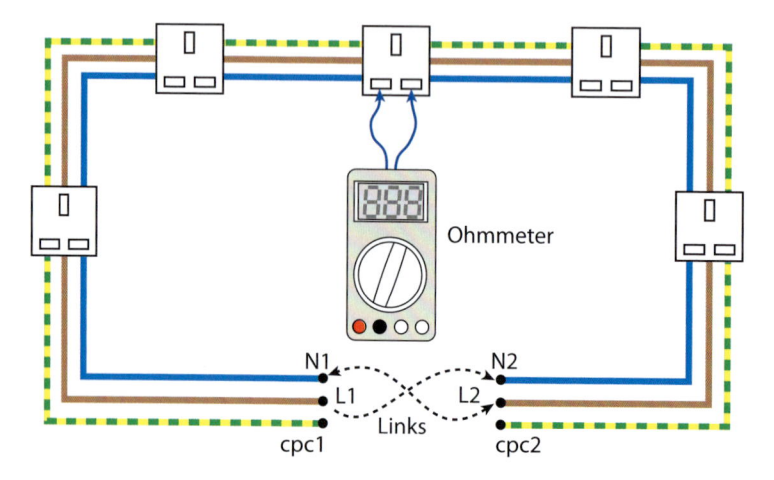

Figure 6.08 Step 2 Line and neutral conductors connected together

The resistance between the line and neutral conductors is then measured at each socket outlet.

The readings obtained at each socket will be substantially the same provided they are connected to the ring (the distance around a circle is the same no matter where you measure it from), and the value will be approximately half the resistance of the line or the neutral loop resistance. Any sockets wired as spurs will have higher resistance value due to the extra length of the spur cable.

Step 3

We then repeat Step 2, but this time with the line and cpc cross-connected as shown in Figure 6.09.

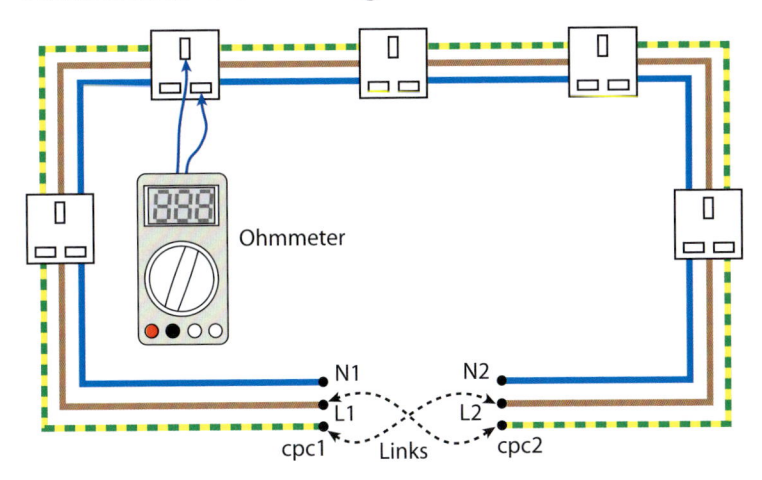

Figure 6.09 Step 3 Line and cpc cross-connected

The resistance between the line and earth is then measured at each socket. Again, as they are connected on a ring, the readings should be basically the same and the value at each socket, with the value being about ¼ of the line plus cpc loop resistances, namely $\frac{R_1 + R_2}{4}$.

The highest value recorded will represent the maximum $R_1 + R_2$ of the circuit and is recorded on the test schedule. This can also be used to determine the earth-loop impedance (Z_s) of the circuit to verify compliance with the loop impedance requirements of the Regulations.

Insulation resistance

For compliance with BS 7671, insulation resistance tests verify that the insulation of conductors, electrical accessories and equipment is satisfactory and that electrical conductors and protective conductors are not short-circuited, or do not show a low insulation resistance (which would indicate faulty insulation).

In other words, we are testing to see whether the insulation of a conductor is so poor as to allow any conductor to 'leak' to earth or to another conductor.

Before testing we need to ensure that:

- Pilot or indicator lamps and capacitors are disconnected from circuits to avoid an inaccurate test value being obtained.
- Voltage sensitive electronic equipment such as dimmer switches, delay timers, power controllers, electronic starters for fluorescent lamps, emergency lighting, RCDs etc. are disconnected so that they are not subjected to the test voltage.
- There is no electrical connection between any line and neutral conductor (e.g. lamps left in).

To show why we remove lamps, consider Figure 6.10. Should we leave the lamp in? No, because the lamp filament is effectively creating a short circuit between the line and neutral conductors. This gives the same effect as if there were no insulation there at all.

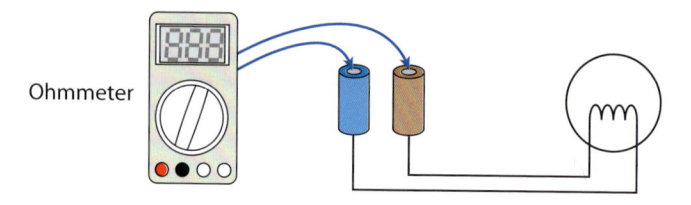

Figure 6.10 Insulation resistance test

The test equipment we use is an insulation resistance tester meeting the criteria laid down in BS 7671, with insulation resistance tests carried out using the appropriate d.c. test voltage as specified in Table 61 of BS 7671.

The installation can be said to conform to the Regulations if firstly the main switchboard and then each distribution circuit (tested separately, with all its final circuits connected but with current using equipment disconnected), have an insulation resistance not less than that specified in Table 61 of BS 7671 as reproduced in Table 6.01 opposite.

For a basic installation, i.e. one with only one DB, the test would normally be carried out on the whole installation, with the main switch off, all fuses in place, switches and circuit breakers closed, lamps removed, and fluorescent and discharge luminaires and other equipment disconnected.

Circuit nominal voltage	Test voltage d.c. (V)	Minimum insulation resistance
SELV and PELV	250	0.5 MΩ
Up to and including 500 V with the exception of the above systems	500	1.0 MΩ
Above 500 V	1000	1.0 MΩ

Table 6.01 Minimum value of insulation resistance (BS 7671 Table 61)

Where the removal of lamps and/or the disconnection of current using equipment is impracticable, the local switches controlling such lamps and/or equipment should be open.

On any two-way/intermediate circuits you will have to operate the two-way switch and re-test the circuit to make sure that you have tested all of the 'strappers'.

Although an insulation resistance value of not less than 1 MΩ complies with BS 7671, if an insulation resistance value of less than 2 MΩ is recorded, there is the possibility of a defect and then each circuit should be separately tested.

We are now checking two things: conductors under test leaking to another conductor and then any conductor under test leaking to earth. Apart from any need to test individual circuits, most electricians prefer, or find it quicker, to test between individual conductors rather than to group them together. As an example, Figure 6.11 shows the ten readings that would need to be taken for a three-phase circuit.

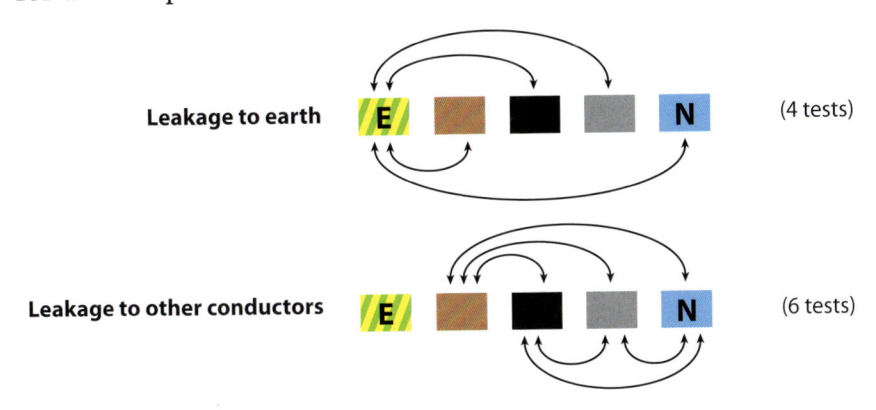

Figure 6.11 Three-phase circuit test

Protection by SELV, PELV or electrical separation (not often required)

Where the protective measures of SELV, PELV or electrical separation have been used then an insulation resistance test must be carried out the with the value being no less than that specified in Table 61 of BS 7671 for the circuit with the highest voltage present in the location.

For example, if an area is being protected by SELV and that area also has low voltage circuits in it, then the minimum insulation resistance value for the SELV circuit must be no lower than that of the low voltage circuit, namely no less than 1 MΩ.

For electrical separation, the source of supply should be inspected to prove that it does not exceed 500 V. The insulation between live parts of the separated circuit and any adjacent conductor (in the same enclosure or touching) and/or to earth must be tested at 500 V d.c. with insulation resistance being no less than 1 MΩ.

Protection by barriers/enclosures provided during erection (not often required)

This test does not apply to the barriers or enclosures of factory-built equipment, but only to those provided and fabricated on the site during the course of assembly. If this is the case then any barrier or enclosure must offer protection to no less than IP2X or IPXXB with readily accessible top surfaces having protection of no less than IP4X or IPXXD.

The **IP Codes** consists of the letters IP followed by two digits and an optional letter. As defined in international standard IEC 60529 (Degrees of protection provided by enclosures – IP Code), it classifies the degree of protection provided against the intrusion of solid objects and water in electrical enclosures.

The first digit indicates the level of protection that the enclosure provides against access by solid objects to hazardous parts (e.g., electrical conductors, moving parts). The second digit refers to protection of the equipment inside the enclosure against the harmful ingress of water. Where there is no protection rating with regard to one of the criteria, the digit is replaced with the letter X.

For example, an electrical socket rated IP22 is protected against insertion of fingers and will not be damaged or become unsafe during a specified test in which it is exposed to vertically or nearly vertically dripping water. IP22 or IP2X are typical minimum requirements for the design of electrical accessories for indoor use.

The various digits and their application are shown in Tables 6.02–6.03.

> **Key term**
>
> **IP Codes** – International Protection, more commonly interpreted as Ingress Protection

Level	Object size protected against	Effective against
0	—	No protection against contact and ingress of objects
1	>50 mm	Any large surface of the body, such as the back of a hand, but no protection against deliberate contact with a body part
2	>12.5 mm	Fingers or similar objects
3	>2.5 mm	Tools, thick wires, etc.
4	>1 mm	Most wires, screws, etc.
5	dust protected	Ingress of dust is not entirely prevented, but it must not enter in sufficient quantity to interfere with the satisfactory operation of the equipment; complete protection against contact
6	dust tight	No ingress of dust; complete protection against contact

Table 6.02 First digit

Level	Protected against	Details
0	not protected	—
1	dripping water	Dripping water (vertically falling drops) shall have no harmful effect.
2	dripping water when tilted up to 15°	Vertically dripping water shall have no harmful effect when the enclosure is tilted at an angle up to 15° from its normal position.
3	spraying water	Water falling as a spray at any angle up to 60° from the vertical shall have no harmful effects.
4	splashing water	Water splashing against the enclosure from any direction shall have no harmful effects.
5	water jets	Water projected by a nozzle against enclosure from any direction shall have no harmful effects.
6	powerful water jets	Water projected in powerful jets against the enclosure from any direction shall have no harmful effects.
7	immersion up to 1 m	Ingress of water in harmful quantity shall not be possible when the enclosure is immersed in water under defined conditions of pressure and time (up to 1 m of submersion).
8	immersion beyond 1 m	The equipment is suitable for continuous immersion in water under conditions which shall be specified by the manufacturer. NOTE: Normally, this will mean that the equipment is hermetically sealed. However, with certain types of equipment, it can mean that water can enter but only in such a manner that produces no harmful effects.

Table 6.03 Second digit

Additional letters

The standard defines additional letters that can be appended to classify only the level of protection against access to hazardous parts by persons.

Level	Protected against access to hazardous parts with
A	back of hand
B	finger
C	tool
D	wire

Table 6.04 Additional letters

The use of these letters can be shown in our BS 7671 requirement that any barrier or enclosure must offer protection to no less than IP2X or IPXXB with readily accessible top surfaces having protection of no less than IP4X or IPXXD. We can now see that IPXXB means only protection against finger contact and that IPXXD means only protection against wire (1 mm with its end at right angles).

Insulation resistance/impedance of non-conducting floors and walls (not often required)

This is a specialist test using a magneto-ohmmeter and test electrodes, consequently no further detail will be given in this book. However, methods of measurement are given in Appendix 13 of BS 7671.

Polarity

A test needs to be performed to check the polarity of all circuits and this is now a two stage process in that it should de-classed as one of the dead tests and done before connection to the supply (with either an ohmmeter or the continuity range of an insulation and continuity tester) and then confirmed once the supply has been switched on using an approved voltage indicator.

The purpose of a polarity test is to verify that:

- each fuse and single-pole control and protective device is connected in the line conductor only
- with the exception of ES14 and ES27 lampholders to BS EN 60238, in circuits that have an earthed neutral, centre contact

bayonet and ES lampholders have the outer or screwed contacts connected to the neutral conductor
- wiring has been correctly connected to socket outlets and similar accessories.

As can be seen from Figure 6.12, having established the continuity of the cpc in an earlier test, we now use this as a long test lead, temporarily linking it out with the circuit line conductor at the distribution board and then making our test across the line and earth terminals at each item in the circuit under test. Remember to close lighting switches before carrying out the test.

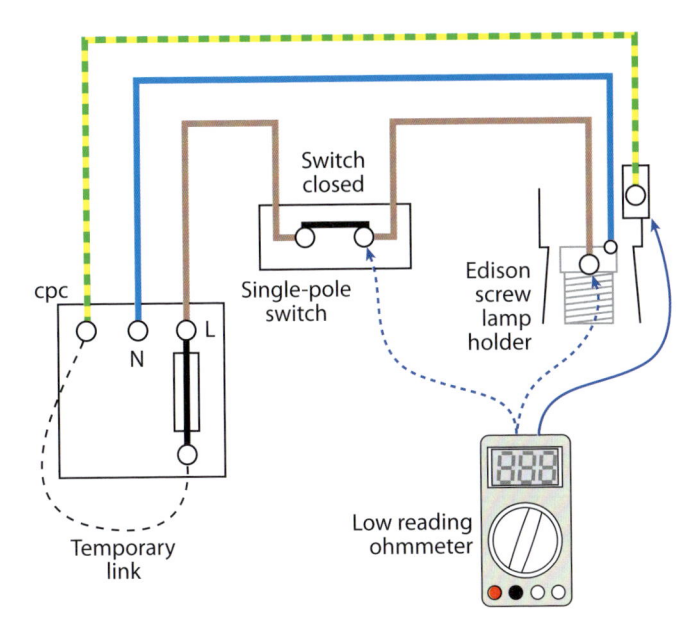

Figure 6.12 Polarity test of lighting circuit using continuity tester

For ring circuits, if the tests required by Regulation 713-03 ring circuit continuity have been carried out, the correct connections of line, neutral and cpc conductors will have already been verified and no further testing is required. However, for radial circuits the $R_1 + R_2$ measurements should also be made at each point, using this method.

Earth electrode resistance

Where an earthing system incorporates an earth electrode as part of the system, such as in rural areas, the electrode resistance to earth needs to be measured. In previous years the metal pipes of the water mains were used, but this practice can no longer be relied upon as pipes are now often made from plastic.

Some of the types of accepted earth electrode are:

- earth rods or pipes
- earth tapes or wires
- earth plates
- underground structural metalwork embedded in foundations
- lead sheaths or other metallic coverings of cables
- metal pipes.

The resistance to earth will depend upon the size and type of electrode used, remember that we want the best connection to earth possible and that the connection to the electrode must be made above ground level.

Measurement by standard method

When measuring earth electrode resistances to earth where low values are required, as in the earthing of the neutral point of a transformer or generator, test method 1 below may be used, using an earth electrode resistance tester.

Test method 1

Before starting the test, disconnect the earthing conductor to the earth electrode either at the electrode or at the main earthing terminal. This will make sure that all the test current passes through the earth electrode. However, as this will leave the installation unprotected against earth-faults, switch off the supply before disconnecting the earth.

The test should be carried out when the ground conditions are at their least favourable, i.e. during a period of dry weather, as this will produce the highest resistance value. The test requires the use of two temporary test electrodes (spikes) and is carried out as shown in Figure 6.13.

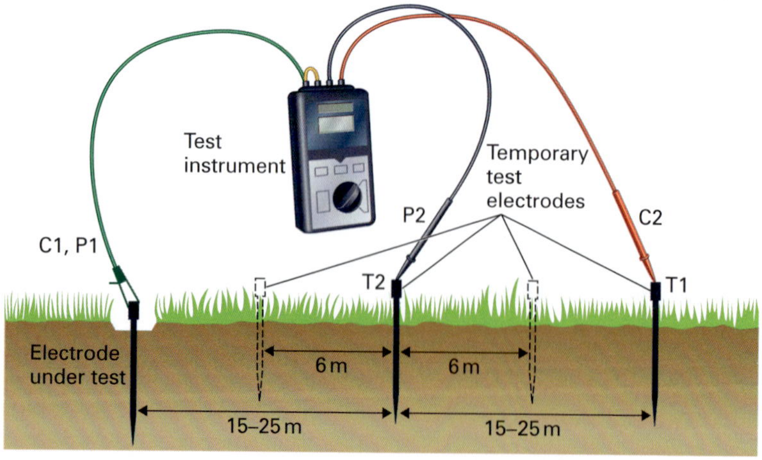

Figure 6.13 Test method 1

Connect the earth electrode to terminals C1 and P1 of a four-terminal earth tester. To exclude the resistance of these test leads from the resistance reading, individual leads should be taken from these terminals and connected separately to the electrode. However, if the test lead resistance is insignificant, the two terminals may be short-circuited at the tester and connection made with a single test lead, the same being true if you are using a three-terminal tester. Connection to the temporary 'spikes' is now made as shown in our diagram.

The distance between the test spikes is important, as if they are too close together their resistance areas will overlap. In general, you can gain a reliable result if the distance between the electrode under test and the current spike is at least ten times the maximum length of the electrode under test, in other words, 30 m away from a 3 m long rod electrode.

We then take three readings:

- first with the potential spike initially midway between the electrode and current spike
- second at a position 10% of the electrode-to-current spike distance back towards the electrode
- third at a position 10% of the distance towards the current spike.

By comparing these three readings a percentage deviation can be determined. We do this calculation by taking the average of the three readings; finding the maximum deviation of the readings from this average in ohms, and then expressing this as a percentage of the average.

The accuracy of the measurement using this technique is on average about 1.2 times the percentage deviation of the readings.

It is difficult to achieve an accuracy better than 2%, and you should not accept readings that differ by more than 5%. To improve the accuracy of the measurement to acceptable levels, the test must be repeated with larger separation between the electrode and the current spike.

Once the test is completed, make sure that the earthing conductor is reconnected.

Test method 2

Guidance Note 3 to BS 7671 lists a test method 2 that uses an earth-fault loop impedance tester. However, this is an alternative

> **Safety tip**
>
> The resistance area is important where livestock are concerned, as an animal may have its front legs outside, and back legs inside the resistance area, thus creating a potential difference. As little as 25 V can be lethal, so it is important to ensure that all of the electrode is well below ground level and that RCD protection is used.

method for use only on RCD protective TT installations and if this is impractical, the measured value of external earth-fault loop impedance may be used.

Protection by automatic disconnection of supply (ADS)

This is an unusual test in the sequence as compliance will come from other required tests. However, we must check the worth of this protection. This can be done as follows.

For TN systems:

- by measurement of earth-fault loop impedance
- confirm by visual inspection that overcurrent devices are the right type and set correctly.
- check that for an RCD the disconnection times of BS 7671 can be met.

For TT systems:

- by measurement of the resistance of the earth arrangement of the exposed-conductive parts of the equipment for the related circuit
- confirm by visual inspection that overcurrent devices are the right type and set correctly.
- check that for an RCD the disconnection times of BS 7671 can be met.

Earth-fault loop impedance

When designing an installation, it is the designer's responsibility to ensure that, should a phase-to-earth fault develop, the protection device will operate safely and within the time specified by BS 7671. Although the designer can calculate this theoretically, it is not until the installation is complete that the calculations can be checked.

It is therefore necessary to determine the earth-fault loop impedance (Z_s) at the furthest point in each circuit and to compare the readings obtained with either the designer's calculated values or the values tabulated in BS 7671 or the Unite Guide to Good Electrical Practice.

Figure 6.14 shows that, starting at the point of the fault, the earth-fault loop is made up of the following elements:

- the circuit protective conductor

- the main earthing terminal and earthing conductor
- the earth return path (dependent on the nature of the supply, TN-S, TN-C-S etc.)
- the path through the earthed neutral of the supply transformer
- the secondary winding (line) of the supply transformer
- the phase conductor from the source of the supply to the point of the fault.

We can then determine the value of earth-fault loop impedance (Z_s) by one of three methods as follows:

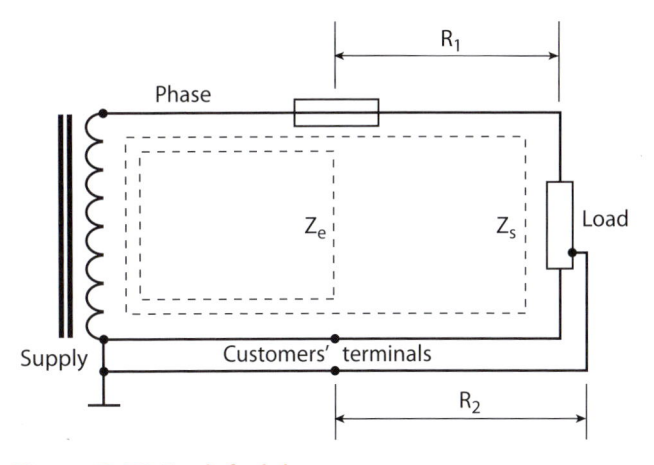

Figure 6.14 Earth-fault loop

- Direct measurement of Z_s.
- Direct measurement of Z_e at the origin of the circuit and adding to this the value of (R_1+R_2) measured during continuity tests: $Z_s = Z_e + (R_1+R_2)$.
- Obtaining the value of Z_e from the electricity supplier and adding to this the value of (R_1+R_2) as above. However, where the value of Z_e is obtained from the electricity supplier and is not actually measured, a test must be carried out to ensure that the main earthing terminal is in fact connected to earth using an earth-fault loop impedance tester or an approved test lamp.

Direct measurement of Z_s

Direct measurement of earth-fault loop impedance is achieved by use of an earth-fault loop impedance tester, which is an instrument designed specifically for this purpose. The instrument operates from the mains supply and therefore this measurement can only be taken on a live installation. The instrument is usually fitted with a standard 13 A plug for connecting to the installation directly through a normal socket outlet, although test leads and probes are also provided for taking measurements at other points on the installation.

IET GN3 states that neither the connections with earth or bonding conductors are to be disconnected, but consequently readings may be less than $Z_e + (R_1+R_2)$ because of parallel paths. This should be taken into account when comparing with design readings.

For the very reason of eliminating any parallel earth return paths, a second school of opinion would argue that as the test is being conducted by a skilled, competent person, the main equipotential bonding conductors should be disconnected for the duration of the test. This would then ensure that readings are not distorted by the presence of gas or water service pipes acting as part of the earth return path. Precautions must be taken, however, to ensure that the main equipotential bonding conductors are reconnected after the test has taken place.

Earth-fault loop impedance testers are connected directly to the circuit being tested and care must be taken to avoid danger. Should a break have occurred anywhere in the protective conductor under test then the whole of the earthing system could become live. It is essential therefore that protective conductor continuity tests be carried out prior to the testing of earth-fault loop impedance. Communication with other users of the building and the use of warning notices and barriers is essential.

Measurement of Z_e

The value of Z_e can be measured using an earth-fault loop impedance tester at the origin of the installation. However, as this requires the removal of covers and the exposure of live parts, extreme care must be taken and the operation must be supervised at all times. The instrument is connected using approved leads and probes between the phase terminal of the supply and the means of earthing with the main switch open or with all sub-circuits isolated. In order to remove the possibility of parallel paths, the means of earthing must be disconnected from the equipotential bonding conductors for the duration of the test. With the instrument correctly connected and the test button pressed, the instrument will give a direct reading of the value of Z_e. It must be remembered to re-connect all earthing connections on completion of the test.

Verification of test results

The measured values of earth-fault loop impedance obtained (Z_s) should be less than the values stated in BS 7671 Chapter 4, Tables 41.2, 41.3 and 41.4, reproduced here as Tables 6.05, 6.06 and 6.07. Please note that these values are for conductors at their normal operating temperature and if the conductors are at a different temperature at the point of test then a de-rating factor should be applied in line with the guidance given in Appendix 14 of BS 7671.

(a) General purpose (gG) fuses to BS 88-2.2 and BS 88.6						
Rating (amperes)	6	10	16	20	25	32
Z_s (ohms)	8.52	5.11	2.70	1.77	1.44	1.04
(b) Fuses to BS 1361						
Rating (amperes)	5	15	20	30		
Z_s (ohms)	10.45	3.28	1.70	1.15		
(c) Fuses to BS 3036						
Rating (amperes)	5	15	20	30		
Z_s (ohms)	9.58	2.55	1.77	1.09		
(d) Fuses to BS 1362						
Rating (amperes)	3	13				
Z_s (ohms)	16.4	2.42				

Table 6.05 Table 41.2 of BS 7671

(a) Type B circuit-breakers to BS EN 60898 and the overcurrent characteristics of RCBOs to BS EN 61009-1													
Rating (amperes)	3	6	10	16	20	25	32	40	50	63	80	100	125	I_n
Z_s (ohms)	15.33	7.67	4.60	2.87	2.30	1.84	1.44	1.15	0.92	0.73	0.57	0.46	0.37	46/I_n
(b) Type C circuit-breakers to BS EN 60898 and the overcurrent characteristics of RCBOs to BS EN 61009-1													
Rating (amperes)		6	10	16	20	25	32	40	50	63	80	100	125	I_n
Z_s (ohms)		3.83	2.30	1.44	1.15	0.92	0.72	0.57	0.46	0.36	0.29	0.23	0.18	23/I_n
(c) Type D circuit-breakers to BS EN 60898 and the overcurrent characteristics of RCBOs to BS EN 61009-14													
Rating (amperes)		6	10	16	20	25	32	40	50	63	80	100	125	I_n
Z_s (ohms)		1.92	1.15	0.72	0.57	0.46	0.36	0.29	0.23	0.18	0.14	0.11	0.09	11.5/I_n

Table 6.06 Table 41.3 of BS 7671

(a) General purpose (gG) fuses to BS-88.2 and BS 88-6								
Rating (amperes)	6	10	16	20	25	32	40	50
Z_s (ohms)	13.5	7.42	4.18	2.91	2.30	1.84	1.35	1.04
Rating (amperes)		63	80	100	125	160	200	
Z_s (ohms)		0.82	0.57	0.42	0.33	0.25	0.19	
(b) Fuses to BS 1361								
Rating (amperes)	5	15	20	30	45	60	80	100
Z_s (ohms)	16.4	5.00	2.80	1.84	0.96	0.70	0.50	0.36
(c) Fuses to BS 3036								
Rating (amperes)	5	15	20	30	45	60	100	
Z_s (ohms)	17.7	5.35	3.83	2.64	1.59	1.12	0.53	
(d) Fuses to BS 1362								
Rating (amperes)	3	13						
Z_s (ohms)	23.2	3.83						

Table 6.07 Table 41.4 of BS 7671

Additional protection

Where an RCD (≤30 mA) is being used to provide additional protection against contact with live parts, then the operating time must not exceed 40 ms when tested at 5 × 30 mA. The maximum test time must be no longer than 40 ms unless the potential on the protective conductor rises by less than 50 V.

Prospective fault current

If a short circuit occurs, current levels are created that are likely to produce considerable thermal and mechanical stresses in electrical equipment, and consequently are potentially life threatening. It is therefore essential to protect personnel and equipment by calculating short circuit currents while designing (or altering) an electrical system, thus allowing the electrical system to be protected using suitable protective devices.

> **Key term**
>
> **Prospective fault current** – The value of overcurrent that would flow at a given point in a circuit if a fault (a short circuit or an earth-fault) were to occur at that point. The prospective fault current tends to be higher than the fault current likely to occur in practice

BS 7671 Regulation 612.11 states that the prospective short circuit current and prospective earth-fault current shall be measured, calculated or determined by another method, both at the origin and other relevant points around the installation; the IET definition of 'relevant points' being any point where a protective device is required to operate under fault conditions. BS 7671 Regulation 434.5.1 additionally requires that where it is installed, the breaking capacity of each protective device is to be no less than the prospective fault current.

This is classed as a live test using the prospective fault scale of an earth loop impedance tester and with the power on and it is the greater of the two fault currents (short circuit or earth-fault) that is recorded on the results schedule.

Except in London (and one or two other major cities) the maximum fault current for 230 V single-phase supplies up to 100 A is unlikely to be greater than 16 kA, and therefore this is taken to be the case where:

- the current ratings of devices are 50 A
- the consumer unit complies with BS 5486-13/BS EN 60439-3
- the consumer unit is supplied through a type 2 fuse 100 A to BS 1361.

Phase sequence

BS 7671 Regulation 612.12 requires that the phase sequence is maintained for multiphase circuits. This is important in terms of checking correct load balancing or motor rotation.

To check phase sequence, the usual method is to use a phase rotation tester sequence indicator. There are typically two types using either a rotating disc system (basically a mini-induction motor) or by digital display as shown in Figure 6.15.

Functional testing

BS 7671 Regulation 612.13.1 requires that where fault and/or additional protection are provided by a Residual Current Device (RCD), the effectiveness of any test facility incorporated in the device shall be verified. Regulation 612.13.2 then requires that all assemblies are to be functionally tested to show that they operate as intended and are mounted and adjusted correctly.

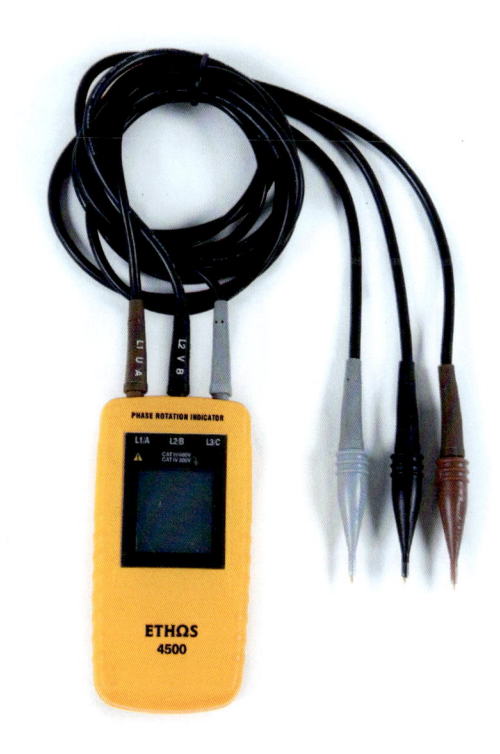

Figure 6.15 A phase rotation tester sequence indicator with a digital display

This includes:

- switches
- motors
- lights
- heaters
- PIRs and similar detectors
- dimmers and other such controls.

RCD testing

Most RCDs have an integral test button, but even a successful test of this does not necessarily confirm that the RCD is working correctly. Basic testing of RCDs therefore involves determining the tripping time (in milliseconds) by inducing a fault current in the circuit. This test can be performed at DBs with test leads or at socket outlets.

The test is performed on a live circuit with all known loads disconnected and because some RCDs are more sensitive in one half cycle of the mains supply waveform than the other, the test must be carried out for both zero and 180 degree phase settings, and the longest time should be recorded. The measured tripping time is then compared with the maximum time permitted in the IEE GN3.

The term residual current device is defined in BS 7671 Part 2 as 'a mechanical switching device or association of devices intended to cause the opening of the contacts when the residual current attains a given value under specified conditions'.

RCDs are now manufactured to harmonised standards and can be identified by their BS EN numbers. An RCD found in an older installation may not provide protection in accordance with current standards. The following list identifies the applicable current standards.

- BS 7071 (PRCD)
- BS 7288 (SRCD)
- BS EN 61008-1 (RCCB)
- BS EN 61009-1 (RCBO)

The most common variations of an RCD are shown in Table 6.08 opposite.

Type of RCD		Description	Usage
RCCB	Residual current operated circuit-breaker without integral overcurrent protection	Device that operates when the residual current attains a given value under specific conditions	Consumer units Distribution boards
RCBO	Residual current operated circuit-breaker (RCCB) with integral overcurrent protection	Device that operates when the residual current attains a given value under specific conditions and incorporates overcurrent protection	Consumer units Distribution boards
CBR	Circuit-breaker incorporating residual current protection	Overcurrent protective device incorporating residual current protection	Distribution boards in larger installations
SRCD	Socket outlet incorporating an RCD	A socket outlet or fused connection unit incorporating a built-in RCD	Often installed to provide supplementary protection against direct contact for portable equipment used out of doors
PRCD	Portable residual current device	A PRCD is a device that provides RCD protection for any item of equipment connected by a plug or socket. Often incorporates overcurrent protection	Plugged into an existing socket outlet. PRCDs are not part of the final installation
SRCBO	Socket outlet incorporating an RCBO	Socket outlet or fused connection unit incorporating an RCBO	Often installed to provide supplementary protection against direct contact for portable equipment used out of doors

Table 6.08 Variations of RCD

There are essentially two RCD tests that must be carried out.

- With a test of half the rated tripping current (e.g. 15 mA for a 30 mA rated RCD) the device should not operate.
- With a test of the full rated tripping current (e.g. 30 mA for a 30 mA rated RCD) the device should trip in less than 300 ms unless it is Type 'S' (or selective) incorporating a time delay, in which case it must trip between 130 ms to 500 ms.

Remember

If the device was to BS 4293 then the full rated test should cause tripping within 200 ms.

We carry out the half-rated test because certain equipment has a small leakage to earth as part of normal operation. IT equipment, for example, uses capacitors within their mains input filters and has a small leakage to earth at mains frequencies. Equally heating elements that are either immersed in liquid or exist in damp conditions, may absorb moisture if damaged, causing leakage through the moisture to the earthed outer case. The half rated test replicates low levels of earth leakage to check there is no 'nuisance' tripping. However, be aware if there is a constant low level of normal leakage, it may create a level close to the rated tripping current and therefore any minor changes in the circuit could cause the RCD to trip.

Additional Protection

If an RCD is affording additional protection to either sockets up to 20 A or for sockets up to 32 A for mobile equipment in use outdoors, then a test of 5 times (5 ×) the full rated tripping current (e.g. 150 mA for a 30 mA RCD) means the device must trip within 40 ms.

The integral test button

As we said earlier, most RCDs have an integral test button, but this only tests the operation of the mechanical parts and even a successful test of this does not necessarily confirm that the RCD is working correctly. Such a button only works if there is a supply present.

Safety

Under certain circumstances these tests can result in potentially dangerous voltages appearing on exposed and extraneous-conductive parts within the installation and therefore suitable precautions must be taken to prevent contact by persons or livestock with any such part. Other building users should be made aware that tests are being carried out and warning notices should be posted as necessary.

Verification of voltage drop

This is not normally part of the initial testing process of a new installation as it would have been part of the initial design process when taking in to account the effect of length of cable runs. However, paragraph 6.4 within Appendix 4 of BS 7671 gives the permitted values of voltage drop as:

- 3% for lighting circuits
- 5% for other uses on LV public systems.

Using an approved voltage indicator, be aware that as voltage drop is determined against the demand of the current using equipment, a greater voltage drop may be experienced for a motor circuit during starting.

Safety tip

Prior to RCD tests being carried out, it is essential for safety reasons that the earth loop impedance of the installation has been tested to check that the earth-return path is sound and that all the necessary requirements have been met. BS 7671 stipulates the order in which tests should be carried out.

Working life

A colleague suggests it does not matter in what sequence the tests are done on an electrical installation. How would you explain to him the importance of sequences? Think about:

1. What reference to the sequence would you use?
2. What is the key reason for the sequence?
3. How would you explain why the sequence must be followed?

Actions to take in event of unsatisfactory test results

Before we look at this topic, we must establish a basic concept when using a continuity meter or insulation resistance tester. This is illustrated using a basic scale on an analogue insulation resistance tester, as shown in Figures 6.16 and 6.17.

When we touch the leads of our meter together we are creating a short circuit. Remember a short circuit means '...negligible impedance between live conductors'. In other words there is almost no resistance between the two leads as they are touching. The needle on the display therefore shows zero to reflect this finite reading.

However, when we separate the leads, we have clearly created an open circuit. This is the same as having a broken conductor and the needle on the display therefore shows infinity to reflect this infinite resistance.

Real tests require more detailed readings, but the principle is the same.

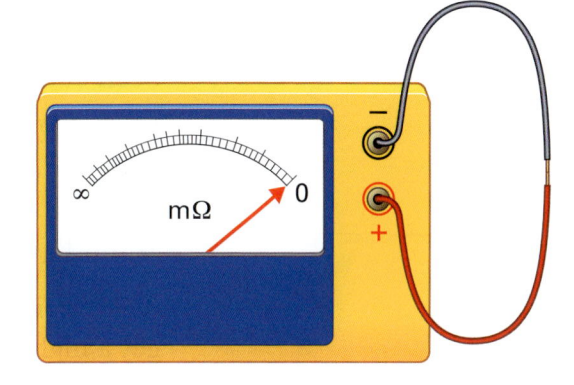

Figure 6.16 Creating a short circuit

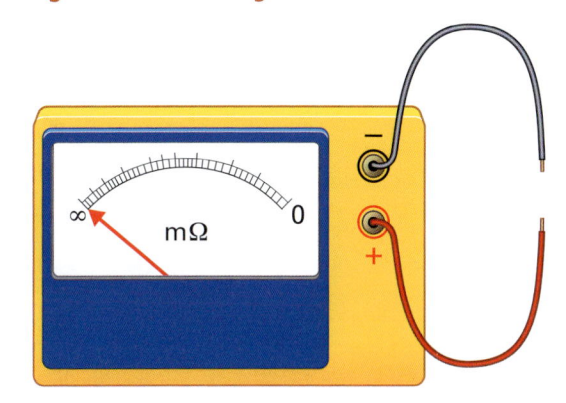

Figure 6.17 Creating an open circuit

Continuity

When testing the continuity of circuit protective conductors or bonding conductors we should always expect a very low reading (near to zero), which is why we must always use a low reading ohmmeter.

Main and supplementary bonding conductors should have a reading of not more than 0.05 ohms while the maximum resistance of circuit protective conductors can be estimated from the value of ($R_1 + R_2$) given in the Unite Guide to Good Electrical Practice. These values will depend upon the cross-sectional area of the conductor, the conductor material and its length.

A very high (near to infinity end of scale) reading would indicate a break in the conductor itself or a disconnected termination that must be investigated. A mid-range reading may be caused by the poor termination of an earthing clamp to the service pipe, e.g. a service pipe which is not cleaned correctly before fitting the clamp or corrosion of the metal service pipe due to its age and damp conditions.

> **Remember**
>
> When measuring the end to end resistance of each of the conductors, a finite reading shows there is no break in the conductor, and as long as the conductors are the same size, then the readings should be the same. If the cpc is smaller, then its resistance will be higher.

When testing continuity of a ring final circuit, remember that the purpose of the test is to establish that a ring exists and that it has been correctly connected.

In Step 2 of the test process (see pages 104–105) we said that the readings obtained at each socket will be substantially the same provided they are connected to the ring (because the distance around a circle is the same no matter where you measure it from).

As we can see from Figure 6.18, should you test at a socket and get a higher reading, then this has probably been connected as a spur, and any sockets wired as spurs will have higher resistance value due to the extra length of the spur cable.

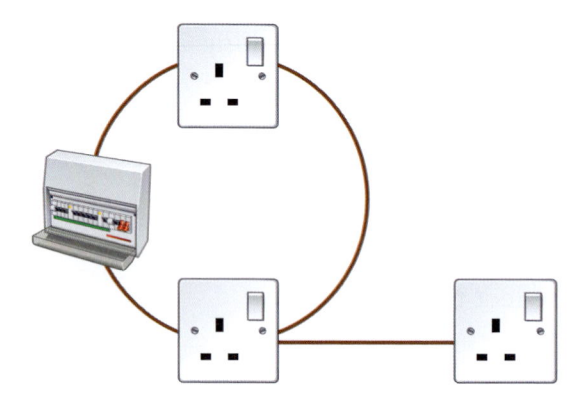

Figure 6.18 Testing at a socket

However, if while testing at each socket you found that your readings were increasing as you moved away from the DB, then it is likely that instead of having a ring, you have the ends incorrectly identified and are not cross-connected between the outgoing leg live to the incoming leg neutral. Instead you may be 'linked out' across the live and neutral of the same leg and are therefore measuring more cable at each socket. As shown in Figure 6.19, the reading taken at socket C will therefore include more cable than that taken at socket A and therefore it will have a higher resistance reading.

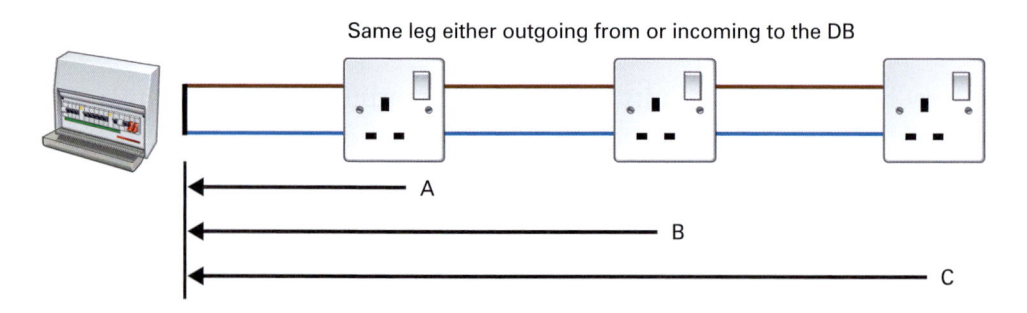

Same leg either outgoing from or incoming to the DB

Figure 6.19 Testing at each socket with readings increasing

Insulation resistance and effects on values

The value of the insulation resistance of an installation will depend upon the size and complexity of the installation and the number of circuits connected to it. When testing a small domestic installation you may expect an insulation resistance reading in excess of 200 MΩ while a large industrial or commercial installation with many sub-circuits, each providing a parallel path, will give a much smaller reading if tested as a whole. Bear in mind that if you double the length of a cable, you halve its insulation resistance. Length and insulation resistance are inversely proportional and longer cables will have more parallel paths and therefore a lower insulation resistance.

It is recommended that, where the insulation resistance reading is less than 2 MΩ, individual distribution boards or even individual sub-circuits be tested separately in order to identify any possible cause of poor insulation values.

An extremely low value of insulation resistance would indicate a possible short circuit between line conductors or a bare conductor in contact with earth at some point in the installation, either of which must be investigated. A reading below 1.0 MΩ would suggest a weakness in the insulation, possibly due to the ingress of dampness or dirt in such items as distribution boards, joint boxes or lighting fittings.

Although PVC-insulated cables are not generally subject to a deterioration of insulation resistance due to dampness (unless the insulation or sheath is damaged), mineral-insulated cables can be affected if dampness has entered the end of a cable before the seal has been applied properly. Other causes of low insulation resistance can be the infestation of equipment by rats, mice or insects.

Polarity

Correct polarity is achieved by the correct termination of conductors to the terminals of all equipment. This may be main intake equipment such as isolators, main switches and distribution boards or accessories such as socket outlets, switches or lighting fittings.

Polarity is either correct or incorrect. Incorrect polarity is caused by the termination of live conductors to the wrong terminals and is corrected by reconnecting all conductors correctly.

Earth-fault loop impedance

As explained previously the earth-fault loop path is made up of those parts of the supply system external to the premises being tested (Z_e) and the phase conductor and circuit protective conductor within the installation ($R_1 + R_2$), the total earth-fault loop impedance being $Z_s = Z_e + (R_1 + R_2)$.

The path followed by a fault current as the result of a low impedance occurring between the phase conductor and earthed metal is called the earth-fault loop and the current is 'driven' through the loop by the supply voltage.

The loop for a TT system is shown in red in Figure 6.20.

> **Remember**
>
> The loop through the other earthing systems will be similar and these systems are represented in Unit ELTK 04a of this book.

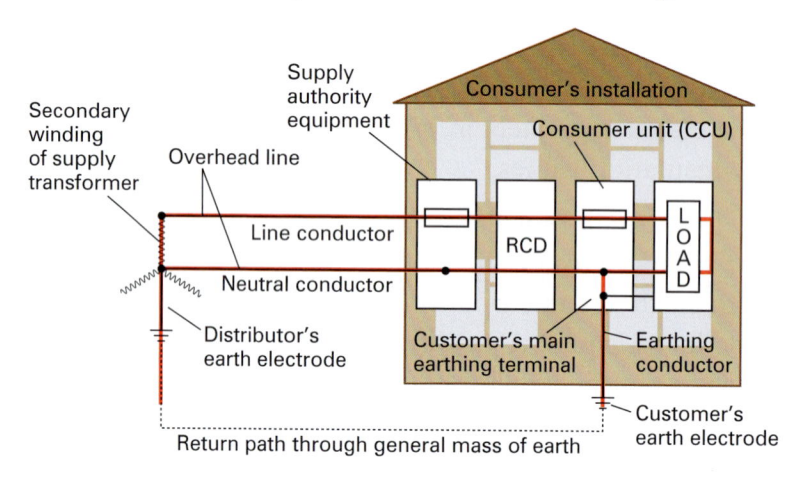

Figure 6.20 Earth-fault loop in a TT system

As we have no influence on the external value of impedance (Z_e), then if the value of impedance measured is higher than that required by the design of the installation, then we can only reduce the value of Z_s by installing circuit protective conductors of a larger cross-sectional area or, if aluminium conductors have been used, by changing these to copper.

If the value is still too high to guarantee operation of the circuit protective device in the time required by BS 7671, then consideration will have to be given to changing the type of protective device (i.e. fuses to circuit breakers).

Residual current devices (RCDs)

Where a residual current device (RCD) fails to trip out when pressing the integral test button, this would indicate a fault within the device itself, which should therefore be replaced. Should an RCD fail to trip out when being tested by an RCD tester, it would suggest a break in the earth return path, which must be investigated. If the RCD does trip out but not within the time specified then a check should be made that the test instrument is set correctly for the nominal tripping current of the device under test.

Periodic inspection and testing

All electrical installations must be maintained in a safe condition, and regular inspection and testing (periodic inspection) is an essential part of any such preventative maintenance programme. In addition to any statutory requirements, other bodies such as licensing authorities, insurance companies and mortgage lenders may also require periodic inspection and testing to be carried out.

However, in the case of an installation that is under constant supervision while in normal use, such as a factory or other industrial premises, periodic inspection and testing may be replaced by a system of continuous monitoring and maintenance of the installation by skilled persons, provided that adequate records of such maintenance are kept.

Periodic inspection and testing is required:

- to confirm compliance with BS 7671
- on a change of ownership of the premises
- on a change of use of the premises
- on a change of tenancy of the premises
- on completion of alterations or additions to the original installation
- following any significant increase in the electrical loading of the installation
- where there is reason to believe that damage may have been caused to the installation.

In summary, periodic inspection and testing is generally necessary because any installation can deteriorate due to wear and tear, age, overload or damage. Inspection and testing is also needed, however, if there is an addition or alteration to an installation, as obviously these must comply with BS 7671 and must not compromise, or be compromised by, the existing installation.

Frequency of inspection and testing

In establishing the frequency of periodic inspection and testing we should consider the following:

- type of installation
- use and operation of installation
- external influences affecting the installation
- frequency and quality of any maintenance activity.

> **Remember**
> With periodic inspection and testing, inspection is the vital operation and testing is carried out in support of that inspection.

> **Remember**
> For domestic or commercial premises, a change in occupancy might require an additional visit.

Type of installation	Routine check	Maximum period between inspections and testing as necessary
1	2	3
General installation		
Domestic	–	Change of occupancy/10 years
Commercial	1 year	Change of occupancy/5 years
Educational establishments	4 months	5 years
Hospitals	1 year	5 years
Industrial	1 year	3 years
Residential accommodation	at change of occupancy/1 year	5 years
Offices	1 year	5 years
Shops	1 year	5 years
Laboratories	1 year	5 years
Buildings open to the public		
Cinemas	1 year	1 to 3 years
Church installations	1 year	5 years
Leisure complexes (excluding swimming pools)	1 year	3 years
Places of public entertainment	1 year	3 years
Restaurants and hotels	1 year	5 years
Theatres	1 year	3 years
Public houses	1 year	5 years
Village halls/community centres	1 year	5 years
Special installations		
Agricultural and horticultural	1 year	3 years
Caravans	1 year	3 years
Caravan parks	6 months	1 year
Highway power supplies	as convenient	6 years
Marinas	4 months	1 year
Fish farms	4 months	1 year
Swimming pools	4 months	1 year
Emergency lighting	Daily/monthly	3 years
Fire alarms	Daily/weekly/monthly	1 year
Launderettes	1 year	1 year
Petrol filling stations	1 year	1 year
Construction site installations	3 months	3 months

Table 6.09 Frequency of testing (according to IET Guidance Note 3)

Table 6.09 gives guidance regarding frequency of periodic inspection and testing; it is contained within IET Guidance Note 3. However, the competent person carrying out the inspection and testing must recommend the interval between future visits, increasing or decreasing the interval based on their findings. The intervals given in the table should therefore only be seen as guidance to help the competent person decide the frequency of visits.

Routine checks

In Table 6.09 there is a column titled 'Routine Checks'. This is in place so no installation is left without attention for the periods indicated in the table. A system of routine checks should be set up to take place between formal periodic inspections. The frequency of these checks will depend entirely on the nature of the premises and the usage of the installation but this column aims to set out some suggested intervals.

For example, in domestic premises it is likely that the occupier will soon notice any damage or breakages to electrical equipment and will take steps to have repairs carried out. In commercial or industrial installations a suitable reporting system should be available for users of the installation to report any potential danger from deteriorating or damaged equipment.

The level of routine check being suggested at these intervals is shown in Table 6.10 and it should be noted that they are not intended to be carried out by a skilled person, but instead by a safe user of the installation and someone who would therefore recognise any defects.

General procedure

In old installations where information such as drawings, distribution board schedules, charts etc. are not available, some exploratory work may be necessary to enable inspection and testing to be carried out safely and without damage to existing equipment.

A survey should therefore be carried out to identify all items of switchgear and control gear and their associated circuits and during the survey a note should be made of any known changes in either the structure of the building, the environmental conditions or of any alterations or additions to the installation which may affect the suitability of the wiring or the method of installation.

Activity	Check
Defect reports	Check that all reported defects have been rectified and that the installation is safe
Inspection	Look for: • breakages • wear or deterioration • signs of overheating • missing parts (covers/screws) • switchgear still accessible • enclosure doors secure • labels still adequate (readable) • loose fittings
Operation	Check operation of: • switchgear (where reasonable) • equipment (switch off and on) • RCD (using test button)

Table 6.10 Routine checks

A careful check should be made of the type of equipment on site so that, where required, electronic or other equipment that may be damaged by high-test voltages can be either disconnected or short-circuited to prevent damage. If computer equipment is to be disconnected for the purpose of testing then the user of the equipment must be informed in order that data within the computer can be backed up and stored on disc if necessary.

Many commercial computer installations have an emergency back-up electrical supply that will automatically energise if the mains supply is disconnected. This means that circuits that you assume are isolated may well be kept live from a different source of supply.

If the inspection and testing cannot be carried out safely without the use of drawings, diagrams or schedules etc. then the person ordering the work should be informed and Section 6 of the Health and Safety at Work Act can be used to call for their preparation.

Sampling

Care and judgement by the person carrying out the inspection and testing are essential when deciding the extent of the inspection and testing.

Relative information such as installation certificates, minor works certificates, maintenance records etc. should be obtained to assist this process. The less information available then logically the higher the percentage of the installation to inspect and test, and any percentage of sample must be agreed with the person ordering the work before commencing.

The sampling should be flexible. This means if a low sample finds serious defects, then the sample should be increased, following agreement with the person ordering the work.

Large installations may be difficult to inspect and test at one time and therefore different sections should be inspected at different visits. Obviously it would not be good practice to keep seeing the same section of an installation at each inspection visit.

Scope

The requirements of BS 7671 for periodic inspection and testing are for a thorough (or partial if required) inspection of the installation without dismantling; supplemented by testing as felt appropriate by the competent person carrying out the inspection and testing to show that the requirements for disconnection times in BS 7671 Chapter 41 are complied with.

The degree of work should be agreed with the client prior to commencing. This is essential to agree periods of isolation and disconnection.

For safety reasons a visual inspection must be carried out before opening any enclosures, removing covers or carrying out any tests. To comply with the Electricity at Work Regulations 1989 the inspection should be carried out wherever possible with the supply de-energised.

A thorough visual inspection should be carried out of all electrical equipment that is not concealed and, where damage or deterioration has occurred, this must be recorded on the inspection schedule.

Carrying out periodic inspections

The inspection should include a check on the condition of all equipment and materials used in the installation with regard to the following.

Joints and connections

It may be impossible to inspect every joint and termination in an electrical installation. Therefore, where necessary, a sample inspection should be made. Provided the switchgear and distribution boards are accessible as required by the Regulations, then a full inspection of all conductor terminations should be carried out and any signs of overheating or loose connections should be investigated and included in the report. For lighting points and socket outlets a suitable sample should be inspected in the same way.

Conductors

The means of identification of every conductor including protective conductors should be checked and any damage or deterioration to the conductors, their insulation, protective sheathing or armour should be recorded. This inspection should include each conductor at every distribution board within the installation and a suitable sample of lighting points, switching points and socket outlets.

Flexible cables and cords

Where flexible cables or cords form part of the fixed installation the inspection should include:

- examination of the cable or cord for damage or deterioration
- examination of the terminations and anchor points for any defects
- checking the correctness of the installation with regard to additional mechanical protection or the application of heat resistant sleeving where necessary.

Switches

Guidance Note 3 recommends that a random sample of all switching devices be given a thorough internal visual inspection to assess their electrical and mechanical condition. Should the inspection reveal excessive wear and tear or signs of damage due to arcing or overheating then, unless it is obvious that the problem is associated only with that particular switch, the inspection should be extended to include all remaining switches associated with the installation.

Protection against thermal effects

Although it is sometimes difficult due to the structure of the building, the presence of fire barriers and seals should be checked wherever reasonably practicable.

Basic and fault protection

SELV is used as a means of protection against both basic and fault protection. When inspecting this type of system, the points to be checked include the use of a safety isolating transformer, the need to keep the primary and secondary circuits separate and the segregation of exposed-conductive parts of the SELV system from any connection with the earthing of the primary circuit or from any other connection with earth.

Basic protection

Inspection of the installation should confirm that all the requirements of the Regulations have been met with regard to protection against direct contact with live conductors. This means checking to ensure there has been no damage or deterioration of any of the insulation within the installation, no removal of barriers or obstacles and no alterations to enclosures that may allow access to live conductors.

Fault protection

The protective measure must be established and recorded on the inspection schedule. Where ADS is used, the requirements of BS 7671 must be met in terms of protective earthing (Regulation 411.3.1.1) and protective equipotential bonding (Regulation 411.3.1.2).

> **Safety tip**
>
> An RCD must not be used as the sole means of protection against direct contact with live parts.

Protective devices

A check must be made that each circuit is adequately protected with the correct type, size and rating of fuse or circuit-breaker. A check should also be made that each protective device is suitable for the type of circuit it is protecting and the earthing system employed. For example, will the protective device operate within the disconnection time allowed by the regulations and is the rating of the protective device suitable for the maximum prospective short circuit current likely to flow under fault conditions?

Enclosures and mechanical protection

The enclosures of all electrical equipment and accessories should be inspected to ensure that they provide protection not less than IP2X or IPXXB.

Marking and labelling

Labels should be applied adjacent to every fuse or circuit-breaker indicating the size and type of fuse or the nominal current rating of the circuit-breaker and details of the circuit they protect.

Other notices and labels required by the Regulations are as follows; the Regulations also state minimum physical sizes for these:

a) At the origin of every installation:

> **IMPORTANT**
> This installation should be periodically inspected and tested and a report on its condition obtained as prescribed in BS 7671 (formally the IET Wiring Regulations for Electrical Installations) published by the Institute of Electrical Engineers.
>
> Date of last inspection : ....
>
> Recommended date of next inspection : ....

b) Where different voltages are present:

- in equipment or enclosures within which a voltage exceeding 250 volts exists but where such a voltage would not be expected
- terminals between which a voltage exceeding 250 volts exists, which although contained in separate enclosures are within arm's reach of the same person
- the means of access to all live parts of switchgear or other live parts where different nominal voltages exist.

c) At earthing and bonding connections a label as shown below to BS 951 shall be permanently fixed in a visible position at or near the following points:

- the point of connection of every earthing conductor to an earth electrode
- the point of connection of every bonding conductor to an extraneous-conductive-part
- the main earth terminal of the installation where separate from the main switchgear.

> **Safety Electrical Connection
> – Do Not Remove**

d) Where RCDs are fitted within an installation, a suitable permanent durable notice as shown on page 135 shall be permanently fixed in a prominent position at or near the main distribution board.

> **Important**
>
> This installation, or part of it, is protected by a device that automatically switches off the supply if an earth-fault develops. Test quarterly by pressing the button marked 'T' or 'Test'. The device should switch off the supply and should then be switched on to restore the supply. If the device does not switch off the supply when the button is pressed, seek expert advice.

e) If alterations/additions are made such that an installation contains both the current harmonised wiring colours and the old colours of previous regulations, then the following sign should be fixed to the DB:

> **CAUTION**
>
> This installation has wiring colours to two versions of BS 7671.
>
> Great care should be taken before undertaking extension, alteration or repair that all conductors are correctly identified.

f) All caravans and motor caravans should have the following durable and permanently fixed label near the main switch giving instructions on the connection and disconnection of the caravan installation to the electricity supply (see Figure 6.21 on page 136).

Carrying out periodic testing

Periodic testing is supplementary to periodic inspection and in this case the level of testing needed for a new installation is not necessarily required or perhaps even possible. Equally, previously tested installations for which there are comprehensive test results do not require the same level of testing as an installation with no records of testing.

Periodic testing can also cause danger and therefore the person carrying out the work must be competent and capable of carrying out sample testing. If a sample of at least 10% produces no satisfactory results then further investigation will be required and the sample size increased if necessary.

GN3 recommends the following testing takes place, as shown in Table 6.11 on page 137.

> **INSTRUCTIONS FOR ELECTRICITY SUPPLY**
>
> **TO CONNECT**
>
> 1. Before connecting the caravan installation to the mains supply, check that:
>
>   a) the supply available at the caravan pitch supply point is suitable for the caravan electrical installation and appliances, and
>
>   b) the voltage and frequency and current ratings are suitable, and
>
>   c) the caravan main switch is in the OFF position.
>
>   Also, prior to use, examine the supply flexible cable to ensure there is not visible damage or deterioration.
>
> 2. Open the cover to the appliance inlet provided at the caravan supply point, if any, and insert the connector of the supply flexible cable.
>
> 3. Raise the cover of the electricity outlet provided on the pitch supply point and insert the plug of the supply cable.
>
>   **THE CARAVAN SUPPLY FLEXIBLE CABLE MUST BE FULLY UNCOILED TO AVOID DAMAGE BY OVERHEATING**
>
> 4. Switch on at the caravan main isolating switch.
>
>   **IN CASE OF DOUBT, OR IF CARRYING OUT THE ABOVE PROCEDURE THE SUPPLY DOES NOT BECOME AVAILABLE, OR IF THE SUPPLY FAILS, CONSULT THE CARAVAN PARK OPERATOR OR THE OPERATOR AGENT OR A QUALIFIED ELECTRICIAN**
>
> 5. Check the operation of residual current devices (RCDs) fitted in the caravan by depressing the test button(s) and reset.
>
> **TO DISCONNECT**
>
> 6. Switch off at the caravan main isolating switch, unplug the cable first from the caravan pitch supply point and then from the caravan inlet connector.
>
> **PERIODIC INSPECTION**
>
> Preferably, not less than once every three years and annually if the caravan is used frequently, the caravan electrical installation and supply cable should be inspected and tested and a report on their condition obtained as prescribed in BS 7671 Requirements for Electrical Installation published by the Institution of Engineering and Technology and BSI.

Figure 6.21 Fixed label for caravan installation

It should be noted that there is no requirement to carry out the tests in the same sequence as initial testing. For example, the earth-fault loop impedance test may be used to confirm the continuity of protective conductors at socket outlets and at accessible exposed-conductive parts of current using equipment and accessories. However GN3 gives the following detail regarding periodic testing.

> **Remember**
>
> The tests themselves were described under initial testing, on pages 99–122.

Test	Recommendation
Protective conductors continuity	Between the earth terminal of distribution boards to the following exposed-conductive parts: • socket outlet earth connections (note 4) • accessible exposed-conductive parts of current-using equipment and accessories (notes 4 and 5)
Bonding conductors continuity	• all protective bonding conductors • all necessary supplementary bonding conductors
Ring circuit continuity	Where there are proper records of previous tests, this test may not be necessary. This test should be carried out where inspection/documentation indicate that there may have been changes made to the ring final circuit.
Insulation resistance	If tests are to be made: • between live conductors, with line(s) and neutral connected together, and earth at all final distribution boards • at main and sub-main distribution panels, with final circuit distribution boards isolated from mains (note 6)
Polarity	At the following positions: • origin of the installation • distribution boards • accessible socket outlets • extremity of radial circuits (note 8)
Functional tests RCDs	Tests as required by Regulation 612.13.1, followed by operation of the functional test button.
Circuit-breakers, isolators and switching devices	Manual operation to prove that the devices disconnect the supply

Table 6.11 Summary of periodic testing

Continuity of protective conductors and equipotential bonding conductors

Where the installation can be safely isolated from the supply, then the circuit protective conductors and equipotential bonding conductors can be disconnected from the main earthing terminal in order to verify their continuity.

If this cannot be done, the circuit protective conductors and the equipotential bonding conductors must not be disconnected from the main earthing terminal as under fault conditions extraneous metalwork could become live. Under these circumstances a combination of inspection, continuity testing and earth-fault loop impedance testing should establish the integrity of the circuit protective conductors.

When testing the effectiveness of the main bonding conductors or supplementary bonds, the resistance value between any service pipe or extraneous metalwork and the main earthing terminal should be about 0.05 Ω or less.

Insulation resistance

Insulation resistance tests can only be carried out where it is possible to safely isolate the supply. All electronic equipment susceptible to damage should be disconnected or alternatively the insulation resistance test should be made between line and neutral conductors connected together and earth.

Where practicable the tests should be carried out on the whole of the installation with all switches closed and all fuse links in place. Where this is not possible, the installation may be subdivided by testing each separate distribution board one at a time. BS 7671 Table 61 states a minimum acceptable resistance value of 1 MΩ. However, if the measured value is less than 2 MΩ then further investigation is required to determine the cause of the low reading.

Where equipment is disconnected for these tests and it has exposed-conductive parts connected to protective conductors, then the insulation resistance of each item of equipment should be checked. In the absence of any other requirements the minimum value of insulation resistance between the live components and the exposed metal frame of the equipment should be not less than 1 MΩ.

Polarity

Polarity tests should be carried out to check that:

- polarity is correct at the intake position and the consumer unit or distribution board. Single-pole switches or control devices are connected in the line conductor only
- socket outlets and other accessories are connected correctly
- with the exception of ES14 and ES27 lampholders to BS EN 60238, in circuits that have an earthed neutral, centre contract bayonet and ES lampholders have the outer or screwed contacts connected to the neutral conductor
- all multi-pole devices are correctly installed.

Where it is known that no alterations or additions have been made to the installation since its last inspection and test, then the number of items to be tested can be reduced by sampling. It is recommended that at least 10% of all switches, control devices and centre contact lamp holders should be tested and 100% of socket outlets.

However if any cases of incorrect polarity are found then a full test should be made in that part of the installation and the sample of the remaining installation should be increased to 25%.

Earth-fault loop impedance

Earth-fault loop impedance tests should be carried out at:

- the origin of each installation and at each distribution board
- all socket outlets
- the furthest point of each radial circuit.

Results obtained should be compared with the values documented during previous tests, and where an increase in values has occurred these must be investigated.

Operation of residual current devices

A check should also be made that the tripping current should not exceed 30 mA when the RCD provides additional protection for sockets, ≤ 20 A for use by ordinary persons or for mobile equipment ≤ 32 A being used outdoors.

Operation of circuit-breakers

The manual operating mechanism of each circuit-breaker should be tested to see that it opens and closes correctly. Circuit-breakers with facility for injection testing should be so tested. All circuit-breakers should also be inspected for visible signs of damage or damage caused by overheating.

Operation of devices for isolation and switching

All such devices should be operated to verify their correct operation and checked for clear labelling and also to check that access to them or their operation has not been obstructed.

Where it is a requirement that the device interrupts all the supply conductors, the use of a test lamp connected between each line and the neutral on the load side of the device may be required to ensure that all supply conductors have been broken.

If devices have lockable or detachable handles, then it must be checked that keys etc. are not interchangeable and the means of isolation checked for integrity.

Condition report

BS 7671 requires that the results and extent of periodic inspection and testing are recorded on an Electrical Installation Condition Report (formerly called the Periodic Inspection Report) and provided to the person ordering the inspection.

The report must include the following:

- a description of the extent of the inspection and tests and what parts of the installation were covered
- any limitations (e.g. portable appliances not covered)
- details of any damage, deterioration or dangerous conditions which were found
- any non-compliance with BS 7671
- schedule of test results
- a recommendation for the interval until the next periodic inspection.

If any items were found which may cause immediate danger these should be rectified immediately. If this is not possible then they must be reported to a responsible person without delay and if necessary made safe.

When inspecting older installations, that may have been installed in accordance with a previous edition of the IET Wiring Regulations, any items that differ from or any other departures from the current BS 7671 must be recorded. Guidance on the action to be taken is then given a reference (C1 to C3) that must be recorded in the 'Observations and recommendations' section of the report.

Explanation of this referencing and the actual document will be seen in pages 155–160.

Types of test instrument and their safe and correct use

Instrument standards

BS EN 61010 covers basic safety requirements for electrical test instruments, and all instruments should be checked for conformance with this standard before use. Older instruments may have been manufactured in accordance with BS 5458 but, provided these are in good condition and have been recently calibrated, there is no reason why they cannot be used. Lead sets should conform to HSE GS38.

Instruments may be analogue (i.e. fitted with a needle that gives a direct reading on a fixed scale) or digital, where the instrument provides a numeric digital visual display of the actual measurement being taken. Insulation and continuity testers can be obtained in either format while earth-fault loop impedance testers and RCD testers are digital only.

> **Did you know?**
>
> The basic standard covering the performance and accuracy standards of electrical test instruments is BS EN 61557, which incidentally also requires compliance with BS EN 61010.

Calibration and instrument accuracy

To ensure that the reading being taken is reasonably accurate, all instruments should have a basic measurement accuracy of at least 5%. In the case of analogue instruments a basic accuracy of 2% of full-scale deflection should ensure the required accuracy of measured values over most of the scale.

All electrical test instruments should be calibrated on a regular basis. The time between calibrations will depend on the amount of usage that the instrument receives, although this should not exceed 12 months in any circumstances.

Instruments have to be calibrated in laboratory conditions against standards that can be traced back to national standards; therefore this usually means returning the instrument to a specialist test laboratory.

On being calibrated the instrument will have a calibration label attached to it stating the date the calibration took place and the date the next calibration is due. An example is shown in Figure 6.22.

> **Remember**
>
> Although older instruments often generated their own operating voltage by use of an in-built hand-cranked generator, almost all modern instruments have internal batteries for this purpose. It is essential therefore that the condition of the batteries is checked regularly, especially ensuring the absence of corrosion at the battery terminals.

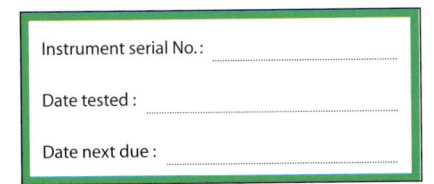

Instrument serial No.: _____
Date tested: _____
Date next due: _____

Figure 6.22 Adhesive calibration label

It will also be issued with a calibration certificate detailing the tests that have been carried out and a reference to the equipment used. The user of the instrument should always check to ensure that the instrument is within calibration before being put to use.

A further adhesive label (see Figure 6.23) is often placed over the joint in the instrument casing stating that the calibration is void should the seal be broken. A broken seal will indicate whether anyone has deliberately opened the instrument and possibly tampered with the internal circuitry.

> **Calibration void if seal is broken**

Figure 6.23 Calibration seal

Records of calibration and copies of all calibration certificates should all be retained in a safe and secure location. It is also best practice that any instrument should be checked for accuracy/re-calibrated after instances of rough electrical or mechanical handing.

Types of instrument

Low resistance ohmmeters

The instrument used for low resistance tests may be either a specialised low resistance ohmmeter, or the continuity range of an insulation and continuity tester. The test current may be d.c. or a.c. It is recommended that it be derived from a source with no-load voltage between 4 V and 24 V, and a short-circuit current not less than 200 mA.

The measuring range should cover the span 0.2 Ω to 2 Ω with a resolution of at least 0.01 Ω for digital instruments (instruments to BS EN 61557-4 will meet these requirements).

> **Remember**
> BS 7671 requires that the instruments used for measuring insulation resistance must be capable of providing the test voltages stated above while maintaining a test current of 1 mA. Instruments that are manufactured to BS EN 61557 will satisfy the above requirements.

Errors in the reading obtained can be introduced by contact resistance or by lead resistance. Although the effects of contact resistance cannot be eliminated entirely and may introduce errors of 0.01 Ω or greater, lead resistance can be eliminated either by clipping the leads together and zeroing the instrument before use (where this facility is provided) or alternatively measuring the resistance of the leads and subtracting this from the reading obtained.

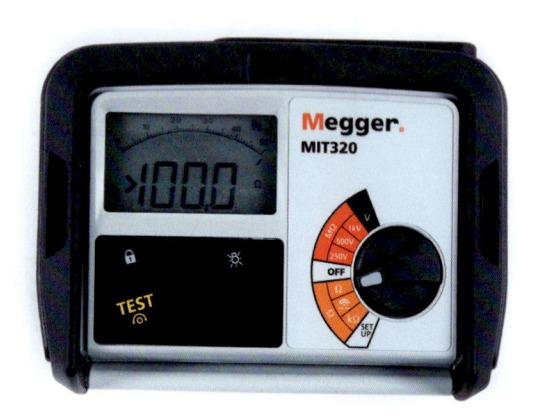

Figure 6.24 Low resistance ohmmeter (set to orange for low resistance test)

Thermocouple effects can be eliminated by reversing the test probes and averaging the resistance readings taken in each direction.

Modern machines combine low resistance ohmmeters with insulation resistance ohmmeters in one device.

Insulation resistance ohmmeters

The instrument used should be capable of developing the test voltage required across the load. Modern ohmmeters have insulation resistance and low resistance testers in one device. The d.c. test voltages required by Table 61 of BS 7671 are:

- 250 V for SELV and PELV
- 500 V for all circuits rated up to and including 500 V, but excluding extra-low voltage circuits mentioned above
- 1000 V d.c. for circuits rated above 500 V up to 1000 V.

Instruments conforming to BS EN 61557-2 will fulfil all the above instrument requirements.

Other features of this particular type of instrument are the ability to lock the instrument in the 'on' position for hands-free operation and an automatic nulling device for taking account of the resistance of the test leads.

Factors affecting in-service reading accuracy include 50 Hz currents induced into cables under test, and capacitance in the test object. These errors cannot be eliminated by test procedures. Capacitance may be as high as 5 μF, and the instrument should have an automatic discharge facility capable of safely discharging such a capacitance. Following an insulation resistance test, the instrument should be left connected until the capacitance within the installation has fully discharged.

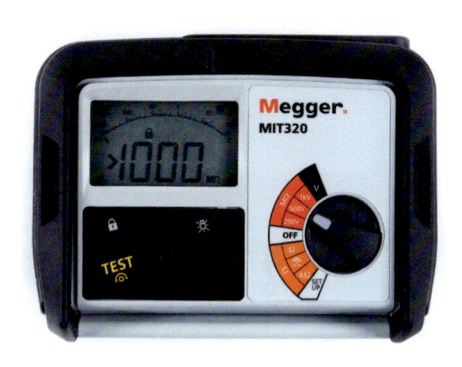

Figure 6.25 Insulation resistance ohmmeter (set to red for insulation resistance)

Earth-fault loop impedance testers

These instruments operate by circulating a current from the line conductor into the protective earth. This will raise the potential of the protective earth system. To minimise electric shock hazard from the potential of the protective conductor, the test duration should be within safe limits.

Instrument accuracy decreases as the scale reading reduces. Aspects affecting in-service reading accuracy include transient variations of mains voltage during the test period, mains interference, test lead resistance and errors in impedance measurement as a result of the test method. To allow for the effect of transient voltages the test should be repeated at least once. The other effects cannot be eliminated by test procedures.

For circuits rated up to 50 A, a line-earth loop tester with a resolution of 0.01 Ω should be adequate. In general, such instruments can be relied upon to be accurate down to values of around 0.2 Ω.

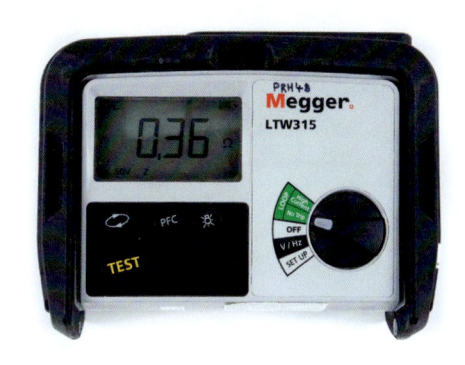

Figure 6.26 Earth-fault loop impedance tester

Instruments conforming to BS EN 61557-3 will fulfil the above requirements.

These instruments often offer additional facilities for deriving prospective short circuit current. The basic measuring principle is generally the same as for earth-fault loop impedance testers. The current is calculated by dividing the earth-fault loop impedance value into the mains voltage. Instrument accuracy is determined by the same factors as for loop testers. In this case, instrument accuracy decreases as scale reading increases, because the loop value is divided into the mains voltage. It is important to note these aspects, and the manufacturer's documentation should be referred to.

Earth electrode resistance testers

This may be a four-terminal instrument (or a three-terminal one where a combined lead to the earth electrode would not have a significant resistance compared with the electrode resistance) so that the resistance of the test leads and temporary spike resistance can be eliminated from the test result. One new development is that some meters now offer 'stakeless' testing.

Aspects affecting in-service reading accuracy include the effects of temporary spike resistance, interference currents and the layout of the test electrodes. The instrument should carry some facility to check that the resistance to earth of the temporary potential and current spikes are within the operating limits of the instrument. It may be helpful to note that instruments complying with BS EN 61557-5 incorporate this facility. Care should be exercised to ensure that temporary spikes are positioned with reasonable accuracy.

Figure 6.27 Earth electrode resistance tester

RCD testers

The test instrument should be capable of applying the full range of test current to an in-service accuracy as given in BS EN 61557-6. This in-service reading accuracy will include the effects of voltage variations around the nominal voltage of the tester.

To check RCD operation and to minimise danger during the test, the test current should be applied for no longer than 2 s. Instruments conforming to BS EN 61557-6 will fulfil the above requirements.

Figure 6.28 RCD tester

Phase rotation instruments

BS EN 61557-7 gives the requirements for measuring equipment for testing the phase sequence in three-phase distribution systems whether indication is given by mechanical, visual and/or audible means.

Indication should be unambiguous between 85% and 110% of the nominal system voltage or within the range of the nominal voltage and between 95% and 105% of the nominal system frequency.

The measuring equipment should be suitable for continuous operation.

'All in one' test instruments

A 'modern' innovation by manufacturers is the production of an 'all in one' instrument that has the ability to carry all of the tests required by BS 7671.

An example of this type of instrument is shown in Figure 6.30. The tests it can perform include:

- continuity test
- insulation resistance test
- RCD test
- loop impedance with no trip feature
- prospective short circuit current (L-N, L-E)
- polarity test
- phase rotation
- earth electrode resistance and resistivity testing
- power
- energy
- power factor
- harmonic monitoring
- light/lux measurements
- cable/fuse location.

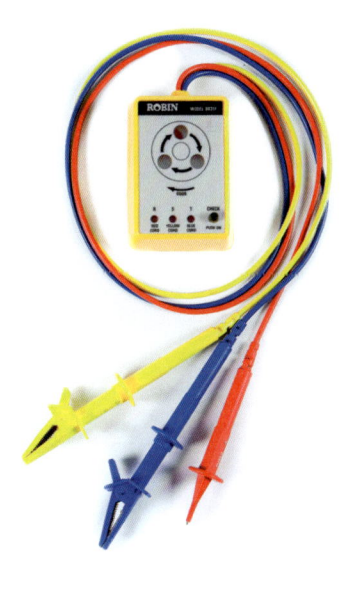

Figure 6.29 Phase rotation instrument

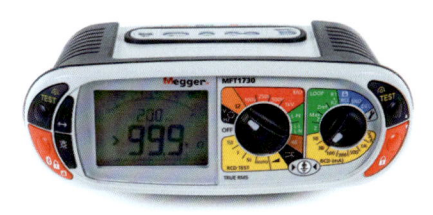

Figure 6.30 All-in-one tester

Working life

Your site supervisor asks you, to explain the need for carrying out electrical condition reporting to the apprentice working with you.

1. Where are the requirements made in BS 7671?
2. What is the purpose of an Electrical Condition Report?
3. What would determine a classification of unsatisfactory on the report?
4. Can an Electrical Condition Report be used to certify an additional circuit?

Progress check

1. What is the sequence of tests to be carried out on a new installation?
2. What precautions would be taken before conducting an insulation-resistance test?
3. Describe how a measurement of Z_e is made and explain the significance of this test.
4. What considerations should be made when deciding upon the frequency of inspection and testing?

K5. Understand the procedures and requirements for the completion of electrical installation certificates and related documentation

Purpose and information contained within certificates and documentation

BS 7671 Chapter 631 requires the completion of certificates upon completion of a new installation or changes to an existing installation; upon completion of a periodic inspection and upon completion of any minor electrical works. BS 7671 Appendix 6 then provides examples of model forms to be used for certification and reporting as follows.

The Electrical Installation Certificate and certification process

The Electrical Installation Certificate (see Figure 6.31) is designed for use when inspecting and testing a new installation, a major alteration or an addition to an existing installation. Where the design, construction, and inspection and testing of the installation are the responsibility of one person a single signature declaration should be made in lieu of signing every box.

BS 7671 Appendix 6 goes on to give the following guidance regarding the Electrical Installation Certificate. The first section is aimed at the person completing the certificate and the second section at the person receiving the certificate.

Electrical Installation Certificate Section 1

1. The Electrical Installation Certificate is to be used only for the initial certification of a new installation or for an addition or alteration to an existing installation where new circuits have been introduced.

   It is not to be used for a Periodic Inspection, for which an Electrical Installation Condition Report form should be used. For an addition or alteration which does not extend to the introduction of new circuits, a Minor Electrical Installation Works Certificate may be used.

   The 'original' certificate is to be given to the person ordering the work (Regulation 632.31). A duplicate should be retained by the contractor.

ELECTRICAL INSTALLATION CERTIFICATE
(REQUIREMENTS FOR ELECTRICAL INSTALLATIONS - BS 7671 [IET WIRING REGULATIONS])

DETAILS OF THE CLIENT
....
....

INSTALLATION ADDRESS
....
....

DESCRIPTION AND EXTENT OF THE INSTALLATION Tick boxes as appropriate

Description of installation: ....

Extent of installation covered by this Certificate: ....
....
....
....
....

(Use continuation sheet if necessary)   see continuation sheet No: .....

- New installation ☐
- Addition to an existing installation ☐
- Alteration to an existing installation ☐

FOR DESIGN
I/We being the person(s) responsible for the design of the electrical installation (as indicated by my/our signature(s) below), particulars of which are described above, having exercised reasonable skill and care when carrying out the design hereby CERTIFY that the design work for which I/we have been responsible is to the best of my/our knowledge and belief in accordance with BS 7671:2008 amended to .... (date) except for the departures, if any, detailed as follows:

Details of departures from BS 7671 (Regulations 120.3 and 134.1.8):

The extent of liability of the signatory or the signatories is limited to the work described above as the subject of this Certificate.

For the DESIGN of the installation: **(Where there is mutual responsibility for the design)

Signature: .... Date: .... Name (IN BLOCK LETTERS): .... Designer No. ...

Signature: .... Date: .... Name (IN BLOCK LETTERS): .... Designer No. ...

FOR CONSTRUCTION
I/We being the person(s) responsible for the construction of the electrical installation (as indicated by my/our signature(s) below), particulars of which are described above, having exercised reasonable skill and care when carrying out the construction hereby CERTIFY that the construction work for which I/we have been responsible is to the best of my/our knowledge and belief in accordance with BS 7671:2008 amended to .... (date) except for the departures, if any, detailed as follows:

Details of departures from BS 7671 (Regulations 120.3 and 134.1.8):

The extent of liability of the signatory is limited to the work described above as the subject of this Certificate.

For the CONSTRUCTION of the installation:

Signature: .... Date: .... Name (IN BLOCK LETTERS): .... Constructor No. ...

FOR INSPECTION & TESTING
I/We being the person(s) responsible for the inspection & testing of the electrical installation (as indicated by my/our signature(s) below), particulars of which are described above, having exercised reasonable skill and care when carrying out the inspection & testing hereby CERTIFY that the work for which I/we have been responsible is to the best of my/our knowledge and belief in accordance with BS 7671:2008 amended to .... (date) except for the departures, if any, detailed as follows:

Details of departures from BS 7671 (Regulations 120.3 and 134.1.8):

The extent of liability of the signatory is limited to the work described above as the subject of this Certificate.

For INSPECTION AND TESTING of the installation:

Signature: .... Date: .... Name (IN BLOCK LETTERS): .... Inspector No. ...

NEXT INSPECTION
I/We the designer(s), recommend that this installation is further inspected and tested after an interval of not more than .... years/months

Figure 6.31 Electrical Installation Certificate (page 1 of 2)

PARTICULARS OF SIGNATORIES TO THE ELECTRICAL INSTALLATION CERTIFICATE

Designer (No 1)
Name: .... Company: ....
Address: ....
.... Postcode: .... Tel No: ....

Designer (No 2) (if applicable)
Name: .... Company: ....
Address: ....
.... Postcode: .... Tel No: ....

Constructor
Name: .... Company: ....
Address: ....
.... Postcode: .... Tel No: ....

Inspector
Name: .... Company: ....
Address: ....
.... Postcode: .... Tel No: ....

SUPPLY CHARACTERISTICS AND EARTHING ARRANGEMENTS *Tick boxes and enter details, as appropriate*

Earthing arrangements	Number and Type of Live Conductors	Nature of Supply Parameters	Supply Protective Device Characteristics
TN-C ☐	a.c. ☐ d.c. ☐	Nominal voltage, U/U_o (1) V	Type: ....
TN-S ☐	1-phase, 2-wire ☐ 2-pole ☐	Nominal frequency, f (1) .... Hz	Rated current .... A
TN-C-S ☐	2-phase, 3-wire ☐ 3-pole ☐	Prospective fault current, I_{pf} (2) .... kA	
TT ☐	3-phase, 3-wire ☐ other ☐	External loop impedance, Z_e (2) .... Ω	
IT ☐	3-phase, 4-wire ☐	*(Note: (1) by enquiry, (2) by enquiry or by measurement)*	
Alternative source ☐ of supply (to be detailed on attached schedules)			

PARTICULARS OF INSTALLATION REFERRED TO IN THE CERTIFICATE *Tick boxes and enter details, as appropriate*

Means of Earthing
Distributor's facility ☐

Installation earth electrode ☐

Maximum Demand
Maximum demand (load) .... kVA / Amps *Delete as appropriate*

Details of Installation Earth Electrode (*where applicable*)

Type (e.g. rod(s), tape etc)	Location	Electrode resistance to Earth
....	....	.... Ω

Main Protective Conductors

Earthing conductor: material .... CSA ....mm² Continuity and connection verified ☐

Main protective bonding conductors material .... CSA ....mm² Continuity and connection verified ☐

To incoming water and/or gas service ☐   To other elements ....

Main Switch or Circuit-breaker

BS, Type and No. of poles ....   Current rating ....A   Voltage rating V

Location ....   Fuse rating or setting ....A

Rated residual operating current $I_{\Delta n}$ = ... mA, and operating time of .... ms (at $I_{\Delta n}$) *(Applicable only where an RCD is suitable and is used as a main circuit-breaker.)*

COMMENTS ON EXISTING INSTALLATION: *(In the case of an addition or alteration see Section 633)*

....
....
....
....

SCHEDULES
The attached Schedules are part of this document and this Certificate is valid only when they are attached to it.
.... Schedules of Inspections and .... Schedules of Test Results are attached.
(Enter quantities of schedules attached.)

Figure 6.31 Electrical Installation Certificate (page 2 of 2)

2. This certificate is only valid if accompanied by the Schedule of Inspections and the Schedule(s) of Test Results.

3. The signatures appended are those of the persons authorised by the companies executing the work of design, construction, inspection and testing respectively. A signatory authorised to certify more than one category of work should sign in each of the appropriate places.

4. The time interval recommended before the first periodic inspection must be inserted.

5. The page numbers for each of the Schedules of Test Results should be indicated, together with the total number of sheets involved.

6. The maximum prospective fault current recorded should be the greater of either the short circuit current or the earth-fault current.

7. The proposed date for the next inspection should take into consideration the frequency and quality of maintenance that the installation can reasonably be expected to receive during its intended life, and the period should be agreed between the designer, installer and other relevant parties.

Electrical Installation Certificate Section 2: Guidance for recipients (to be appended to the Certificate)

This safety certificate has been issued to confirm that the electrical installation work to which it relates has been designed, constructed, inspected and tested in accordance with British Standard 7671 (the IET Wiring Regulations).

You should have received an 'original' certificate and the contractor should have retained a duplicate certificate. If you were the person ordering the work, but not the owner of the installation, you should pass this certificate, or a full copy of it including the schedules, immediately to the owner.

The 'original' certificate should be retained in a safe place and be shown to any person inspecting or undertaking further work on the electrical installation in the future. If you later vacate the property, this certificate will demonstrate to the new owner that the electrical installation complied with the requirements of British Standard 7671 at the time the certificate was issued.

The Construction (Design and Management) Regulations require that, for a project covered by those Regulations, a copy of this certificate, together with schedules, is included in the project health and safety documentation.

For safety reasons, the electrical installation will need to be inspected at appropriate intervals by a competent person. The maximum time interval recommended before the next inspection is stated on Page 1 under 'Next Inspection'.

This certificate is intended to be issued only for a new electrical installation or for new work associated with an addition or alteration to an existing installation. It should not have been issued for the inspection of an existing electrical installation. An 'Electrical Installation Condition Report' should be issued for such an inspection.

The inspection schedule

A requirement of Regulation 631, the inspection schedule provides confirmation that a visual inspection has been carried out as required by Part 6 of BS 7671.

As stated, the completed inspection schedule is attached to and forms part of the Electrical Installation Certificate.

Each item on the inspection schedule (see page 88) should be checked and either ticked as satisfactory or ruled out if not applicable. On completion of the inspection schedule it should be signed and dated by the person responsible for carrying out the inspection.

Schedule of Test Results

Also a requirement of Regulation 631, the Schedule of Test Results (see Figure 6.32) is a written record of the results obtained when carrying out the electrical tests required by Part 6 of BS 7671 and must be attached to the Electrical Installation Certificate.

The Minor Electrical Installation Works Certificate

The Minor Works Certificate (see Figure 6.33) is intended to be used for additions and alterations to an installation that do not extend to the provision of a new circuit.

Examples include the addition of a socket outlet or a lighting point to an existing circuit, the relocation of a light switch etc.

This certificate may also be used for the replacement of equipment such as accessories or luminaires, but not for the

Unit ELTK 06

DB reference no ....
Location ....
Zs at DB (I) ....
I_{pf} at DB (kA) ....
Correct polarity of supply confirmed   yes/no
Phase rotation confirmed (where appropriate) ☐

Details of circuits and/or installed equipment vulnerable to damage when testing ....

Details of test instruments used (state serial and/or asset numbers
Continuity ....
Insulation resistance ....
Earth fault loop impedance ....
Earth electrode resistance ....
RCD ....

Circuit details																Test results							
		Overcurrent device				Conductor details			Ring final Circuit continuity (Ω)					Continuity (Ω) $(R_1 + R_2)$ or R_2		Insulation Resistance (MΩ)		Correct polarity (yes/no)	Z_s (Ω)	RCD (ms)			Remarks (continue on a separate sheet if necessary)
Circuit number	Circuit Description	BS (EN)	type	Rating (A)	PFC (kA)	Reference Method	Live (mm²)	cpc (mm²)	r_1 (line)	r_n (neutral)	r_2 (cpc)	Cross-connected line and neutral	Cross-connected line and cpc*	$(R_1 + R_2)$	R_2	Live-Live	Live-E			@IΔn	@5IΔn	Test button operation	
1	2	3	4	5	6	7	8	9	10	11	12	13	14	15	16	17	18	19	20	21	22	23	24

*Where there are no spurs connected to a ring circuit this value is also the $(R_1 + R_2)$ of the circuit

Figure 6.32 Schedule of Test Results

MINOR ELECTRICAL INSTALLATION WORKS CERTIFICATE
REQUIREMENTS FOR ELECTRICAL INSTALLATIONS - BS 7671 [IET WIRING REGULATIONS]
To be used only for minor electrical work which does not include the provision of a new circuit.

PART 1: Description of minor works
1. Description of the minor works
2. Location/Address
3. Date minor works completed
4. Details of departures, if any, from BS 7671:2008

PART 2: Installation details
1. System earthing arrangement   TN-C-S ☐   TN-S ☐   TT ☐
2. Method of fault protection:
3. Protective device for the modified circuit: Type ....   Rating .... A
4. Omission of additional protection by 30 mA RCD
   for socket outlets (see 411.3.3) ☐
   for cables concealed in walls (see 522.6.102) ☐

Comments on existing installation, including adequacy of earthing and bonding arrangements (see Regulation 134.1.9):

PART 3: Essential tests
Earth continuity satisfactory ☐
Insulation resistance:
   Line/neutral.................... MΩ
   Line/earth...................... MΩ
   Neutral/earth................... MΩ
Earth-fault loop impedance .... Ω
Polarity satisfactory ☐
RCD operation (if applicable). Rated residual operating current $I_{\Delta n}$ .... mA and operating time of ...... ms (at $I_{\Delta n}$)

PART 4: Declaration
I/we CERTIFY that the said works do not impair the safety of the existing installation, that the said works have been designed, constructed, inspected and tested in accordance with BS 7671:2008 (IET Wiring Regulations), amended to .... (date) and that the said works, to the best of my/our knowledge and belief, at the time of my/our inspection, complied with BS 7671 except as detailed in Part 1 above.

Name: ....   Signature: ....
For and on behalf of: ....   Position: ....
Address: ....

Date: ....

Figure 6.33 Minor Electrical Installation Works Certificate

replacement of distribution boards or similar items. Appropriate inspection and testing, however, should always be carried out irrespective of the extent of the work undertaken.

The component parts of the certificate are described as follows:

Part 1 Description of minor works

1, 2 The minor works must be so described that the work that is the subject of the certification can be readily identified.

4 See Regulations 120.3 and 120.4. No departures are to be expected except in most unusual circumstances. See also Regulation 633.1.

Part 2 Installation details

2 The method of fault protection must be clearly identified, e.g. automatic disconnection of supply (ADS).

Part 3 Essential tests

The relevant provisions of Part 6 'Inspection and testing' of BS 7671 must be applied in full to all minor works. For example, where a socket outlet is added to an existing circuit it is necessary to:

1. Establish that the earthing contact of the socket outlet is connected to the main earthing terminal.
2. Measure the insulation resistance of the circuit that has been added to, and establish that it complies with Table 61 of BS 7671.
3. Measure the earth-fault loop impedance to establish that the maximum permitted disconnection time is not exceeded.
4. Check that the polarity of the socket outlet is correct.
5. (If the work is protected by an RCD) verify the effectiveness of the RCD.

Part 4 Declaration

1, 3 The Certificate shall be made out and signed by a competent person in respect of the design, construction, inspection and testing of the work.

As with the Electrical Installation Certificate, the Minor Works Certificate goes on to give guidance to the person receiving the certificate.

Minor Electrical Installation Works Certificate: Guidance for Recipients (to be appended to the certificate)

This certificate has been issued to confirm that the electrical installation work to which it relates has been designed, constructed, inspected and tested in accordance with British Standard 7671 (the IET Wiring Regulations).

You should have received an 'original' certificate and the contractor should have retained a duplicate. If you were the person ordering the work, but not the owner of the installation, you should pass this certificate, or a copy of it, to the owner.

A separate certificate should have been received for each existing circuit on which minor works have been carried out. This certificate is not appropriate if you requested the contractor to undertake more extensive installation work, for which you should have received an Electrical Installation Certificate.

The certificate should be retained in a safe place and be shown to any person inspecting or undertaking further work on the electrical installation in the future. If you later vacate the property, this certificate will demonstrate to the new owner that the minor electrical installation work carried out complied with the requirements of British Standard 7671 at the time the certificate was issued.

Note for installer

With respect to Part 2 item 4 of this Certificate, the omission of additional protection by 30 mA RCD for socket outlets (Regulation 411.3.3) and cables concealed in walls (Regulation 522.6.102), the person issuing this Certificate should be aware that they are the designer of the work. Additional protection by RCD should only be omitted where the designer is satisfied that this will not result in an increased risk of electric shock and that there is minimal risk of damage to the circuit cable due to penetration by screws, nails and the like.

Note for customer

With respect to Part 2 item 4 of this Certificate, the installer (designer of the installation) may have decided to omit additional protection by 30 mA RCD for socket outlets and cables concealed in walls, where in their opinion this omission will not result in an increased risk of electric shock.

ELECTRICAL INSTALLATION CONDITION REPORT

Section A. Details of the client / person ordering the report
Name ....
Address ....
....

Section B. Reason for producing this report ....
....
Date(s) on which inspection and testing was carried out ....

Section C. Details of the installation which is the subject of this report
Occupier ....
Address ....
....
Description of premises (tick as appropriate)
Domestic ☐   Commercial ☐   Industrial ☐   Other (include brief description) ☐ ....
Estimated age of wiring system .... years
Evidence of additions / alterations   Yes ☐   No ☐   Not apparent ☐   If yes, estimate age .... years
Installation records available?   Yes ☐   No ☐   Date of last inspection .... (date)

Section D. Extent and limitations of inspection and testing
Extent of the electrical installation covered by this report
....
....
Agreed limitations including the reasons (See Regulation 634.2) ....
....
Agreed with: ....
Operational limitations including the reasons (see page no ....) ....
....
The inspection and testing detailed in this report and accompanying schedules have been carried out in accordance with BS 7671: 2008 (IET Wiring Regulations) as amended to ....
It should be noted that cables concealed within trunking and conduits, under floors, in roof spaces, and generally within the fabric of the building or underground, have **not** been inspected unless specifically agreed between the client and inspector prior to the inspection.

Section E. Summary of the condition of the installation
General condition of the installation (in terms of electrical safety) ....
....
....
Overall assessment of the installation in terms of its suitability for continued use
                    SATISFACTORY / UNSATISFACTORY* (Delete as appropriate)
*An unsatisfactory assessment indicates that dangerous and/or potentially dangerous conditions have been identified.

Section F. Recommendations
Where the overall assessment of the suitability of the installation for continued use above is stated as UNSATISFACTORY, I / we recommend that any observations classified as *'Danger present'* (Code C1) or *'Potentially dangerous'* (Code C2) are acted upon as a matter of urgency.
Investigation without delay is recommended for observations identified as *'Requiring further investigation'*.
Observations classified as *'Improvement recommended'* (Code C3) should be given due consideration.

Subject to the necessary remedial action being taken, I / we recommend that the installation is further inspected and tested by .... (date)

Section G. Declaration
I/We, being the person(s) responsible for the inspection and testing of the electrical installation (as indicated by my/our signatures below), particulars of which are described above, having exercised reasonable skill and care when carrying out the inspection and testing, hereby declare that the information in this report, including the observations and the attached schedules, provides an accurate assessment of the condition of the electrical installation taking into account the stated extent and limitations in section D of this report.

INSPECTED AND TESTED BY:	REPORT AUTHORISED FOR ISSUE BY:
Name (Capitals)	Name (Capitals) ....
Signature	Signature ....
For/on behalf of	For/on behalf of ....
Position	Position ....
Address	Address ....
Date	Date ....

Section H. Schedule(s)
.... schedule(s) of inspection and .... schedule(s) of test results are attached.
The attached schedule(s) are part of this document and this report is valid only when they are attached to it.

Figure 6.34 Electrical Installation Condition Report (page 1 of 2)

Section I. Supply characteristics and earthing arrangements				
EARTHING ARRANGEMENTS	NUMBER AND TYPE OF LIVE CONDUCTORS	NATURE OF SUPPLY PARAMETERS	SUPPLY PROTECTIVE DEVICE
TN-S ☐ TN-C-S ☐ TT ☐ TN-C ☐ IT ☐	a.c. ☐ d.c. ☐ 1-phase, 2-wire ☐ 2-wire ☐ 2-phase, 3-wire ☐ 3-wire ☐ 3-phase, 4-wire ☐ Confirmation of supply polarity ☐	Nominal voltage, $U / U_0.$ [1] .... V Nominal frequency, $f.$ [1] .... Hz Prospective fault current, I_{pf} [2] .... kA External loop impedance Z_e [2] .... Ω Note: (1) by enquiry (2) by enquiry or by measurement	BS (EN) .... Type .... Rated Current .... A

Alternative source of supply (as detailed on attached schedule) ☐

Section J. Particulars of installation referred to in report

MEANS OF EARTHING	DETAILS OF INSTALLATION EARTH ELECTRODE (WHERE APPLICABLE)
Distributor's facility ☐ Installation earth electrode ☐	Type .... Location .... Resistance to Earth .... Ω

Section K. Main protective conductors

Earthing conductor	Material	Csa .... mm²	Connection / continuity verified ☐
Main protective bonding conductors	Material	Csa .... mm²	Connection / continuity verified ☐
To incoming water service ☐	To incoming gas service ☐	To incoming oil service ☐	To structural steel ☐
To lightning protection ☐	To other incoming service(s) ☐ Specify		

Section L. Main switch / switch-fuse / circuit-breaker / RCD

Location .... BS(EN) .... No of poles	Current rating .... A Fuse / device rating or setting ...... A Voltage rating V	**If RCD main switch** Rated residual operating current ($I_{\Delta n}$) .... mA Rated time delay .... ms Measured operating time (at $I_{\Delta n}$) .... ms

Section M. Observations

Referring to the attached schedules of inspection and test results, and subject to the limitations specified at the *Extent and Limitations of the Inspection and testing* section

No remedial action is required. ☐   The following observations are made

OBSERVATION(S)	CLASSIFICATION CODE	FURTHER INVESTIGATION REQUIRED (YES / NO)
....	....	....
....	....	....
....	....	....
....	....	....
....	....	....
....	....	....
....	....	....
....	....	....
....	....	....
....	....	....
....	....	....
....	....	....
....	....	....
....	....	....

One of the following codes, as appropriate, has been allocated to each of the observations made above to indicate to the person(s) responsible for the installation the degree of urgency for remedial action.

C1 – Danger present. Risk of injury. Immediate remedial action required
C2 – Potentially dangerous - urgent remedial action required
C3 – Improvement recommended

Figure 6.34 Electrical Installation Condition Report (page 2 of 2)

In this case, an RCD is an automatic switching device installed to disconnect the supply when a person makes contact with a live part. The installer should have discussed the omission of the RCD with you before carrying out the work.

The Electrical Installation Condition Report

Previously known as the 'Periodic Inspection & Testing Report', BS 7671 requires that the results of any periodic inspection and test should be recorded on an Electrical Installation Condition Report as shown in Figure 6.34.

When inspecting older installations that may have been installed in accordance with a previous edition of the IET Wiring Regulations, the installation may still be acceptable provided that all items which do not conform to the present edition of BS 7671 are reported and provided that no risk of shock, fire or burns exists. BS 7671 gives the following guidance for the person producing the report:

5. This report should only be used for the reporting on the condition of an existing electrical installation.

6. The report, normally comprising at least six pages, should include schedules of both the inspection and the test results. Additional pages may be necessary for other than a simple installation. The number of each page should be indicated, together with the total number of pages involved.

7. The reason for producing this report, such as change of occupancy or landlord's periodic maintenance, should be identified in Section B.

8. The maximum prospective fault current (I_{pf}) recorded should be the greater of either the short-circuit current or the earth-fault current.

9. Those elements of the installation that are covered by the report and those that are not should be identified in Section D (Extent and Limitations). These aspects should have been agreed with the person ordering the report and other interested parties before the inspection and testing is carried out. Any operational limitations, such as inability to gain access to parts of the installation or an item of equipment, should also be recorded in Section D.

10. The summary of condition of the installation in terms of safety should be clearly indicated in Section E. Observation(s), if any, should be categorised in Section M using the coding

C1 to C3 as appropriate. Any observation given a C1 or C2 classification should result in the overall condition of the installation being reported as unsatisfactory.

11. Where an installation has an alternative source of supply a further schedule of supply characteristics and earthing details based upon Section I of this report should be provided.

12. Where an observation requires further investigation because the inspection has revealed an apparent deficiency which could not, owing to the extent or limitations of this inspection, be fully identified, this should be indicated in the column headed 'Further investigation required' within Section M.

13. The date by which the next electrical installation condition report is required should be given in Section F. The interval between inspections should take into account the type and usage of the installation and its overall condition.

14. If the space available for observations in Section M is insufficient, additional pages should be provided as necessary.

15. Wherever practicable, items classified as 'Danger present' (C1) should be made safe on discovery. Where this is not practical the owner or user should be given written notification as a matter of urgency.

Electrical Installation Condition Report: Guidance for recipients (to be appended to the Report)

This report is an important and valuable document which should be retained for future reference.

This report form is for reporting on the condition of an existing electrical installation.

1. The purpose of this condition report is to confirm, so far as reasonably practicable, whether or not the electrical installation is in a satisfactory condition for continued service (see Section E). The report should identify any damage, deterioration, defects and/or conditions which may give rise to danger (see Section M).

2. The person ordering the report should have received the original report and the inspector should have retained a duplicate.

3. The original report should be retained in a safe place and be made available to any person inspecting or undertaking work on the electrical installation in the future. If the property is

vacated, this report will provide the new owner/occupier with details of the condition of the electrical installation at the time the report was issued.

4. Where the installation incorporates residual current devices (RCDs) there should be a notice at or near the devices stating that they should be tested quarterly. For safety reasons it is important that these instructions are followed.

5. Section D (Extent and Limitations) should identify fully the extent of the installation covered by this report and any limitations on the inspection and testing. The inspector should have agreed these aspects with the person ordering the report and with other interested parties (licensing authority, insurance company, mortgage provider and the like) before the inspection was carried out.

6. Some operational limitations such as such as inability to gain access to parts of the installation or an item of equipment may have been encountered during the inspection. The inspector should have noted these in Section D.

7. For items classified in Section M as C1 ('Danger Present'), the safety of those using the installation is at risk, and it is recommended that a competent person undertakes the necessary remedial work immediately.

8. For items classified in Section M as C2 ('Potentially Dangerous'), the safety of those using the installation may be at risk and it is recommended that a competent person undertakes the necessary remedial work as a matter of urgency.

9. Where it has been stated in Section M that an observation requires further investigation the inspection has revealed an apparent deficiency which could not, due to the extent or limitations of this inspection, be fully identified. Such observations should be investigated as soon as possible. A further examination of the installation will be necessary, to determine the nature and extent of the apparent deficiency (see Section F).

10. For safety reasons, the electrical installation will need to be re-inspected at appropriate intervals by a competent person. The recommended date by which the next inspection is due is stated in Section F of the report under 'Recommendations' and on a label near to the consumer unit/distribution board.

The Electrical Installation Condition Report must be accompanied by a schedule of test results and an inspection schedule. The actual report shown earlier finished at Section M. However, classed as Section N, BS 7671 then gives a set of model inspection forms applicable to various installation conditions:

- Schedule of inspections for a single distribution board installation for domestic and similar premises for use with Electrical Installation Condition Report
- Schedule of inspections for the main intake and associated circuits for use with Electrical Installation Condition Report (Main Intake and Associated Circuits Inspection Schedule)
- Schedule of inspections for a distribution board and its associated circuits for use with Electrical Installation Condition Report (Distribution Board Inspection Schedule for Multiple Distribution Board Installations).

Progress check

1. For what purpose should an Electrical Installation Certificate be issued and who to?
2. Explain briefly when a Minor Electrical Installation Certificate should be issued.
3. What is the main function of an Electrical Condition Report and could it be issued to cover the addition of lighting to a property?

Getting ready for assessment

All electrical installations can be potentially dangerous and a complete record of their testing, inspection and commissioning must be kept. This will allow the final users of the installation to be sure that what they are using is safe and that they are using it correctly.

For this unit you will need to be familiar with:

- the principles, regulatory requirements and procedures for completing the safe isolation of an electrical circuit and complete electrical installations in preparation for inspection, testing and commissioning
- principles and regulatory requirements for inspecting, testing and commissioning electrical systems, equipment and components
- regulatory requirements and procedures for completing the inspection of electrical installations
- regulatory requirements and procedures for the safe testing and commissioning of electrical installations
- procedures and requirements for the completion of electrical installation certificates and related documentation

For each learning outcome, there are several skills you will need to acquire, so you must make sure you are familiar with the assessment criteria for each outcome. For example, for Learning Outcome 3 you will need to be able to explain the cable termination techniques for a range of different cables, and explain the advantages, limitations and application of the different connection methods. You will also need to describe the procedures for proving terminations and connections are electrically and mechanically sound and the consequences of this not being the case. You will also need to name the health and safety requirements for terminating and conducting cables, conductors and flexible cords.

It is important to read each question carefully and take your time. Try to complete both progress checks and multiple choice questions, without assistance, to see how much you have understood. Refer to the relevant pages in the book for subsequent checks. Always use correct terminology as used in BS 7671. There are some simple tips to follow when writing answers to exam questions:

- **Explain briefly** – usually a sentence or two to cover the topic. The word to note is 'briefly' meaning do not ramble on. Keep to the point.
- **Identify** – refer to reference material, showing which the correct answers are.
- **List** – a simple bullet list is all that is required. An example could include, listing the installation tests required in the correct order.
- **Describe** – a reasonably detailed explanation to cover the subject in the question.

This unit has a large number of practical skills in it, and you will need to be sure that you have carried out sufficient practice and that you feel you are capable of passing any practical assessment. It is best to have a plan of action and a method statement to help you work. It is also wise to check your work at regular intervals. This will help to ensure that you are working correctly and help you to avoid any problems developing as you work. Remember, don't rush the job as speed will come with practice and it is important that you get the quality of workmanship right.

Good luck!

CHECK YOUR KNOWLEDGE

1. What is the minimum value of insulation resistance on a lighting circuit?
   a) 0.5 Ω
   b) 0.5 MΩ
   c) 1 Ω
   d) 1 MΩ

2. An item of equipment has an IP code of IP56, what does the second digit (6) apply to?
   a) Size of test finger to apply
   b) Protection against the ingress of solids
   c) Protection against the ingress of water
   d) Type of coating on the equipment surface

3. Which of the following is unacceptable as an earth electrode?
   a) Earth rod
   b) Water service pipe
   c) Earth plate
   d) Underground structural metalwork

4. Where a 30 mA RCD is being used to provide additional protection against contact with live parts, what must the maximum operating time not exceed?
   a) 40 ms when tested at 1 × 30 mA
   b) 40 ms when tested at 5 × 30 mA
   c) 100 ms when tested at 1 × 30 mA
   d) 100 ms when tested at 5 × 30 mA

5. What is the maximum period between inspection and testing of a domestic property as shown in IET Guidance Note 3?
   a) 3 months
   b) 1 year
   c) 5 years
   d) 10 years

6. What is the maximum period between inspection and testing of a construction site as shown in IET Guidance Note 3?
   a) 3 months
   b) 1 year
   c) 5 years
   d) 10 years

7. Which document lays down the safety requirements of test equipment leads?
   a) HSG 85
   b) HSE GS38
   c) BS7671
   d) BS5266

8. Which instrument is used for carrying out continuity tests?
   a) RCD tester
   b) High resistance ohmmeter
   c) Low resistance ohmmeter
   d) Earth-fault loop impedance tester

9. Who should an Electrical Installation Certificate be issued to?
   a) Architect
   b) Electrician
   c) Site Engineer
   d) Person ordering the work

10. Which of these should a Minor Electrical Installation Works Form **not** be issued for?
    a) New circuit
    b) Moving a light position
    c) Adding a new lighting point to an existing circuit
    d) Adding an extra socket outlet to an existing circuit

UNIT ELTK 07

Understanding the principles, practices and legislation for diagnosing and correcting electrical faults in electrotechnical systems and equipment

Electricians should have the ability to recognise when something is not up to standard or is not functioning correctly, such as when a piece of equipment is not suitable (e.g. a metal-clad switch that has been installed in a damp environment) or a circuit is not operating correctly (e.g. a two-way lighting circuit that is only operating from one switch).

Not all faults are easily visible; some are concealed and some may develop over a period of time. However, regular inspection, tests and maintenance checks will limit such faults.

This unit will cover the following learning outcomes:

- Understand the principles, regulatory requirements and procedures for completing the safe isolation of electrical circuits and complete electrical installations
- Understand how to complete the reporting and recording of electrical fault diagnosis and correction work
- Understand how to complete the preparatory work prior to fault diagnosis and correction work
- Understand the procedures and techniques for diagnosing electrical faults
- Understand the procedures and techniques for correcting electrical faults.

K1. Understand the principles, regulatory requirements and procedures for completing the safe isolation of electrical circuits and complete electrical installations

Any electrical work can be dangerous. Fault diagnosis and rectification has the potential to be most dangerous, as you will often be trying to repair something installed in a location that is probably now occupied by people.

Remember as well that it's not just about your own safety. For example, if you are asked to find a fault on a ward in a hospital, you cannot just switch off the supply to the whole ward without taking into account supplies to life-support machines or similar.

Equally, if a faulty ceiling-mounted lighting fitting in the busy main corridor of a leisure centre was reported, you would need to consider issues such as using barriers to ensure your safety from passers-by and ensuring that temporary lighting is used to keep the corridor safely lit. You may have to get agreement to close the corridor. As with all aspects of health and safety, risk assessment is crucial to the work process.

It may be that following risk assessment you will need to isolate supplies, erect barriers or place notices and signs. One thing you must certainly do is liaise with everyone concerned to make sure that they fully understand the extent of work to be carried out, whether that be investigative or corrective.

Details of how we carry out risk assessment, use work permits, use any relevant PPE and a full explanation of safe isolation procedures and other related issues are included in Unit ELTK 01 Health and Safety in Book A.

Understanding the health and safety issues in the electrical industry are vital to gaining a full understanding of the safe methods of working and working within health and safety requirements for tools, PPE and when in unsafe situations. ELTK 01 also covers the dangers of working with electricity.

K2. Understand how to complete the reporting and recording of electrical fault diagnosis and correction work

Experience and understanding of electrical installations takes years to acquire and a vast amount of knowledge is gathered during your apprenticeship. However, putting that knowledge into practice only really occurs when you take up your responsibilities as an electrician and when analysing faults on electrical installations even the trained electrician will need to ask questions to help in rectifying faults.

By now you should have come across many different types of wiring systems, equipment, enclosures and protective devices. Equally, you should have now used the procedures for safe isolation and locking off procedures many times. However, when a fault occurs, the art of finding information and asking questions can be a daunting task.

Understanding the electrical system, installation and equipment

These can generally be categorised as follows:

- Voltage – 230 volt single-phase or 400 volt three-phase
- Installation type – domestic, industrial or commercial etc.
- System type – lighting, power, fire alarm system, emergency lighting, heating etc.

If a fault occurs on a system which is in the process of being installed then the information and data for the system should be readily available. However, the main problem starts when you are asked to rectify a fault on a system that you have had no experience or knowledge of. In this situation you will need to access drawings and data to familiarise yourself with layout, distribution boards etc. The type of information that you will require when called out to a fault can be listed as:

- type of supply – single-phase or three-phase
- nominal voltage – 230 volt or 400 volt (remember this when using your voltage indicator)
- type of earthing supply system – TT, TN-S, TN-C-S
- type of protective measure – ADS (automatic disconnection of supply)

> **Remember**
>
> An electrician's knowledge and training can be a great advantage in recognising and tracing faults, but the task will be made easier and the fault will be remedied quicker if the information listed is available. If this information is not readily available, you should seek advice or try to obtain it.

- types of protective device – HRC, MCB, RCD etc.
- ratings of devices
- location of incoming supply services
- location of electrical services
- distribution board schedules
- location drawings
- design and manufacturer's data
- nature of the fault.

Optimum use of personal and other people's experience of systems and equipment

Your own knowledge of an installation can be an asset not only in the fault finding process, but also to your company for future business. If your company can solve problems quickly and be relied upon by a client when a problem occurs, it will be your company that will be the first to be called.

The fault may not exist on the installation wiring but on auxiliary equipment such as refrigeration and air conditioning and unless you have specialist-trained knowledge of such equipment, it is not safe to attempt to rectify the faults and your company may not be insured to work on such systems. However, if the fault is on the circuit supplying these systems, the fault may be investigated and rectified.

You may also have to work alongside other specialists, assisting each other in commissioning and testing such systems. On no account should you attempt to investigate or rectify faults on their systems without knowledge or experience and training.

Equally, you will often be asked to repair faults on wiring and equipment that have been caused by inexperienced personnel attempting to install or repair circuits or equipment.

When you first arrive at any installation to investigate and rectify a fault, you should always use the personnel present to help you to obtain any background to the fault and relevant information, and if possible seek the specialist knowledge of the person responsible for the electrical installation or general maintenance.

The different people that you will be dealing with to obtain special and essential knowledge and information could include:

- the electrician who previously worked on the system
- design engineer
- works engineer

- shift engineer
- maintenance electrician
- machine operator
- home owner
- site foreman
- shop manager
- school caretaker.

These persons may have access to:
- operating manuals
- wiring and connection diagrams
- manufacturer's product data/information
- maintenance records
- inspection and test results
- installation specifications
- drawings
- design data
- site diary.

Their experience of the installation and day-to-day knowledge is essential as it can help you to solve, replace or repair the fault more efficiently.

Providing relevant persons with information

Procedures vary in terms of being informed of a fault. It may be a colleague, a supervisor or a customer who reports it and their description may range from 'it's not working' to a complex detailed report. Remember to be courteous at all times when dealing with people, particularly clients and customers who may not have much technical knowledge or are uncertain when describing faults. It is important for your reputation and future business that you are able to put people at their ease and inspire their confidence.

Irrespective of this, the details of the fault should at least be written down and this information passed to the person who will be repairing the fault, ideally on a written job card. Remember, the more information that can be received and given about the problem, the easier it may be to find and repair the fault.

Customer relations

Good communication is essential throughout the entire diagnosis and correction process. Involving the customer throughout can assist the process and enable them to work around you.

The same is true when handing the installation back to the customer once any faults have been rectified. Take the time to explain what has been done and hand over all relevant paperwork such as test results or manufacturer's instructions if equipment has been replaced.

Remember that the time spent developing a good relationship will be useful in helping to find the fault and also may gain your employer future business, if the customer feels their problem was dealt with in a caring and efficient manner.

> **Progress check**
>
> 1. You have been asked to attend one of your customer's factory complexes to repair a fault. Who are the possible personnel you may need to refer to when checking the technical details of the installation?
>
> 2. List the usual sources of information held, that may assist you when diagnosing faults on site.
>
> 3. What are the main considerations when completing a handover to a customer following the rectification of faults?

K3/K4. Understand how to complete the preparatory work prior to fault diagnosis and correction work and the procedures and techniques for diagnosing electrical faults

To be able to competently investigate, diagnose and find faults on electrical installations and equipment is one of the most difficult jobs for an electrician. Therefore, if the electrician is to be successful in fault finding and diagnosis, a thorough working knowledge of the installation or equipment involved is essential.

The person carrying out the procedure should take a reasoned approach and apply logical steps or stages to the investigation and subsequent remedy. The electrician should also understand the limitations of their own knowledge and seek expert advice and support where necessary.

In an ideal world an electrician should not embark on any testing or fault finding without some forward planning. However, because of their very nature, an emergency or dangerous fault may allow very little time for planning the remedy or repair. That said, some faults that are visible and straightforward can be easily repaired, and some careful planning and liaison with the client will limit disruption.

Inspection and testing prior to energising a new installation can be important and can save embarrassment by rectifying problems before circuits are energised. As an example, pre-energising tests can be extremely important when checking the function of switching circuits, particularly two-way and intermediate switching.

Each electrician should therefore fully understand inspection and testing in order to find faults competently – in particular the correct use of test instruments, the choice of instrument for testing and knowledge of each instrument's range and limitations.

> **Remember**
>
> The IET Wiring Regulations on testing can be a useful guide, but the Electricity at World Regulations and safe use of test equipment should be followed during the process.

Logical stages of diagnosis and rectification

Some faults can be rectified very easily, especially when the electrician has a working knowledge of the installation. But there are many occasions when you may be called out to a repair where there is no information available. It is on these occasions that

> **Remember**
>
> Information is a necessity and no fault finding should commence without the relevant information and background to the fault being made available to the person carrying out the work. Some faults are recognisable by simply analysing test results and can be remedied simply.

your training and knowledge of wiring systems have to be used. This knowledge may not always be enough and a sequential and logical approach to rectifying a fault and gathering information is needed.

One popular approach is called the 'half-split technique', which simply means if you are told that there is a fault on a circuit, then split the circuit in to two parts (halves) and see which half has the fault in it. You can then repeat the process inside the half that you know has the fault in it to narrow down the area where the fault is.

We normally say that a sensible approach to finding the fault would be as follows.

- **Identify the symptom:** This can be done by establishing the events that led up to the problem or fault on the installation or equipment.
- **Gather information:** This is achieved by talking to people and obtaining and looking at any available information. Such information could include manufacturer's data, circuit diagrams, drawings, design data, distribution board schedules and previous test results and certification.
- **Analyse the evidence:** Carry out a visual inspection of the location of the fault and cross-reference with the available information. Interpret the collected information and decide what action or tests need to be carried out. Then determine the remedy.
- **Check supply:** Confirm supply status at origin and locally using an approved voltage indicator. Confirm circuit or equipment when fault is isolated from supply.
- **Check protective devices:** Check status of protective devices. If they have operated this would determine location of fault on circuit or equipment.
- **Isolate and test:** Confirm isolation prior to carrying out sequence of tests.
- **Interpret information and test results:** By interpreting the test results, the status of protective devices and other information, the fault may be identified and remedied/rectified.
- **Rectify the fault:** This may be done quite simply or parts or replacement may be needed.
- **Carry out functional tests:** Before restoring the supply, the circuit or equipment will need to be tested not only electrically but also for functionality.

- **Restore the supply:** Care must be taken that the device has been reset or repaired to correct current rating. Make sure the circuit or equipment is switched off locally before restoring supply.
- **Carry out live and functional tests:** Once the supply has been restored, it may be prudent to complete live tests to make sure the supply is stable, and also carry out any live functional tests such as operator switches etc.

In summary, the logical sequence of events would be:

1. Identify the symptoms
2. Gather information
3. Analyse the evidence
4. Check supply
5. Check protective devices
6. Isolation and test
7. Interpret information and test results
8. Rectify the fault
9. Carry out functional tests
10. Restore the supply
11. Carry out live and functional tests.

> **Remember**
>
> Some information regarding the fault will be gathered by asking people questions relating to the events leading up to the fault. If the fault occurred on the installation when the electrician was working on it then the information will be to hand. More often the situation regarding faults is when we are called out to a breakdown and asked to carry out repairs. In this situation, we need the information listed on pages 165–166.

Procedures for carrying out tests and interpretation of test results

Some faults can be recognised at the installation stage when the testing and inspection process is carried out. We should be able to recognise typical test results for each non-live test and interpret the type of fault that may exist.

BS 7671 Part 6 (612) lists non-live and live tests and as the non-live tests are carried out on the installation wiring circuits before they are ever energised, they will confirm the integrity of the circuit.

The non-live tests are:

- continuity of protective conductors including main and supplementary bonding
- continuity of ring final circuit conductors
- insulation resistance
- polarity.

> **Remember**
>
> Low voltage present at the equipment may also be causing operating problems, e.g. reduced levels of lighting.

Full details of the various tests plus related results, instruments and documentation are contained in Unit ELTK 06 of this book, on the following pages:

- Tests (pages 99–140)
- Instruments (pages 140–145)
- Documentation (pages 146–160).

> **Working life**
>
> You have been called to a property with the following reported fault: 'All the lights suddenly dimmed and the electric is now off.'
> 1. What should you do when you arrive?
> 2. What precautions should you take before commencing work?
> 3. Identify the steps necessary to identify the likely fault.

Non-live tests

Continuity of protective conductors including main and supplementary bonding

BS 7671 Regulation 411.3.1.1 requires that installations providing protection against electric shock using automatic disconnection of supply (ADS) must have a circuit protective conductor run to and terminated at each point in the wiring and at each accessory. An exception is made for a lamp holder having no exposed-conductive parts and suspended from such a point.

Regulation 612.2.1 then requires that every protective conductor, including circuit protective conductors, the earthing conductor and main and supplementary equipotential bonding conductors, should be tested to verify that they are electrically sound and correctly connected.

Test methods 1 (*linking the circuit line to the protective conductor at the DB*) and 2 (*using a long wandering lead*) are alternative ways of testing the continuity of protective conductors and should be performed using a low resistance ohmmeter. Remember that the resistance readings obtained include the resistance of the test leads and therefore the resistance of the test leads should be measured and deducted from all resistance readings obtained unless the instrument can auto-null.

Test method 1 as well as checking the continuity of the protective conductor, also measures ($R_1 + R_2$) which, when added to the external impedance (Z_e), enables the earth-fault loop impedance (Z_s) to be checked against the design. However, as ($R_1 + R_2$) is the resistance of the line conductor R_1 and the circuit protective conductor R_2 the reading may be affected by parallel paths through exposed-conductive parts and/or extraneous-conductive parts.

Installations incorporating steel conduit, steel trunking, MI and SWA cables will introduce parallel paths to protective conductors and similarly, luminaires fitted in grid ceilings and suspended from steel structures in buildings may create parallel paths.

In such situations, unless a plug and socket arrangement has been incorporated in the lighting system by the designer, the $(R_1 + R_2)$ test will need to be carried out prior to fixing accessories and bonding straps to the metal enclosures and finally connecting protective conductors to luminaires. Under these circumstances, some of the requirements may have to be visually inspected after the test has been completed. This consideration requires tests to be performed during the erection of an installation, in addition to tests at the completion stage.

When testing the effectiveness of main equipotential bonding conductors, the resistance between a service pipe/other extraneous-conductive part and the main earthing terminal should be about 0.05 Ω.

Similarly, supplementary bonding conductors should have a reading of 0.05 Ω or less.

Continuity of ring final circuit conductors

This test, carried out on the circuit wiring, is designed to highlight open circuits and interconnections within the wiring, and a low resistance ohmmeter is used to carry out this test. Typical readings will be 0.01 to 0.1 ohms depending on the conductor size and length of circuit wiring.

If the circuit is wired correctly, then the readings on the instrument will be the same at each point of test. If variable readings are found at each point of the test, this may indicate an open circuit or interconnections. If these types of fault exist, the consequence can be an overload on part of the circuit wiring.

Insulation resistance

This test is designed to confirm the integrity of the insulation resistance of all live conductors, between each other and earth. The type of instrument is known as an insulation resistance tester (not a megger) and typical test results would be around 50 to 100 megohms, i.e. 50 to 100 million ohms.

Depending on the circuit, Table 61 of BS 7671 indicates minimum values between 0.5 MΩ and 1 MΩ. These are minimum values and, if these values exist then the circuit may need further investigation. Generally, if a reading of less than 2 MΩ is recorded for an LV circuit then the circuit should be investigated.

Depending on where the test takes place, variable values of resistance may be recorded. In the case of a test on the supply side of a distribution board (Figure 7.01), a group value reading will indicate circuits connected in parallel (Figure 7.02), which invariably most are.

From Figures 7.01 and 7.02, it can be seen that when testing grouped circuit conductors, a parallel reading would indicate a poor reading, but if each circuit were tested individually then the actual reading is likely to be acceptable.

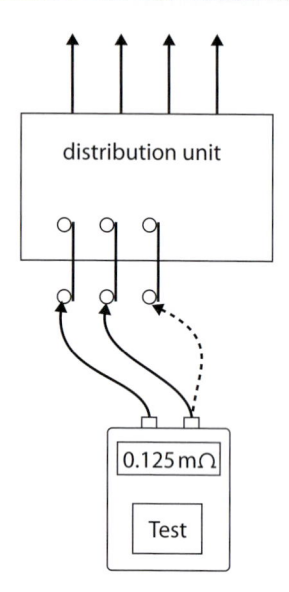

Figure 7.01 Test on supply side of distribution board

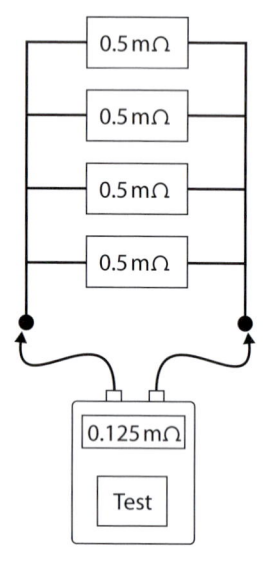

Figure 7.02 Test on circuits connected in parallel

Polarity

The polarity test can be carried out separately using a low resistance ohmmeter as used for continuity testing. However, as polarity of a ring final circuit is confirmed in the ring circuit test mentioned earlier, polarity of switches and lighting can equally be confirmed during the test on the continuity of protective conductors including main and supplementary bonding.

Typical test results will depend on the resistance of the conductors under test but will generally be low, around 0.01 to 0.1 ohms. Values in excess of this may indicate an open circuit or incorrect polarity.

Live tests

Earth-fault loop impedance testing

Overcurrent protective devices must, under earth-fault conditions, disconnect fast enough to reduce the risk of electric shock. This can be achieved if the actual value of the earth loop impedance does not exceed the tabulated values given in BS 7671. The purpose of the test, therefore, is to determine the actual value of the loop impedance (Z_s) for comparison with those values.

The test procedure requires that all main equipotential bonding is in place. The test instrument is then connected by the use of 'flying' leads to the phase, neutral and earth terminals at the remote end of the circuit under test. Press to test and record the results.

Once Z_s and the voltage have been established at the remote point in the circuit, you can divide the voltage by Z_s to give you the fault current to earth. Apply this calculation to the time current characteristic graphs shown in BS 7671 and you will be able to determine the actual disconnection time of the circuit.

RCD test

Regulation 612.13.1 states that where fault protection and/or additional protection is to be provided by an RCD, the effectiveness of any test facility incorporated in the device shall be verified.

Such a test would use an appropriate RCD test instrument and the test should confirm the operation of the device independently of the device's integral test button. Residual current circuit breakers or devices are rated in milliamperes and the test should show that the device operates within the milliampere ratings of the device and within the time constraints. Typically if the rated current of the device was 30 mA then the instrument should prove operation at this value within 200 milliseconds.

Functional testing

Regulation 612.13.2 states that assemblies such as switchgear and control gear assemblies, drives, controls and interlocks shall be subjected to a functional test, to show that they are properly mounted, adjusted and installed in accordance with relevant requirements of BS 7671.

Typical functional tests should be applied to:

- lighting controllers, switches etc.
- motors, fixings, drives, pulleys etc.
- motor controllers
- controls and interlocks
- main switches
- isolators.

Selecting test instruments and confirming their operation

As mentioned above, the uses, functions and calibration of test instruments was covered in Unit ELTK 06, pages 140–145. This section will give a quick reminder of the key points.

Voltage indicating devices

Instruments used solely for detecting a voltage fall into two categories:

- **Detectors that rely on an illuminated lamp (test lamp) or a meter scale (test meter)** – test lamps are fitted with a 15 watt lamp and should not give rise to danger if the lamp is broken. A guard should also protect it.
- **Detectors that use two or more independent indicating systems (one of which may be audible) and limit energy input to the detector by the circuitry used** – an example is a two-pole voltage detector, i.e. a detector unit with an integral test probe, an interconnecting lead and a second test probe.

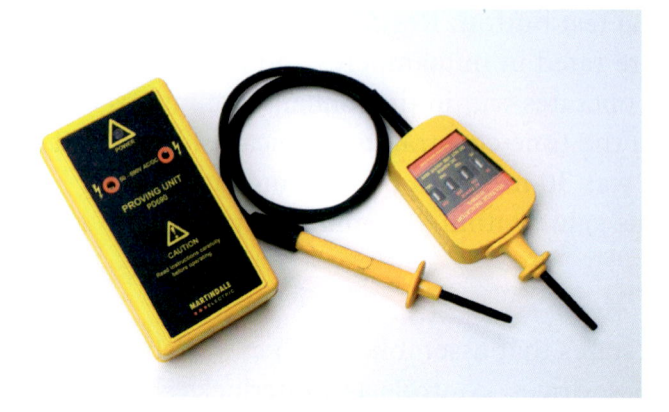

Figure 7.03 Voltage indicator device

Both these detectors are designed and constructed to limit the current and energy that can flow into the detector. This limitation is usually provided by a combination of the circuit design, using the concept of protective impedance, and current-limiting resistors built into the test probes.

The detectors are also provided with in-built test features to check their functioning before and after use. The interconnecting lead and second test probes are not detachable components. These types of detector do not require additional current-limiting resistors or fuses to be fitted, provided that they are made to an acceptable standard and the contact electrodes are shrouded.

Test lamps and voltage indicators are recommended to be clearly marked with the maximum voltage which may be tested by the device and any short time rating for the device if applicable. This rating is the recommended maximum current that should pass through the device for a few seconds. This is because these devices are generally not designed to be connected for more than a few seconds.

Safety tip

Care must be taken when using the external earth probe on this tester because the probe tip is at mains potential.

The limitation and range of instruments

The correct choice of instrument for each particular test should now be known, but if you are still unsure refer to Table 7.01.

Test	Instrument	Range
Continuity of protective conductors	Low resistance ohmmeter	0.005 to 2 ohms or less
Continuity of ring final circuit conductors	Low resistance ohmmeter	0.05 to 0.8 ohms
Insulation resistance	High reading ohmmeter	1 MΩ to greater than 200 MΩ
Polarity	Ohmmeter Bell/Buzzer	Low resistance None

Table 7.01 Limitation and range of test instruments

Some instruments can provide more than one facility and many manufacturers provide such instruments. Inspectors and testers should know the range of operation of their instruments in order to be conversant with their operation and understand the instructions for using the instrument.

When carrying out test procedures, and before any instrument is used, some simple checks should be carried out to ensure that the instrument is operating within its range and calibration. A suitable checklist for use prior to testing is given in Figure 7.04.

- ✔ Check the instrument and leads for damage or defects.
- ✔ Zero the instrument.
- ✔ Check the battery level.
- ✔ Select the correct scale for testing. If in doubt, ask or select the highest range available.
- ✔ Check the calibration date and record the serial number.
- ✔ Check leads for open and closed circuit prior to test.
- ✔ Record test results.
- ✔ After test, leave selector switches in off position. Some analogue instruments turn off automatically to save battery life.
- ✔ Always store instruments in their cases in secure, dry locations when not in use.

Figure 7.04 Checklist for testing

Common categories of electrical faults

In this section you will recognise different types of faults and understand that the consequences of the fault depend on its location within the installation or in a specific circuit. You should also understand the need for care when installing electrical systems and how poor, or a lack of, maintenance could lead to such faults. The essential points to be covered are:

- position of faults (complete loss of supply at the origin of the installation and localised loss of supply)
- operation of overload and fault current devices
- transient voltages
- insulation failure
- plant, equipment or component failure
- faults caused by abuse, misuse and negligence
- prevention of faults by regular maintenance.

Position of faults

Your knowledge of fault finding and the diagnosis of faults can never be complete because no two situations are exactly the same. Also, as technology advances and systems are being constantly improved, the faults that develop become more difficult to solve. An understanding of the electrical installation and the equipment we install becomes essential.

To diagnose faults an electrician will adopt basic techniques and these usually solve the most common faults. However, there are occasions when it may be impractical to rectify a fault.

This could be due to costs involved in down-time, or the cost of repair being more than the cost of replacement, in which case it could mean the renewal of wiring and/or equipment or components. Such situations and outcomes must be monitored, and the client should be kept informed and be made party to the decision to repair or replace.

Almost all faults should never exist. A fault can be compared to an accident and careful planning and thought can prevent many accidents. It is therefore part of a designer's job to build fault protection and damage limitation into the design of the electrical installation, such as:

- Installing more than one circuit, so that when a fault occurs on one of the circuits the fault is limited to affecting only that circuit.

Remember

Always keep the client informed.

- Installing fuses and protective devices to disable the fault and limit its effect.
- Ensuring the ability to access, maintain and repair. Good maintenance will limit faults, and access to the installation allows for maintenance and repair.

It is easier to find faults on installations where there are plenty of circuits.

Regulation 314.1 of BS 7671 states the following.

> Every installation shall be divided into circuits as necessary to:
> 1. avoid hazards and minimise inconvenience in the event of a fault
> 2. facilitate safe inspection, testing and maintenance (see also section 357)
> 3. take account of hazards that may arise from the failure of a single circuit such as a lighting circuit
> 4. reduce the possibility of unwanted tripping of RCDs due to excessive protective conductor (PE) currents not due to a fault
> 5. mitigate the effects of electromagnetic interferences (EMI)
> 6. prevent the indirect energising of a circuit intended to be isolated.

We can see that compliance with this regulation will help us to locate faults more easily, as usually by a process of elimination (operating each fuse individually) or simply looking at the device to see which one has operated we can identify a defective circuit or device.

Protective devices and simple fuses are designed to operate when they detect large currents due to excess temperature. In the case of a short circuit fault, high levels of fault current can develop, causing high temperatures and breakdown of insulation (see Figures 7.05 and 7.06). Such faults, if allowed to develop, can cause fires.

Figure 7.05 Meltdown of insulation due to overcurrent

When considering the consequences of a fault, it is important to know that the location of the fault can limit its severity with regard to disruption and inconvenience.

Figure 7.07 shows a typical layout of equipment for a small industrial or commercial installation where a three-phase supply is required and the load is divided on the three final consumer units, and will assist you in understanding simple installation layouts and the consequence of a located fault.

Figure 7.06 Meltdown of plastic connector due to overcurrent

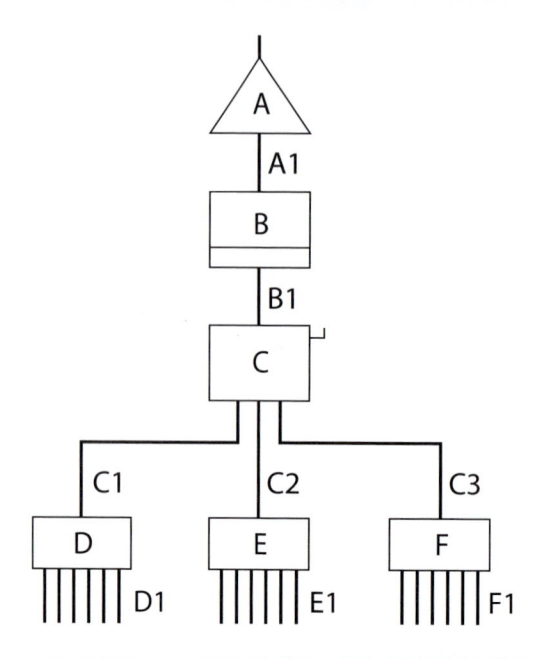

Key to Figure 7.07	
A	Supply company service with incoming supply cable and fuses.
A1	Supply company cables feeding KWh meter.
B	Supply company metering equipment KWh meter.
B1	Consumer's meter tails.
C	Consumer's main switch – typically 100 A TP&N.
C1	Sub-main circuit conductors (brown phase).
C2	Sub-main circuit conductors (black phase).
C3	Sub-main circuit conductors (grey phase).
D	Consumer unit with DP main switch and protective devices.
E	Consumer unit with DP main switch and protective devices.
F	Consumer unit with DP main switch and protective devices.
D1	Consumer's final circuits i.e. sockets and lighting.
E1	Consumer's final circuits i.e. sockets and lighting.
F1	Consumer's final circuits i.e. sockets and lighting.

Figure 7.07 Typical three-phase supply for small installation

Loss of supply

Complete loss of supply at the origin of the installation

With reference to Figure 7.07, it can be seen that a fault at any point from A to B1 could result in the complete loss of supply to the rest of the installation. The equipment at these points is mainly the property of the supply company and faults that appear are usually the result of problems with the supply, such as an underground cable being severed by workmen.

Faults on the installation wiring could also result in the supply company's protective devices operating but this is unusual and is only a consequence of poor design, misuse of the equipment or overloading of the supply company's equipment.

As an example of this, if the supply company's protective device or fuses were rated at 100 A per phase and the designer loaded each phase up at 200 A, this would result in a 100% overload and the consequence of this would be the operation of the mains protective devices and the whole installation being without power.

Localised loss of supply

Looking again at Figure 7.07, we can recognise what happens when different parts of an installation lose their supply. If, for example, a fault appeared on Cable C1 then the consumer unit (D) and all the final circuits from it would be dead. However, if a fault occurred on any one final circuit fed by that DB then only that circuit's protective device should have operated and only that circuit would be dead.

If a fault occurred on any one final circuit it should not affect any other final circuit or any other cable leading up to that consumer unit and therefore because the system in our previous diagram is three-phase, this should mean that a fault on any one phase should only result in that phase being affected, again limiting the consequence of a fault. This further highlights the reason for Regulation 314.1 and its requirements as mentioned earlier.

Operation of overload and fault current devices

When a fault is noticed, it is often because a circuit or piece of equipment has stopped working and this is usually because the protective device has done its job and operated.

The rating of a protective device should be greater than, or at least equal to, the rating of the circuit or equipment it is protecting, e.g. 10 × 100 watt lamps equate to a total current use of 4.35 A. Therefore a device rated at 5 or 6 A could protect this circuit.

A portable domestic appliance which has a label rating of 2.7 kW equates to a total current of 11.74 A. Therefore a fuse rated at 13 A should be fitted in the plug.

Protective devices are designed to operate when an excess of current (greater than the design current of the circuit) passes through it. The fault current's excess heat can cause a fuse element to rupture or the device mechanism to trip, depending on which type of device is installed. These currents may not necessarily be circuit faults, but short-lived overloads specific to a piece of equipment or outlet and BS 7671 categorises these as **overload current** and **overcurrent**.

For conductors, the rated value is the current-carrying capacity. Most excess currents are however due to faults, either **earth faults** or **short circuits,** which cause excessive currents.

Whichever type of fault occurs, the designer should take account of its effect on the installation wiring and choose a suitable device to disconnect the fault quickly and safely. The fundamental effect of any fault is a rise in current and therefore a rise in temperature.

High temperature destroys the properties of insulation, which in itself could lead to a short circuit. High currents damage equipment, and earth-fault currents present the danger of electric shock. Some examples of each are given below.

Overload faults

- adaptors used in socket outlets exceeding the rated load of the circuit
- extra load being added to an existing circuit or installation
- not accounting for the starting current on a motor circuit (start-up currents on motors are transient, short-lived heavy current, therefore overcurrent but **not** overload unless the motor is on load).

Short circuit faults

- insulation breakdown
- severing of live circuit conductors (e.g. penetration by a nail)
- inappropriate connection method
- wrong termination of conductors energised before being tested.

Figure 7.08 Wrong type of starter in fluorescent tube has melted due to excess power demand

Figure 7.09 Fuse holder has melted due to overloading

Earth-faults

- insulation breakdown
- incorrect polarity
- poor termination of conductors.

Arcing

If someone reports that there is a faulty switch they can hear 'cracking' when they operate it, they are probably reporting arcing. In the case of the light switch, arcing occurs when loose or corroded connections make an intermittent contact. This in turn causes sparking or 'arcing' between the connections. This translates into heat, which will break down the insulation of the wire and can result in fire.

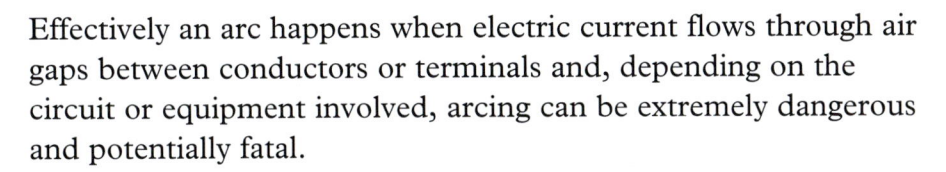

Figure 7.10 Poor termination of circuit protective conductors (cpc)

Effectively an arc happens when electric current flows through air gaps between conductors or terminals and, depending on the circuit or equipment involved, arcing can be extremely dangerous and potentially fatal.

The production of arc-fault current and voltage concentrated in one place, produces enormous energy. It can generate large amounts of heat that can severely burn human skin and set clothing on fire. High arc temperatures can vaporise the conductors in an explosive change in state from solid to vapour. Such a conductive vapour can help sustain the arc.

The duration of the arc is primarily determined by the time it takes for any overcurrent protective devices to open the circuit. For example, fast acting fuses may open the circuit in 8 ms or faster while other devices may take much longer to operate and open.

Arcing can cause metal to be blasted away from the fault location, as well as producing a shock wave that can blow people off their feet. This in turn can create a risk when touching other energised conductors.

Transient voltages

A transient voltage can be defined as a variance or disturbance to the normal voltage state, the normal voltage state being the voltage band in which the equipment is designed to operate.

It is the installation designer's responsibility to stipulate adequate sizes for the circuit conductors, but this sizing procedure will only prevent voltage drop during normal circuit conditions. Designers

Did you know?

Transient voltages are becoming increasingly problematic with the expanding use of IT equipment, not only in commerce and industry, but also in the home.

> **Did you know?**
>
> If transient voltages are thought to exist on installations, monitoring equipment can be installed to record events over periods of time. The results of the monitoring can usually trace the cause.

cannot prevent transient voltages as they are outside their control, but they can compensate by including equipment that will protect against such voltage variations and disturbances.

For example, it is now becoming common practice to install filter systems to IT circuits. These devices provide stabilised voltage levels and can also reduce the effect of voltage spikes by suppressing them. If transient voltages are not recognised they can cause damage to equipment and lead to a loss of data that could prove costly, especially if the user was an organisation such as a bank.

Some common causes of transient voltages are:

- supply company faults
- electronic equipment
- heavy current switching (causing voltage drops)
- earth-fault voltages
- lightning strikes.

Insulation failure

Insulation is designed primarily to separate conductors and to ensure their integrity throughout their life. Insulation is also used to protect the consumer against electric shock by protecting against direct contact and is also often used as a secondary protection against light mechanical damage, such as in the outer sheath of PVC/PVC twin and cpc cables. However, as previously mentioned, the breakdown of electrical insulation can lead to short circuit or earth-faults and when insulation of conductors and cables fail it is usually due to one or more of the following:

- poor installation methods
- poor maintenance
- excessive ambient temperatures
- high fault current levels
- damage by a third party.

Open circuit

We can say a circuit is 'open' when for some reason the circuit is incomplete and a break is preventing the flow of current. This may be because a conductor has come out of a terminal. However, be aware that an open circuit can be deliberate in that we use a switch to control a light and therefore it may be a defective component that is creating the open circuit.

Figure 7.11 Poorly connected 13 A plug top resulting in dangerous working conditions

Plant, equipment and component failure

It is said that nothing lasts forever and this is certainly true of electrical equipment. There will be some faults that you will attend that will be the result of a breakdown simply caused by wear and tear, although planned maintenance systems and regular testing and inspections can extend the life of equipment. Some common failures and their causes on installations and plant are explained below:

- **switches not operating** – damaged mechanism created by excessive use/age
- **motors not running** – probably due to age of brushes
- **lighting not working** – passed useful life of lamp
- **fluorescent luminaire not working** – new lamp or starter needed
- **outside PIR not switching** – ingress of insects/water causing failure
- **corridor socket outlet not working** – worn contacts created by excessive use/age.

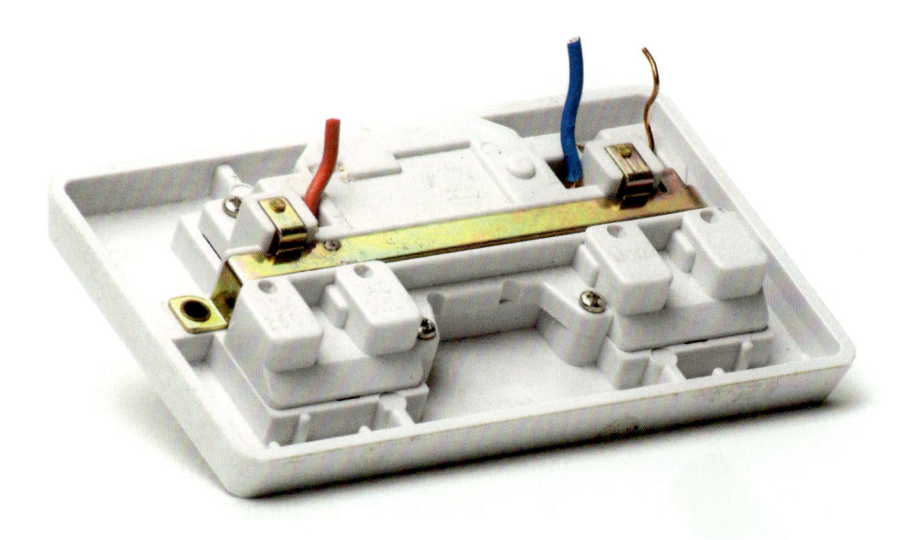

Figure 7.12 Poorly wired socket outlet

Faults caused by misuse, abuse and negligence

A common reason for faults on any electrical system or equipment is misuse, where the system or the equipment is simply not being used as it was designed. Every item of electrical equipment comes with user instructions that usually cover procedures and precautions and such instructions should be read carefully as misuse can lead to invalidation of any guarantees.

> **Remember**
>
> If you are called out to repair/rectify a fault which you find was caused by misuse of the system, then the guarantee for the installation may have been invalidated by the fault. In this case the client may have to pay the repair costs.

When an electrical installation is completed, a manual is normally handed over to the client that includes all product data and installation schedules and test results. This data will help the client when additions to the installation are made, inspections and tests are carried out or to assist in maintaining the installation.

Some faults are caused by carelessness during the installation process and simple errors such as poor termination or stripping of conductors can also lead to serious short circuits or overheating. Therefore good mechanical and electrical processes should always be carried out and every installation should be tested and inspected prior to being energised.

Some examples of faults arising from misuse are as follows:

User misuse

- using an MCB as a switch where constant use could lead to breakdown
- unplugging on-load portable appliances thus damaging socket terminals
- damp accessories, for instance, due to hosing walls in dry areas
- MCBs' nuisance-tripping caused by connecting extra load to circuits.

Installer misuse

- poor termination of conductors – overheating due to poor electrical contact
- loose bushes and couplings – no earth continuity – electric shock risk!
- wrong size conductors used – excessive voltage drop – excess current which could lead to an inefficient circuit and overheating of conductors (remember that low voltage present at the equipment may also be causing operating problems, e.g. reduced levels of lighting)
- not protecting cables when drawing them in to enclosures – causing damage to insulation
- overloading conduit and trunking capacities – causing overheating and insulation breakdown.

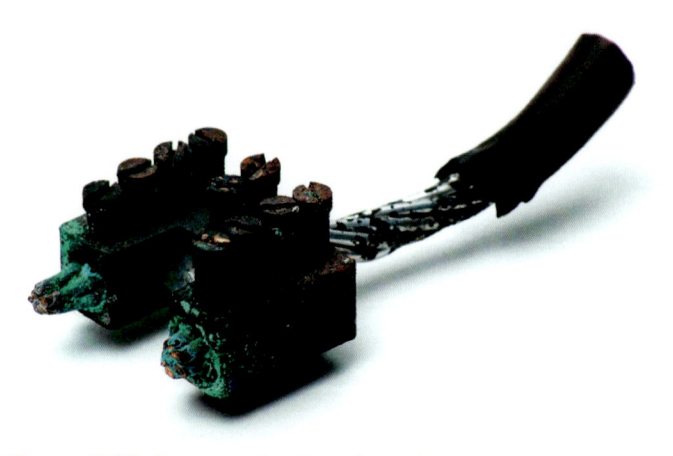

Figure 7.13 Poor termination of conductors causing electric shock risk

> **Working life**
>
> A customer has undertaken some DIY work replacing an electric shower themselves. They have since found that when the shower was in use, the TV flickered and the lights were dim. On your arrival the shower is not working. You are shown the rewireable fuse protecting the circuit and observe the element is melted in the middle.
>
> 1. Where would you start?
> 2. What are the likely problems?
> 3. How would you explain the likely problem to the customer?

Prevention of faults by regular maintenance

A well-designed electrical installation should serve the consumer efficiently and safely for many years, but an installation that is not regularly inspected and tested could lead to potential faults and areas of weakness not being discovered. Even a visual inspection can uncover simple faults which, if left, could lead to major problems in the future.

Large organisations in the commercial sector and industry have realised the need for regular inspections and planned maintenance schedules. Some major supermarkets choose to have regular maintenance take place when the stores are closed to the public, as they realise a store closure or part-closure to carry out corrective electrical work could be costly in terms of lost sales.

A typical example of maintenance in shops is lamp cleaning, changing and repairs and servicing to refrigeration and air conditioning. Large manufacturing companies usually carry out maintenance during planned holiday periods when the workforce is on annual leave and machinery, equipment and the electrical installation wiring can be accessed without loss of production.

Carrying out repairs and servicing in this way gives electricians and engineers the opportunity to test installations, carry out remedial work and to do general servicing on equipment and machinery.

Specific types of fault and their location

The type of equipment we install as electricians is normally chosen by a designer to meet the needs of the client's specification. We therefore often take it for granted that the equipment and wiring we fit is right for the job, and the designer has not only met the needs of the client but also the environment.

> **Remember**
>
> It is part of the electrician's job to understand why items of equipment and wiring types are being installed and to know they are suitable.

The manufacturer and the types of cables and accessories that you install must comply with BS and BS EN Standards and must then be installed in accordance with BS 7671.

Part 5 of BS 7671 covers 'Selection and Erection of Equipment', in particular:

- type of wiring system
- selection and erection of wiring systems in relation to external influences
- current-carrying capacity of cables
- cross-sectional area of conductors of cables
- voltage drop in consumers' installations
- electrical connections
- selection and erection of wiring systems to minimise the spread of fire
- proximity of wiring systems to other services
- selection and erection in relation to maintainability, including cleaning.

The items listed in this area of the Regulations should have been accounted for in the initial design, but it is the responsibility of the installer to enforce them. This is achieved by adopting good practice when installing, and understanding the consequences (faults) that would occur if such good practice and Part 5 of BS 7671 were ignored.

Cable interconnections

Cable interconnections are used on many occasions in electrical installations. Their use is generally seen to be the result of poor planning, and good design would limit their use. That said, they are normally used in one or more of the following ways:

- lighting circuit joint boxes
- power circuits, ring final socket outlet wiring and spurs etc.
- street lighting and underground cables where long runs are needed
- general alterations and extensions to circuits when remedial work is being done
- rectification of faults or damage to wiring.

Where they must be used, they should be mechanically and electrically suitable. They must also be accessible for inspection as laid down by Regulation 526.3, which states that 'every connection and joint shall be accessible for inspection, testing and maintenance'.

There is an exception to Regulation 526.3 when any of the following is used:

- a joint designed to be buried underground
- a compound-filled or encapsulated joint
- a connection between a cold tail and heating element as in ceiling heating, floor heating or a trace heating system
- a joint made by welding, soldering, brazing or appropriate compression tool
- a joint forming part of the equipment complying with the appropriate product standard.

The joints in non-flexible cables shall be made by soldering, brazing, welding, mechanical clamps or of a compression type. The devices used should relate to the size of the cable and be insulated to the voltage of the system used. Any connectors used must be to the appropriate British Standard and the temperature of the environment must be considered when choosing the connector.

Where cables having insulation of dissimilar characteristics are to be jointed, for example XLPE thermosetting to general purpose PVC, the insulation of the joint must meet the highest temperature of the two cable insulating materials. The most common types of terminating devices you will come across are:

- plastic connectors and porcelain connectors
- soldered joints
- screwits (older installations only as these have largely fallen into disuse)
- un-insulated connectors
- compression joints
- junction boxes.

Cable interconnections are usually seen as the first point to investigate in the event of a fault. This is because they are seen as the weak link in the wiring system and because they are usually readily accessible.

Figure 7.14 Screwit

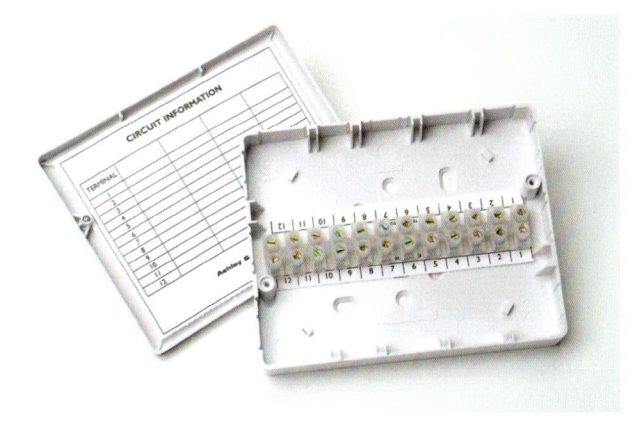

Figure 7.15 RB4 Junction box

> **Remember**
>
> Follow the safe isolation procedures before removing any cover or lid from any termination point.

The mechanical and electrical connection when joining two conductors relies on good practice by the installer to ensure that the connection is made soundly. The terminating device used should:

- be the correct size for the cross-section of the cable conductor
- be at least the same current rating as the circuit conductor
- have the same temperature rating as the circuit conductor
- be suitable for the environment.

The most common fault would probably be due to a poor/loose connection, which would produce a high resistance joint and excessive heat that could lead to insulation breakdown or eventually fire.

When a fault occurs at a cable interconnection it may not necessarily be due to the production of heat, but instead to a lack of support causing strain on the conductors, which may lead to the same outcome as above.

Cable terminations, seals and glands

Terminations

The rules for cable interconnections also apply to the termination of cables, i.e. good mechanical and electrical connections and the same consequences will occur if they are not adhered to. Therefore please note the following:

- Care must be taken not to damage the wires.
- BS 7671 522.8.5 requires that a cable should be supported to avoid any appreciable mechanical stress on the terminations of the conductor.
- BS 7671 526 gives detailed information regarding electrical connections.
- A termination under mechanical stress is liable to disconnect, overheat or spark.
- When current is flowing in a conductor a certain amount of heat is developed and the consequent expansion and contraction may be sufficient to allow a conductor under stress, particularly tension, to be pulled out of the terminal or socket.
- If the power factor (PF) improvement capacitor became disconnected in a luminaire, the circuit current would increase.
- A fault caused at a poorly connected terminal would be known as a high resistance fault.

- One or more strands or wires left out of the terminal or socket will reduce the effective cross-sectional area of the conductor at that point. This may result in increased resistance and likely overheating.
- Poorly terminated conductors in circuits that continue to operate correctly are commonly known as latent defects.

Types of terminals

There is a wide variety of conductor terminations. Typical methods of securing conductors in accessories include pillar terminals, screw heads, nuts and washers, and connectors as respectively are shown in Figures 7.16–7.18.

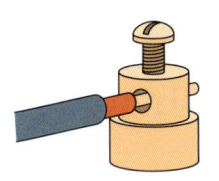

Figure 7.16 Pillar terminal

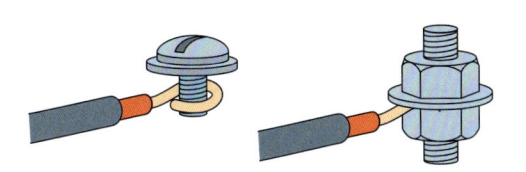

Figure 7.17 Screwhead, nut and washer terminals

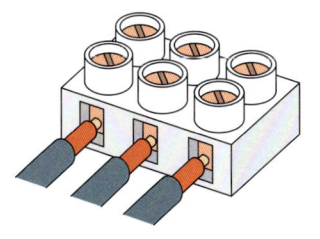

Figure 7.18 Strip connectors

Seals and entries

Where a cable or conductor enters an accessory or piece of equipment, the integrity of the conductor's insulation and sheath, earth protection and of the enclosure or accessory must be maintained. Some cables and wiring systems have integrated mechanical protection, albeit of varying effectiveness, such as:

- PVC/PVC twin and cpc
- PVC/SWA/PVC
- PVC covered mineral insulated copper cable (MICV)
- FP200.

Remember

BS 7671 now refers to thermoplastic and thermo-setting, depending on the main properties of the material.

Figure 7.19 FP 200 cable

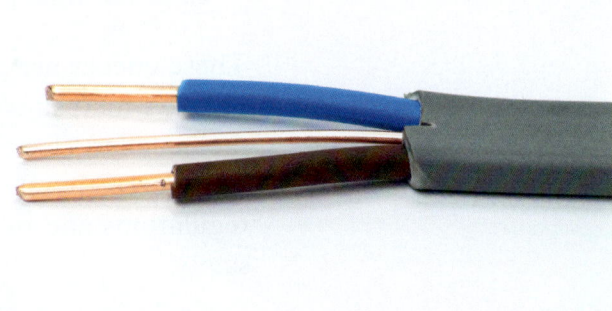

Figure 7.20 Twin and cpc PVC insulated and sheathed cable

Figure 7.21 MICV cable with pot and seal

When these and similar cables are installed, their design capabilities should not be degraded and special glands and seals produced by the makers should be used.

Where PVC/PVC cables enter accessories, the accessory itself should have no loss of integrity and there should be no damage to any part of the cable.

When carrying out a visual inspection of an installation, either at the completion of works or at a periodic inspection, the checking of terminations of cables and conductors is an integral part of the inspection.

BS 7671, Regulation 611.3 states the minimum items for inspection, some of the more pertinent to this section being listed below:

- connection of conductors
- identification of conductors
- routing of cables in safe zones, or protection against mechanical damage, in compliance with section 522
- selection of conductors for current-carrying capacity and voltage drop, in accordance with the design
- correct connection of accessories and equipment
- presence of fire barriers, suitable seals and protection against thermal effects
- methods of protection against electric shock
- selection of equipment and protective measures appropriate to external influences.

These checks are made to comply with BS 7671 Regulation 611.2, which states 'The inspection shall be made to verify that the installed electrical equipment is in compliance with section 511, correctly selected and erected in accordance with the regulations and not visibly damaged or defective so as to impair safety.'

Accessories including switches, control equipment, contactors, electronic and solid state devices

Faults can appear on most items of equipment that we install. The most common fault is due to wear of the terminal contacts as they constantly make and break during normal operation. All items of equipment should be to BS or BSEN and must be type tested by the manufacturer.

Some accessories will break down due to excessive use, and through experience the electrician can usually recognise them.

Consequently, during regular inspections and maintenance of an installation, the inspector will check 10% of accessories visually, which will usually be those items that are constantly being used such as:

- entrance hall switches
- socket outlets in kitchens for kettles
- cleaner's socket
- any item which is regularly being switched on and off.

> **Remember**
>
> Operating mechanisms do not normally break down and where contact points can be serviced and the build-up of carbon removed, this will extend the life of the accessory. However, many items are not serviceable and when a breakdown occurs through constant use, a repair is often not possible or cost-effective.

Control equipment

This is also termed switchgear and is found at the origin of the supply after the metering equipment. It can take many forms; a few examples are:

- domestic installation – double-pole switch in consumer unit
- industrial installation – single-phase double-pole switch fuse or TPN switch fuse
- large industrial installation moulded case circuit-breakers – polyphase.

Switchgear is categorised not only in normal current ratings for load in amperes, but also for fault current ratings in 1000 A or kA. If a fault occurred on the outgoing conductors from these devices then the level of fault current would be dependent on the location.

The basic rule for fault current level is that the nearer to the supply that the fault occurs, then the greater the level of fault current. This is due to there being less impedance (resistance) to impede the current flow at that point, whereas the further away from the origin of the supply the greater the impedance (resistance) to impede the current flow.

The operation of this type of control gear, either manually or automatically, due to a fault would leave all circuits fed from such equipment dead.

The manual operation of such devices is usually for maintenance reasons and it should be noted that permission of the client by prior arrangement would need to be obtained before commencing such work.

Protective devices are usually an integral part of switchgear, which will be one of the following:

- high breaking capacity fuses HBC or HRC
- moulded case circuit-breakers (MCCB).

(Specialist information and manufacturer's instructions are needed when replacing fuses that have operated and when re-setting tripped circuit-breakers.)

In older type installations you may come across switchgear that is prone to fault due to lack of maintenance. For example oil type circuit-breakers have their contact breaker points submerged in a special mineral-based oil. The oil, which quenches arcing, assists in breaking high levels of load and fault currents. If the oil viscosity and level is not regularly checked, this could lead to high levels of fault current circulating, leading to equipment damage and insulation breakdown or even fire.

Contactors

Contactors are commonly used in electrical installations. They can be found in:

- motor control circuits
- control panels
- electronic controllers
- remote switching.

The basic principle of operation of a contactor is the use of the solenoid effect, in which a magnetic coil is used in the energised or de-energised mode and spring-loaded contacts for auxiliary circuits are either made or broken when the coil has been energised or de-energised. Operating coils used in motor control circuits are energised when the start button is pressed, with the most common fault associated with their use being coils burning out. This can be due to age, but can be due to the wrong voltage type being used, e.g. a 230 volt coil being used across a 400 volt supply.

> **Did you know?**
>
> As contacts are regularly being made and broken, a build-up of carbon appears on the contact points. On larger contactors these can simply be cleaned periodically using fine emery paper to remove the residue and make the electrical contact point more efficient.

Remote switching can utilise the contactor to great effect and convenience because the contacts of large, load switching contactors can be used to carry large loads, i.e. distribution boards and large lighting loads such as used in car parks, floodlights etc. The benefit of the system is that a local 5 A switch in an office or reception area can be operated when entering or leaving, thus energising the coil of the contactor and allowing heavy loads such as the building's lighting to be carried through the contactor's contacts.

Electronic and solid state devices

Solid state and electronic equipment work within sensitive voltage and current ranges in millivolts and milliamperes, and are consequently sensitive to mains voltage and heat.

In the case of resistors on circuit boards, excessive heat produces more resistance and eventually an open circuit when the resistor breaks down. We can use a low resistance ohmmeter to check resistor values.

In the case of capacitors, voltages in excess of their working voltage will cause them to break down, which will result in a short circuit. Such equipment obviously requires specialist knowledge, but some basic actions by the electrician when carrying out tests and inspections can prevent damage to this type of equipment. When testing, such equipment should be disconnected prior to applying tests to the circuits. The reason for this is that some test voltages can damage these sensitive components and their inclusion in the circuit would also give an inaccurate reading.

We have discussed transient voltages and these are a major cause of faults on components and equipment of this type. These voltages can arise from supply company variations or lightning strikes, and therefore most large companies protect their equipment from such voltages and employ specialist companies to install lightning protection and filtering equipment.

> **Remember**
>
> On smaller contactors used in electronic circuits it is not possible to clean contact points, and when breakdown occurs it is normal practice to replace rather than repair.

Protective devices

These devices operate in the event of a fault occurring on the circuit or equipment that the device is protecting. Typical faults include:

- short circuits between live conductors
- earth-faults – between any live conductor and earth
- overcurrents.

> **Did you know?**
>
> When a protective device has operated correctly and this device was the nearest device to the fault, **discrimination** is said to occur.

The most common reason for a protective device not operating is that the wrong type or rating of device has been used. Therefore, when replacing or re-setting a device after a fault has occurred, it is important to replace the device or repair it with the same type and rating. This should only be done after the fault has been investigated and corrected.

The consequence of switching on a device with an unrepaired fault will be that the protective device will operate again. This type of activity will cause damage to the circuit wiring, the equipment and possibly the device itself.

Luminaires

The most common fault with luminaires is expiry of the lamp, which obviously only requires the lamp to be replaced, albeit discharge lighting systems employ control gear which will need replacement if they fail as they cannot be repaired.

Discharge type lighting may also have problems with the control circuit, with common items of equipment in the luminaire control circuit being the following.

- The capacitor used for power factor correction. If this had broken down it would not stop the lamp operating but would prevent the luminaire operating efficiently.
- The choke or ballast used to create a high p.d. (potential difference) needed to assist the lamp in starting. A common item which could break down and would need replacement.
- The starter, which is used to assist the discharge across the lamp when switched on at the start. This is the usual part to replace when a lamp fails to light and hence is usually easy to access on the luminaire.

> **Did you know?**
>
> Many fluorescent luminaires have starter-less electronic control gear, which is not only quick-start with increased efficiency, but requires less maintenance. They have a longer lamp life but when the quick-start unit fails will need replacement.

Flexible cables and cords

This type of conductor is used to connect many items within an electrical installation, for example:

- ceiling rose and pendant lamp holders
- flex outlets and fused connection units
- fused plugs to portable appliances
- immersion heaters
- flexible connections to fans and motors.

BS 7671 requires that all flexible cables and cords shall comply with the appropriate British and Harmonised Standard. The most common faults that will occur with this type of conductor are likely to arise from poor choice and suitability for the equipment and the environment.

Common faults relating to flexible cables and cords are:

- poor terminations into accessory – conductors showing etc.
- wrong type installed, e.g. PVC instead of increased temperature type
- incorrect size of conductor – usually too small for load
- incorrectly installed when load bearing in luminaires.

Such problems should be identified during the visual inspection stage.

Portable appliances and equipment

As discussed, flexible cables and cords are used to connect many items of fixed and portable equipment to the installation wiring and most faults on cords and flexible cables usually relate to poor choice or installation.

Nowadays portable appliances come pre-wired with a fused plug to BS 1363. However, when supplying equipment with flexible cables and cords it is important that the requirements of BS 7671 are followed, specifically:

- the correct size conductor used for the load
- the correct type is used for the environment – temperature – moisture – corrosion
- the correct termination method used to ensure good connection and no stress
- the correct type and rating of protective device is used to protect the cable and appliance.

Instrumentation and metering

Switch room panels have integrated metering equipment. In this part of the installation please note that large currents are being monitored. This monitoring allows the consumer to view current and voltage values at particular times of the day. Some panels will have equipment which monitors and records these values, allowing the consumer to plot peak energy times and budget for energy costs.

As these instruments measure large currents, transformers are used to reduce current and voltage values, and this reduces the size of instrument needed. These transformers are known as CTs (current transformers) and VTs (voltage transformers). However, great care must be taken when servicing and repairing such equipment as large voltages can exist across the transformer terminals.

The common remedial work on such equipment is the replacement of faulty instruments and burnt-out transformers and it is essential that safe isolation procedures are followed.

Special precautions that should be taken for special situations

Special precautions and a risk analysis should always be carried out before any work is commenced. It is a requirement of the Health and Safety at Work Act 1974 that precautions should be accounted for and that any risk is removed or persons carrying out any work activity should be protected from harm. Risk assessments are now commonplace and are part of our daily work routine. In this section we shall be looking at the risks involved in carrying out fault diagnosis and repair.

Fibre optic cabling

This type of wiring system is becoming increasingly popular in the data and telecommunications industry, although it is unlikely that you will terminate or fault-find on such systems. However, your company may be asked to install such systems and you may be asked to assist.

As its name suggests the cable is made up of high-quality minute glass fibres, which can carry vast amounts of transmitted signals and data using emitted signals of light. The levels of light transmitted can reach dangerously high levels. Therefore **you should never look into the end of fibre optic cable!**

Antistatic precautions

Antistatic precautions are necessary in areas that are hazardous due to the high risk of ignition or explosion. This is usually where flammable, ignitable liquids, gases or powders are stored or dispensed, for instance:

- petrol filling stations
- chemical works
- offshore installations
- paint stores
- flour mills.

> **Remember**
>
> Many of the precautions for working in hazardous areas and when working alone have been covered in Unit ELTK 01 in Book A.

> **Did you know?**
>
> Some formal risk assessments are recorded and noted in method statements. These can be seen in your company's health and safety statements.

> **Safety tip**
>
> The levels of light carried by fibre optic cable can be dangerously high so you should never look into the end of a fibre optic cable.

In such situations **no naked flames** or equipment which can cause a spark or discharge of energy can be used. There would obviously be a no smoking policy. It is important to adhere to the safety signs and regulations in such areas. When asked to work in such environments you should be given special induction and training so you are aware of the dangers and the risks and almost certainly all work will be carried out via work permits and method statements.

Electrostatic discharge

Figure 7.22 Petrol filling station

Static is an electric charge, caused by an excess or deficiency of electrons collected on a conducting or non-conducting surface, thus creating a potential difference measured in volts. Static charges are caused in several ways, as described below.

Friction

When two surfaces are rubbed together electrons are rubbed off one surface and on to the other. This leaves one surface with a deficiency of electrons, therefore being positively charged and the other surface acquires and retains electrons and so is negatively charged.

Separation

Figure 7.23 Static electricity

When self-adhesive plastic or cellulose tape is rapidly pulled off a reel it generates a static charge by tearing electrons from one surface to another.

Induction

Static is generated by induction when a charged item, for example a folder of paper, is placed on a desk and on top of that is placed a conductor, say a printed circuit board. If the paper was positively charged, the charge from the paper is immediately induced into the printed circuit board, making it negatively charged.

All materials can be involved in the production of static. In the case of non-conductors (insulators such as paper, plastic and textiles) the charge cannot redistribute itself and therefore remains on the surface.

> **Did you know?**
>
> The reality is that every step, even every shuffle on a seat, generates static charges. Picking up, putting down, wrapping or unwrapping, even just handling an object creates static. So you can see that everyday movements, while wearing clothes made from man-made fibres, will generate large amounts of static.

Protection from electrostatic discharge

The main effort must be to protect devices that are sensitive to electrostatic by providing special handling areas where they can be handled, stored, assembled, tested, packed and despatched.

In the special handling area there should be total removal of all untreated wrappings, paper, polystyrene, plastics, non-conductive bins, racks and trays, and soldering irons including tips should be earthed.

Specifications, drawings and work instructions must be in protective bags, folders or bins. Wrist straps should be worn at all times in the static free area. These straps 'earth' the handler and provide a safe path for the removal of static electricity.

Damage to electronic devices due to over voltage

When testing or inspecting faults, we must take account of electronic equipment and their rated voltage. It is usual to isolate or disconnect such equipment to avoid damage due to test voltages exceeding the equipment's rated voltage.

Such voltage levels will cause components, control equipment, or data and telecommunication equipment to be faulty. This could be costly not only in repairs but also to the reputation of your company, and could result in the loss of future contracts. The voltage of an insulation resistance tester can reach levels between 250 to 1000 V.

Avoidance of shut down of IT equipment

On no account should any circuit be isolated where computer equipment is connected. This type of equipment is to be found in many commercial, industrial and even domestic installations. The loss of data is the most common occurrence when systems fail due to faults or power supply problems.

Larger organisations will have data protection in the form of uninterruptible power supplies (UPS systems). Whether UPS systems are in place or not, it is still important to plan the isolation of circuits where IT systems are connected. This will give the client time to arrange data storage and download, which will allow systems to be switched off and repairs to be carried out.

Risk of high frequency on high capacitive circuits

Capacitive circuits could be circuits with capacitors connected to them or some long runs of circuit wiring which may have a capacitive effect. This is usual for example on long runs of mineral-insulated cables.

When working on such circuits, no work should commence until the capacitive effect has been discharged. In some cases it would be practical to discharge capacitors manually by shorting out the capacitor with an approved device.

Danger from storage batteries

Batteries can be found on many installations and provide an energy source in the event of an emergency when there is a power failure. This can be seen where battery panels provide a back-up for the emergency lighting in a building when the power fails.

Batteries can also be found in alarm panels, IT UPS systems and emergency stand-by generators for starting, but wherever you have cause to work with or near batteries special care should be observed, particularly with regard to:

- lead acid cells, which contain dilute sulphuric acid. This acid is harmful to the skin and eyes, can rot clothing and is highly corrosive. If there is any contact with the skin it should be immediately diluted and washed with water
- lead acid cells also emit hydrogen gas when charging and discharging, which when mixed with air is highly explosive
- high voltages applied to cell terminals will damage the battery
- when connecting cells, shorts across the terminals will produce arcing that could cause an explosion
- cells should never be disconnected, as this can also produce arcing that could cause an explosion.

> **Safety tip**
>
> On some systems where high frequencies exist, there is a danger of exceptionally high voltages occurring across terminals. On no account should you, as a trainee, ever attempt to tamper with capacitive circuits and, if in doubt, seek the assistance of your supervisor.

> **Progress check**
>
> 1 Identify the logical sequence of events that should be undertaken during fault diagnosis and rectification.
>
> 2 What are the common symptoms to be considered with electrical faults?
>
> 3 There are a number of common faults in electrical installations attributed to misuse by both users and installation personnel. Identify some of these.
>
> 4 The checking of termination of cables and conductors is an integral part of an inspection. List the main areas to be covered during such an inspection.

K5. Understand the procedures and techniques for correcting electrical faults

Having located the fault it would seem simple enough for the electrician to eliminate the cause and re-energise the circuit so that the equipment or circuit is working again. However, in many cases there are issues that will not allow the power to be switched on to the circuit or equipment under scrutiny.

Factors which can affect fault correction, repair or replacement

In many cases the final judgement about repair or replacement cannot be made by the electrician and may need to be discussed by third parties, e.g. the client, the manufacturer and the person ordering the work. So in this section we will look at the factors which will influence this decision-making process, to decide whether the equipment or circuit is replaced or repaired and whether the fault can be remedied instantly or at a pre-determined future time.

Cost of replacement

When successfully finding a fault you should not repair the fault without first considering the consequences to your company or your client. There are points that need consideration such as: 'What will the cost of replacement be?'

This question can only be answered by liaison and consultation with your engineer and the customer, as an agreement must be reached on how to proceed. It may well be that some electrical contractors have an agreement or service contract in place where remedial works are carried out on a day work and contracted basis. This is where a prior contract agreement is signed and the contractor carries out maintenance and repair work at pre-determined rates.

However, where this is not the case the question may arise: 'Will the replacement cost be too expensive?'

It could be that the price quoted for replacement is high, which may result in the customer using your company's quotation to compare with other companies' prices. This may lead to negotiations where only part of a faulty piece of equipment is replaced or repaired, which may reduce the outlay costs to the customer.

It is often the case that the customer is given a choice of options; they will compare these options and make the decision. The options may be:

- full replacement
- part replacement
- full repair.

In offering such options it can be seen generally that the time taken to carry out a repair or part replacement may in many cases outweigh the time and cost of a full replacement.

Availability of replacement

As already noted, in many cases where a complete replacement of equipment or machinery is needed, it is because the fault has arisen due to the lifespan of the equipment being exceeded. In these cases, the equipment or machinery may have been in operation for so long that a repair may not be possible and the item may even be obsolete. When this happens, a suitable replacement must be found.

The most obvious decision to be made is that of finding a replacement that closely matches the original specification of the faulty item. But this is not always possible and the deciding factor becomes the availability of the equipment. This is important and will influence the decision to order. It is often the manufacturer who has the replacement item in stock and has a fast delivery that will be given the order to supply, as this will enable the electrician to remedy the fault quickly.

> **Remember**
>
> It may be the case that the original manufacturer is no longer in business and therefore the problem becomes finding a suitable alternative manufacturer. Most manufacturers will have their own design team who will help in the process of finding a suitable alternative piece of equipment.

Downtime under fault conditions

It is often the case that the consequence of a fault on a circuit or item of equipment will be to cost the client money. This can be due to:

- lost production where manpower and machinery are not working
- lost business where products or services are not being provided to the customer
- data loss because of power failure or disruption during repair.

Sometimes the fault may not be in the electrical installation but on the supply to the consumer's premises. In such cases the rectification of the fault is out of your company's responsibility and control. However, in such circumstances and where the loss of supply is prolonged, it may make good business sense, where possible, to provide a temporary supply to the customer.

Availability of resources and staff

We have already mentioned the availability of equipment due to choice, manufacturer and delivery time, but the availability of staff is often the cause of any delay in rectifying faults and carrying out repairs.

Some faults and repairs may be minor and need little resource in terms of materials and labour. The response time for small faults could be as low as minutes and hours, but in the case of major industrial and commercial repairs the downtime and repair time could be weeks or even months.

The consequences of some faults may be major contract works such as rewiring or large equipment replacement. In such cases careful planning and decisions must be taken by the contractor's management and design team when deciding to commit the company to carrying out such works.

The following are basic considerations that any company should make before commencing such contracts:

- **Availability of staff** – Has the company enough people employed to carry out and complete the work?
- **Competency** – Does the company have staff with relevant experience?
- **Cost implications** – Has the company the monetary capability for the contract?
- **Special plant and machinery** – Has the company this type of resource available?

Legal responsibility

The legal responsibilities surrounding the replacement or repair of a fault must also be considered when making the decision to repair or replace. It is usual for a contract to be entered into by both parties as this will help in settling any future disputes. Such a contract would itemise:

- costs
- time
- guarantee period/warranty
- omissions from contract.

> **Remember**
> Whatever the cause of the fault, the consequence is often loss of revenue, either directly because of output, or because of poor service to the customer, this can affect the client's reputation and ultimately result in a further loss of business.

> **Remember**
> Repairs must be viable for both client and also for the contractor, who also cannot afford to lose money or reputation on installation and repair works.

The costs should be agreed and adhered to. Most customers will not pay more than the initial agreed costs of repair unless an allowance was made in the contract for unforeseen work. Unforeseen work could result from something found when dismantling or taking out existing items, such as finding faulty or dangerous wiring underneath an item of heavy plant that is to be repaired.

If the initial costing accounted for repair and replacement of parts, then it is essential that you do what you said you were going to do. For instance, if an inspection is subsequently carried out and it is evident that no work has taken place then you may be liable for legal proceedings to be taken against you for a breach of contract terms.

An accurate cost analysis should be considered when giving an estimate or fixed price for repair work, as it is usual practice to carry out repair work on a daily rate where the customer pays only for the time taken and materials used in rectifying the fault.

The customer should also be given a guarantee for the work done, both in respect of quality and also for the materials used.

When deciding whether to repair or replace, it may be necessary to not write a guarantee in the contract when the client insists on re-using existing items in order to reduce costs. It is often the case that old industrial or antiquated items are prone to failure and to avoid legal action such items should not be included in the contract or your guarantee.

Other factors affecting the fault repair process

Careful planning can mean that most work is carried out efficiently with little or no disruption to the client. However, faults on installation equipment and wiring systems are certainly not planned and trained personnel can only minimise the disruption caused by the effects of faults.

In a factory environment, maintenance electricians are on hand to limit the effect of a fault and can usually solve problems and reduce the downtime of a machine or circuit to a minimum. This is because of their day-to-day working knowledge of the electrical services and the equipment connected to it.

When an electrician is called out to a fault they may not have the day-to-day knowledge or experience of the maintenance staff. However, through training, knowledge and experience the electrician will use contingency plans to set up procedures to rectify the fault with the minimum of disruption.

Such contingency plans provide a common format when dealing with fault finding at any customer's premises. Your company will have its own procedures and these may include:

- signing in
- wearing identification badges
- locating supervisory personnel
- locating data drawings etc.
- liaising with the client and office before commencing any work
- following safe isolation procedures.

Special requirements

We have previously looked at the safety procedures and the actions needed for determining causes of faults. If we assume that safety requirements and the location of faults have been met and that the prime objective of the electrical contractor (the electrician) is to locate the fault, rectify and re-commission, then this will be done in liaison with the customer and with minimal disruption. However, the following topics are worthy of note.

Access to the system during normal working hours

There is nothing more costly or embarrassing than arriving at a customer's premises to carry out work of any kind, only to find the premises locked or unmanned because no prior arrangements were made for access.

Access to the premises and the electrical installation or the specific item of equipment may prove a problem, especially if you are unfamiliar with the customer's premises.

When attending a fault or breakdown we must remember the procedures for ascertaining the information and familiarising ourselves with the installation. Remember our logical sequence:

1. Identify the symptoms.
2. Gather information.
3. Analyse the evidence.
4. Check supply.
5. Check protective devices.
6. Isolation and test.

7. Interpret information and test results.
8. Rectify the fault.
9. Carry out functional tests.
10. Restore the supply.
11. Carry out live and functional tests.

If we need to familiarise ourselves with the installation, then access must be readily available not only at the fault repair location but also at the supply intake and isolation point, and this has to be agreed prior to commencing any contracted work, whether it is a repair or a new installation.

Such prior agreement must take into account those personnel who may have their normal day-to-day routine and activities affected. It is not always possible to carry out repair work during normal working hours and so to avoid disruption such work is often done outside of normal working hours. It is therefore important that arrangements for out-of-hours work are pre-arranged for security and access.

Is there a need for building fabric restoration?

Electricians are more often asked to locate and repair faults on existing installations. Consequently, depending on the fault and the method of rectification needed, we may need to disturb the fabric of the building. It is therefore vitally important to fully discuss with the client all of the implications of the fault diagnosis/rectification process.

As part of investigating a fault, we simply have to look inside an item of equipment. However, if the fault has been caused by someone putting a nail through a cable, then the repair will involve replacing the cable and that in turn may involve having to damage the building fabric. You should therefore explain these aspects to the client and agree what repair work (e.g. brickwork, plastering, decorating) will be required and which, if any, of these can be carried out satisfactorily by yourself.

Obviously, as with any aspect of fault finding and rectification, always leave the site in a clean and safe condition, disposing of items such as fluorescent tubes in a safe and approved manner.

> **Remember**
> The expertise of building contractors may be needed and any associated costs need to be agreed before any repairs are carried out.

Whether the system can be isolated section by section

After access to the installation has been agreed, the repairs should be carried out using a logical approach.

> **Remember**
>
> All electrical installation systems should have been designed and installed in accordance with the IET Wiring Regulations, ensuring that the facility for isolating circuits and sections is inherent in the system.

However, if the premises are still occupied and parts of the system are still being used, then it is important that the faulty circuit or plant to be worked on is isolated.

When only a small area or a single circuit or piece of equipment is affected it is not good practice to switch off the mains supply and isolate all the installation. By analysing the installation drawings and data, an area or items can be isolated individually to limit disruption.

Provision of emergency or stand-by supply

Often the disruption to the customer is not having an electrical supply to an area of the installation. If the work cannot be done outside normal working hours, a temporary supply could be arranged. Some installations may have a stand-by service in the form of emergency lighting or a stand-by generator, but this is only usual in hospitals or where computers are being used and data is stored.

> **Remember**
>
> Any temporary supply which is provided should be wired in accordance with the IET Wiring Regulations.

Client demand for a continuous supply

Where the client does not want any interruption in supply and it is not possible to work on the affected area during normal working hours, arrangements should be made for a planned maintenance or shut down in order for the repair to be carried out safely.

Working life

A client has called your company to their offices as they are experiencing a burning smell from the supply intake position. Following investigation you find the damage is due to an overloaded circuit which has burnt away and damaged the fuse carrier inside the distribution board. This will necessitate replacing the board. The IT system is in use all day and cannot be switched off.

1. What recommendations can you make?
2. What are the alternatives?
3. What would you recommend for the IT system, should shut down be required in future?

Progress check

1. What are the main considerations, before deciding upon repair or replacement of faulty equipment?
2. Most companies involved with fault finding at a customer's premises usually have an agreed procedure for staff to follow, to ensure consistency. What would be included in such a procedure?

Getting ready for assessment

As part of your work as an electrician you will need to be able to recognise when an installation has not been carried out correctly, or is not of the correct standard. In many cases, these faults will be repairable once you have successfully diagnosed them. Repairing faults should be seen as being closely related to inspecting and testing – these procedures will often identify the faults, which will need to be repaired.

For this unit you will need to be familiar with:

- principles, regulatory requirements and procedures for completing the safe isolation of electrical circuits and complete electrical installations
- how to complete the reporting and recording of electrical fault diagnosis and correction work
- how to complete the preparatory work prior to fault diagnosis and correction work
- procedures and techniques for diagnosing electrical faults
- procedures and techniques for correcting electrical faults

For each learning outcome, there are several skills you will need to acquire, so you must make sure you are familiar with the assessment criteria for each outcome. For Learning Outcome 3 you will have to be able to specify the safe working procedures that should be adopted for the completion of fault diagnosis and correction work and state the logical stages of each process. You will need to be able to identify and describe the common symptoms of electrical faults and the causes of a range of different types of fault, as well as specifying their likely locations. You will also need to be able to state the special precautions you will need to follow for a range of special locations and electronic components.

It is important to read each question carefully and take your time. Try to complete both progress checks and multiple choice questions, without assistance, to see how much you have understood. Refer to the relevant pages in the book for subsequent checks. Always use correct terminology as used in BS 7671. There are some simple tips to follow when writing answers to exam questions:

- **Explain briefly** – usually a sentence or two to cover the topic. The word to note is 'briefly' meaning do not ramble on. Keep to the point.
- **Identify** – refer to reference material, showing which the correct answers are.
- **List** – a simple bullet list is all that is required. An example could include listing the installation tests required in the correct order.
- **Describe** – a reasonably detailed explanation to cover the subject in the question.

This unit has a large number of practical skills in it, and you will need to be sure that you have carried out sufficient practice and that you feel you are capable of passing any practical assessment. It is best to have a plan of action and a method statement to help you work. It is also wise to check your work at regular intervals. This will help to ensure that you are working correctly and help you to avoid any problems developing as you work. Remember, don't rush the job as speed will come with practice and it is important that you get the quality of workmanship right.

Good luck!

CHECK YOUR KNOWLEDGE

1. Which of the following is it **unnecessary** to familiarise an operative with before fault rectification work commences?
   a) Nominal voltage
   b) Cost of repair
   c) Type of earthing supply system
   d) Distribution board schedules

2. Which one of the following would **not** be included in a fault finding sequence?
   a) Identifying symptoms
   b) Checking the supply
   c) Isolation and testing
   d) Length of time it will take

3. Which of the following tests is carried out during non-live testing?
   a) Insulation resistance
   b) RCD
   c) Functional
   d) Earth-fault loop impedance

4. Identify which instrument is used to carry out an insulation resistance test?
   a) Bell/Buzzer
   b) High resistance ohmmeter
   c) Low resistance ohmmeter
   d) Earth-fault loop impedance tester

5. Which of the following would **not** be regarded as a transient fault?
   a) Heavy current switching
   b) Earth-fault voltages
   c) Lightning strikes
   d) Overload

6. A mechanical and electrical connection when joining two conductors together relies on good practice by the installer. What should a terminating device for use on conductors have?
   a) Current rating lower than that of the conductor
   b) Temperature rating lower than the conductor
   c) Cross-sectional area smaller than that of the conductor
   d) Suitability for the environment and that of the conductor

7. What would the most likely outcome at a loose termination under load conditions be?
   a) Overheating
   b) Normal running
   c) Cooling of the conductor
   d) Protective device operating

8. Which of the following faults would an RCD protect against?
   a) Overcurrent
   b) Excess heat in conductors
   c) Short circuits between live conductors
   d) Earth-faults between live conductors and earth

9. Which of the following would **not** be regarded as a fault in a flexible cable?
   a) Oversized conductor
   b) Incorrect type installed
   c) Incorrect size of conductor
   d) Poor termination into an accessory

10. Antistatic precautions to avoid ignition or explosion are usually required in which of the following installation environments?
    a) Office
    b) School
    c) Flour mill
    d) Shopping mall

UNIT ELTK 08

Understanding the electrical principles associated with the design, building, installation and maintenance of electrical equipment and systems

As well as having a practical competence, electricians also need to know about the operating principles behind the job. The purpose of this unit is to look at the principles of mathematics, electrical science and electronics necessary to support electrical and electronic installations.

This unit covers the following learning outcomes:

- Understand mathematical principles which are appropriate to electrical installation, maintenance and design work
- Understand standard units of measurement used in electrical installation, maintenance and design work
- Understand basic mechanics and the relationship between force, work, energy and power
- Understand the relationship between resistance, resistivity, voltage, current and power
- Understand the fundamental principles which underpin the relationship between magnetism and electricity
- Understand electrical supply and distribution systems
- Understand how different electrical properties can affect circuits, systems and equipment
- Understand the operating principles of d.c. machines and a.c. motors
- Understand the operating principles of different electrical components
- Understand the principles and applications of electrical lighting systems
- Understand the principles and applications of electrical heating
- Understand the types, application and limitations of electronic components in electrotechnical systems and equipment.

K1/K2. Understand standard units of measurement and mathematical principles that are used in electrical installation, maintenance and design work

It is impossible to know what people are talking about unless you understand the language they are speaking. In science and engineering, mathematics is the language that is used to explain how things work. To work with all things electrical, we need to understand and be able to communicate using this universal language and therefore this section will cover at the following topics:

- SI units: how we describe basic measurement quantities
- powers of 10: mega, pico and the decimal system
- basic rules: getting the right answer
- fractions and percentages: working with parts of the whole
- algebra: formulas for all
- indices: powers of anything
- transposition: re-arranging equations
- triangles and trigonometry: angles on reading
- the sines
- statistics: ways of showing data.

SI units

Imagine if each country had a different idea of how long a metre is, or how much beer you get in a pint. What if a kilogram in Leeds was different from one in London?

A common system for defining properties such as length, temperature and time is essential if people in different places are to work together.

Most countries in the world have adopted an agreed international system of units for measuring different properties, and this is known as the SI system, which has seven base units from which all the other units are created.

Table 8.01 shows each base quantity and the name of its base unit. The last column shows the symbol used to represent the unit.

> **Did you know?**
>
> SI stands for Système Internationale.

Base quantity	Base unit	Symbol
Length	metre	m
Mass	kilogram	kg
Time	second	s
Electric current	ampere	A
Temperature	kelvin	K
Amount of substance	mole	mol
Luminous intensity	candela	cd

Table 8.01 SI units

The SI units, or derived units, that you are most likely to come across are shown in Table 8.02.

Length	Obviously useful for all sorts of things, such as measuring the amount of cable you'll need in an installation
Area	Used to measure a surface such as a sports field, e.g. for working out how much floodlighting will be needed
Volume	Used particularly when you are dealing with heating systems, for example, to calculate how much energy is required to heat a hot water cylinder
Mass	Often confused with weight. Mass is 'how much there is' of something. Electricians often need to know how much electrical energy is required to change a mass from one state to another; e.g. to change water into steam, or ice into water
Weight	Related to how gravity has an effect on mass. A mass of 1 kg will weigh about 10 newtons on Earth, i.e. it will exert a *force* of 10 N, but rather less on the Moon, since gravity is lower there. NB It will still have a mass of 1 kg on the Moon since it still contains the same amount of material
Temperature	The SI base unit for temperature is degrees Kelvin (K), but for most purposes we use degrees Celsius (°C). 0 K is −273°C (otherwise known as Absolute Zero – extremely cold!). Sometimes people also use degrees Fahrenheit (°F)

Table 8.02 Base quantities

> **Did you know?**
>
> The boiling point of water at sea level is 100°C or 373 K or 212°F.

SI unit prefixes

Often we want to deal with quantities that are much larger or smaller than the base units. If we could only use the base units, the numbers would become clumsy and it would be easy to make mistakes. For example, the diameter of a human hair is about 0.0009 m and the average distance of the Earth from the Sun is around 150 000 000 000 m.

To make life easier, we can alter the symbols (and the quantities they represent) by adding another symbol in front of them (known as a prefix) and these represent the base units multiplied or divided by one thousand, one million, etc. Table 8.03 shows the most common ones:

Multiplier	Name	Symbol prefix	As a power of 10
1 000 000 000 000	Tera	T	1×10^{12}
1 000 000 000	Giga	G	1×10^{9}
1 000 000	Mega	M	1×10^{6}
1 000	kilo	k	1×10^{3}
1	unit		
0.001	milli	m	1×10^{-3}
0.000 001	micro	μ	1×10^{-6}
0.000 000 001	nano	n	1×10^{-9}
0.000 000 000 001	pico	p	1×10^{-12}

Table 8.03 Common prefixes

Here are some common examples of using the prefixes with the unit symbol:

- km (kilometre = one thousand metres)
- mm (millimetre = one thousandth of a metre)
- MW (megawatt = one million watts).
- μs (microsecond = one millionth of a second)

Later in this unit you will see SI units applied to a wide range of electrical variables:

- resistance
- resistivity
- power
- frequency
- current
- voltage
- energy
- impedance
- inductance and inductive reactance
- capacitance and capacitive reactance
- power factor
- actual power
- reactive power
- apparent power.

Identify and apply appropriate mathematical concepts

Basic rules

Unless we all carry out calculations using the same basic rules, we will all get different answers to the same question. Try working out the sum in the example.

Example

Work out the answer to the following sum.

$(42 \times 4) + (6 \div 3) - 2 = ?$

We asked three apprentices. Andy says it's 56, but Mo says it's 420, and Ali says it's 168.

So who's right? What do you think?

The problem is that each apprentice followed their own set of rules.

Andy simply worked from left to right:	Mo worked from right to left:	Ali did the multiplications first, then the divisions, then the addition and subtraction:
$42 \times 4 = 168$	$3 - 2 = 1$	$42 \times 4 = 168$
$168 + 6 = 174$	$6 \div 1 = 6$	$6 \div 3 = 2$
$174 \div 3 = 58$	$4 + 6 = 10$	$168 + 2 = 170$
$58 - 2 = \mathbf{56}$ ✗	$10 \times 42 = \mathbf{420}$ ✗	$170 - 2 = \mathbf{168}$ ✓

Mathematics needs rules, and we need to know them if we are to get the correct answers. The three main ones are as follows:

Basic Rule 1:
'All numbers are positive unless told otherwise.'

All numbers are either positive or negative (except 0, of course). Since most things we deal with are positive numbers, we don't usually bother to put a positive sign (+) sign in front of them, but we must put a minus sign (–) in front of any negative numbers.

Basic Rule 2:
'Like signs add, unlike signs subtract.'

For example:

4 + (+5) = 4 + 5 =	9	(Like signs add)
4 + (–5) = 4 – 5 =	–1	(Unlike signs subtract)
4 – (+5) = 4 – 5 =	–1	(Unlike signs subtract)
4 – (–5) = 4 + 5 =	9	(Like signs add)

The same applies when multiplying and dividing: 'Like signs give positive results, unlike signs give negative results.'

For example:

+4 × –5 =	–20	(Unlike signs give negative results)
–4 × +5 =	–20	(Unlike signs give negative results)
–4 × –5 =	20	(Like signs give positive results)
20 ÷ +5 =	4	(Like signs give positive results)
–20 ÷ +5 =	–4	(Unlike signs give negative results)
20 ÷ –5 =	–4	(Unlike signs give negative results)
–20 ÷ –5 =	4	(Like signs give positive results)

Basic Rule 3:
'Bodmas rules!'

BODMAS is what we call an 'acronym' and BODMAS represents the order in which we must tackle a calculation.

This is where BODMAS comes in to help you, as it stands for:

Brackets   **O**ther operations   **D**ivision   **M**ultiplication   **A**ddition   **S**ubtraction

> **Remember**
> An acronym is an abbreviation of several words in such a way that the abbreviation itself forms a pronounceable word (e.g. BODMAS) and that this often helps to remember the topic.

So the order you should do your calculations in is:

Brackets (*Always calculate what's in them first*)
Other operations (*Such as powers or square roots*)
Division and **M**ultiplication (*Start on the left and work them out in the order that you find them*)
Addition and **S**ubtraction (*When only addition and subtraction are left in the sum work them out in the order you find them, starting from the left of the sum and working towards the right*)

In BODMAS, division and multiplication have the same priority, as do addition and subtraction. Let's look at some examples.

> **Example**
> Calculate: $2 + 4 \times 4 - 1$
> (× first) $2 + 16 - 1$
> (then +) $18 - 1$
> (then -) $= 17$

> **Example**
> Calculate: $3 \times (4 + 5)$
> (brackets first) 3×9
> (then ×) $= 27$

> **Example**
> Calculate: $4(3 + 2 \times 2)$
> (brackets first) $4(3 + 4)$
> (still brackets) 4×7
> (then ×) $= 28$

We'll show this final example step-by-step to confirm the process.

> **Example**
> Calculate $(4 - 2)^2 \times 2 \times 5$
> (brackets first) $(\mathbf{4 - 2})^2 \times 2 \times 5$
> $2^2 \times 2 \times 5$
> (next operations) $\mathbf{2^2} \times 2 \times 5$
> $4 \times 2 \times 5$
> (next × work L to R) $\mathbf{4 \times 2} \times 5$
> (next × work L to R) 8×5
> (next ×) 8×5
> Answer $\mathbf{40}$

Having looked at these examples, we should note two further things:

When there is a number outside the brackets with no operator (e.g. +, −, ×, ÷), then a multiplication is assumed.

So for example, $5(3 + 9)$ is the same as saying $5 \times (3 + 9)$ where the result is $5 \times 12 = 60$.

In the above example, you could also multiply all the numbers inside the bracket by the number outside, and then do the calculation.

In other words: $5(3 + 9)$
                 $= (5 \times 3) + (5 \times 9)$
                 $= 15 + 45 = \mathbf{60}$

Powers of 10

So far so good, but what happens when you want to do some arithmetic with several quantities?

For instance, speed is calculated by dividing the distance covered by the time taken. So an energetic spider might cover 1 m in 1 s, and its speed would therefore be 1 m/s (metre per second).

But how do we express the speed of a bullet that travels 3 km in 50 ms?

This is where an understanding of powers of 10 will help us.

When we write down any calculation, the position of the figures shows their size relative to the decimal point, e.g. units, tens, hundreds, thousands, etc. The purpose of the decimal point is to separate whole numbers from parts of numbers. Therefore, we can say 1.5 is 1 and 5 tenths, the same as one and a half.

As an example, the number 4123.4 tells us that we have:

Thousands	Hundred	Tens	Units	Decimal point	Tenths
4	1	2	3	.	4

We could also look at it like this.

> **Example**
>
> The number 4123.4 means:
>
> | | |
> |---|---|
> | 4 times one thousand | 4000.0 |
> | Plus 1 times one hundred | 100.0 |
> | Plus 2 times ten | 20.0 |
> | Plus 3 units | 3.0 |
> | Plus 4 times one-tenth | 0.4 |
> | **Total** | **4123.4** |

In both the above examples we can see that going left from the decimal point, each column is 10 times greater than the previous one. Going right from the decimal point, each column is 10 times less or one-tenth of the previous one.

The example has included zeros to make sense of a column of figures, but extra zeros added does not change the number. 4123.4 is exactly the same as 4123.40 or 4123.400.

When adding, always line up the decimal points and each column when writing them down, then you are sure of adding the number of 10s in one number to the number of 10s in another number and so on.

It can help if you draw straight lines, which helps keep your calculations easy to understand when checking answers!

Let's look at adding 24.01 to 110

```
1 1 0 . 0 0
    2 4 . 0 1
  ─────────
1 3 4 . 0 1
```

Here is another example.

> **Example**
>
> | | |
> |---|-----------|
> | | 32 456.24 |
> | + | 123.51 |
> | **Total** | **32 579.75** |

The fact that each column is either ten times bigger or one-tenth of the previous column also makes it easy to multiply and divide by 10, as all you need to do is move all the digits one jump to the right to divide by 10, and one jump to the left to multiply by 10.

To multiply 123.4 by 10 (note the position of the decimal point) we move the decimal point one place to the right.

   123.4 = 1234.0

Therefore, 123.4 × 10 = 1234.0

Division sees the decimal point move in the other direction; so 4321.0 ÷ 10 = 432.1

   4321.0 becomes 432.10

Similarly, multiplying or dividing by 100 means moving the decimal point two places to the left or right. Multiplying or dividing by 1000 then just means moving the decimal point three places to the left or right.

In other words, we are moving the same amount as there are zeros in the number:

- 10 = has one zero, therefore we move the decimal point one place
- 1000 = has three zeros, therefore we move the decimal point three times

6789.345 × 100 (00)	=	6789.345	=	678934.5
6789.345 ÷ 1000 (000)	=	6789.345	=	6.789345

> **Did you know?**
>
> You may also see multiplication by 10 explained as moving the digits to the right, and division as moving it to the left.

Other powers of numbers

We have a shorthand way of showing a number multiplied by itself.

For example 25, which is 5 × 5, can be written as 5^2, which we say as 'five to the power of two', or 'five squared'. (We call it 5 squared because we find the area of a square from its two dimensions L × W or 5 × 5)

We can multiply any number by other powers. So for example, 125 which is 5 × 5 × 5, can be written as 5^3, which we say is '5 to the power of 3' or '5 cubed'. (We call it 5 cubed because we find the volume of a cube from its 3 dimensions L × W × D or 5 × 5 × 5)

The power can take any value, for example

10 000 = 10 × 10 × 10 × 10 = 10^4
1 000 000 = 10 × 10 × 10 × 10 × 10 × 10 = 10^6

We can also have negative powers. If there is a minus sign in front of the power, this represents our number divided by the power.

So 10^{-1} = 1 ÷ 10^1 (1 ÷ 10) = 0.1
And 10^{-3} = 1 ÷ 10^3 (1 ÷ 1000) = 0.001

All scientific calculators have a 'powers' button for working with large powers. Figure 8.01 shows the button that is normally used.

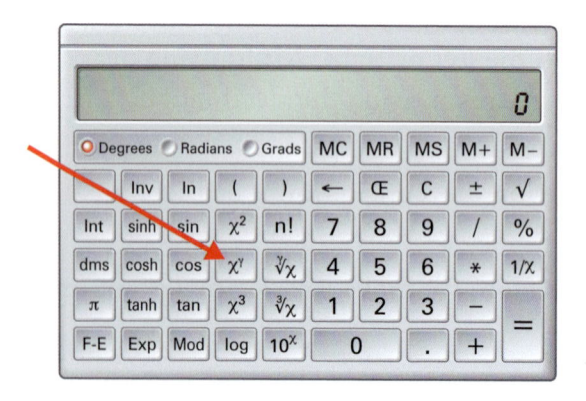

Figure 8.01 Calculator

Try it with a calculation that you know the answer of (e.g. 2^4) to check that you can use **your** calculator correctly.

To find 7^8 we would use the following key sequence:

| 7 | x^y | 8 | = |

The answer should be **5 764 801**

Multiplying and dividing with powers

When we multiply two numbers with powers together, we add the powers.

For example, we know that $5^3 = 5 \times 5 \times 5 = 125$, therefore $5^3 \times 5^3$ will be $125 \times 125 = \mathbf{15\,625}$

Adding the powers we get $5^{3+3} = 5^6$ which also equals $\mathbf{15\,625}$.

Although the powers can be different, i.e. $5^3 \times 5^2$ it must be the same number (in this case 5) that we are applying the powers to. This method would not work for $4^2 \times 5^2$.

When it comes to division we subtract the powers.

For example, we know that 3^3 is 27 and that 3^2 is 9 and that $27 \div 9 = 3$

Therefore, when written as powers this would be $3^{3-2} = 3^1$

When we have a number raised to the power of 1, it is simply the original number; in other words, 3^1 is just 3.

However, when you see any number raised to the power of 0, then the result is always 1.

This can be explained by looking at what happens when we divide a number by itself.

Example

$5 \div 5 = 1$ (as 5 goes into 5 once)

Having established that 5^1 means 5, we could also write this as $5^1 \div 5^1 = 1$

Remember that when dividing we subtract the powers

So we can also write $5^{1-1} = 5^0$

So the value of any number to the power of 0 will be the number 1

Most commonly we see powers of 10 used to express very large numbers.

Example

$$\begin{aligned} 123456.0 &= 12345.6 \times 10 & \text{or} & & 12345.6 \times 10^1 \\ &= 1234.56 \times 100 & \text{or} & & 1234.56 \times 10^2 \\ &= 1.23456 \times 10\,000 & \text{or} & & 1.23456 \times 10^5 \end{aligned}$$

Note: Every time the decimal point moves a place to the left, the power goes up by 1.

The same can be done for very small numbers.

> **Example**
>
> $$0.000123 = 0.00123 \times \frac{1}{10} \text{ or } 0.00123 \times 10^{-1}$$
>
> $$= 1.23 \times \frac{1}{10000} \text{ or } 1.23 \times 10^{-4}$$

It is important to understand that in all the previous examples, the value of the numbers did not change, only the way we wrote them down.

In science we can use any power, but in engineering we normally use only powers of 10 in multiples of 3, e.g. 10^{-3}, 10^3, 10^6 and we do this because they match the prefixes we use such as milli, kilo or Mega.

If you look back to Table 8.03, you should now be able to see how the powers of ten relate to them:

Mega (M) = 1 000 000 times, which is the same as 10^6
micro (µ) = 0.000 001 times, or 10^{-6}

So, just how fast was that speeding bullet?

Well… to recap, we said it was travelling 3 km in 50 ms.

We also said that we calculate speed by using the following formula:

$$\text{Speed} = \frac{\text{Distance travelled}}{\text{Time taken}}$$

3 km = 3 × 10^3 m ..... in other words 3000 m

And 50 ms = 50 × 10^{-3} s .... in other words 0.05 s

So the speed of the bullet is:

$\frac{3 \text{ km}}{50 \text{ ms}}$ which we can write as $\frac{3 \times 10^3}{50 \times 10^{-3}}$ which gives us

$$\frac{3000 \text{ m}}{0.05 \text{ s}} = \mathbf{60\,000 \text{ m/s}}$$

Our answer could also be written as 60 × 10^3 m/s which we could also say as 60 km/s.

It is important to practise using these different ways of expressing numbers as, although a calculator is a great help, it is easy to make a mistake. Being comfortable with numbers and always being able to check that the result looks correct is very useful.

> **Remember**
>
> Don't take your calculator for granted! Always check that the answer looks sensible using approximation.

Fractions

We often need to do calculations with parts of a whole unit and so far we have used the decimal system for this. However, another method is to use fractions.

A fraction is simply the result of dividing something (the whole thing) into smaller, equal parts.

Consider a cake that hasn't been cut up and so we have the whole thing. There is one part available and we've got it.

However, you wish to cut it into 4 equal pieces and so divide the whole thing into 4 parts.

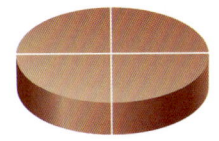

If we do this and then put one piece aside, that piece (shown in blue) will be one quarter of the original cake. That leaves three pieces remaining of the original 4 that made up the whole cake.

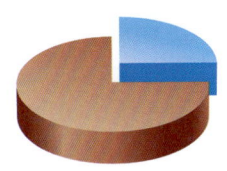

Our small piece of cake is one bit out of the 4 bits that were available, and we would write it as:

$\frac{1}{4}$ one part out of four available. This leaves 3 parts out of four left, in other words $\frac{3}{4}$

The same blue piece of cake could also be described with decimals as 0.25 of the cake; it doesn't make any difference to the size of the piece. Sometimes we use fractions, sometimes decimals, it often depends on which is easiest.

Adding and subtracting fractions

To become an electrician you will need to know how to handle arithmetic involving different fractions.

For example, how do you add

$\frac{1}{4} + \frac{3}{7}$?

So, how do we add quarters to sevenths?

We need to express each of the two fractions with the same common denominator (*the number under the line*) so we can add like to like.

So, to add or subtract fractions, follow these steps:

1. Choose a new common denominator (*the number under the line*) for both fractions. To do this, we find the lowest possible number that both denominators will go into. Quite often the easiest method is to multiply the two numbers together, in this $4 \times 7 = 28$.

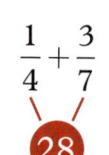

2. Now change each fraction into values of the new common denominator. We do this by multiplying the top part of each fraction (the numerator) by the number of times that the original denominator divides into our new common denominator. So, in our example, in the first fraction (1/4), 4, goes into 28 seven times, so we need to multiply the top of our first fraction (1) by 7. For the second fraction (3/7), because 7 goes into 28 four times, we multiply the top of our fraction (3) by 4.

Now our sum will looks like this:

$\frac{7+12}{28}$ which gives us $\frac{19}{28}$

3. One last step might be needed. If the numerator is bigger than the denominator, this means there's more than one whole thing. It is correctly called an improper fraction.

For example $\frac{3}{2}$ means you have 'three halves' which you would normally say was 'one and a half'.

When this happens, we simply find how many times the denominator will go into the numerator (this gives us the amount of whole numbers we have) and then what's left becomes the numerator (top number) of the fractional part.

For example $\frac{29}{12}$

12 goes into 29 twice, with 5 left over, so this is $2\frac{5}{12}$ as a mixed number

Multiplying fractions

Multiplying fractions is easier than adding and subtracting. You simply multiply all the top numbers together to get a new numerator and then multiply all the bottom ones together to get a new denominator.

So, for example: $\frac{1}{4} \times \frac{3}{7}$ becomes $\frac{1 \times 3}{4 \times 7} = \frac{3}{28}$

and $\frac{3}{4} \times \frac{4}{5}$ becomes $\frac{12}{20} = \frac{3}{5}$

Dividing fractions

Dividing fractions is nearly as easy. First turn the second fraction upside down and then treat the problem as though it was a multiplication.

So for example:

$\frac{1}{4} \div \frac{3}{7}$ becomes $\frac{1}{4} \times \frac{7}{3} = \frac{7}{12}$

And

$\frac{1}{2} \div \frac{1}{4}$ becomes $\frac{1}{2} \times \frac{4}{1} = \frac{4}{2} = 2$

Percentages

Percentages are based upon a fraction with 100 parts. We use the symbol % to show per cent.

If per cent mean 'parts out of a hundred', then 1% must mean one part out of a hundred.

If 80% of the population own a television, we are saying that 80 out of every hundred people own a TV. We can use a grid of squares to show this concept.

Did you know?

The name percentage comes from the Latin word 'centum', meaning 100. So, think about a century break in snooker, or the number of cents in a dollar and you'll see it's still used!

1	2	3	4	5	6	7	8	9	10
11	12	13	14	15	16	17	18	19	20
21	22	23	24	25	26	27	28	29	30
31	32	33	34	35	36	37	38	39	40
41	42	43	44	45	46	47	48	49	50
51	52	53	54	55	56	57	58	59	60
61	62	63	64	65	66	67	68	69	70
71	72	73	74	75	76	77	78	79	80
81	82	83	84	85	86	87	88	89	90
91	92	93	94	95	96	97	98	99	100

Our green square is one out of the 100 available, in other words 1/100 or 1%.

The row of blue has 10 squares, in other words 10 out of the 100 available, or 10/100 or 10%.

1	2	3	4	5	6	7	8	9	10
11	12	13	14	15	16	17	18	19	20
21	22	23	24	25	26	27	28	29	30
31	32	33	34	35	36	37	38	39	40
41	42	43	44	45	46	47	48	49	50
51	52	53	54	55	56	57	58	59	60
61	62	63	64	65	66	67	68	69	70
71	72	73	74	75	76	77	78	79	80
81	82	83	84	85	86	87	88	89	90
91	92	93	94	95	96	97	98	99	100

Here, our orange squares make up 76 of the 100 available, in other words 76%.

These examples show that fractions, decimals and percentages are simply different ways of expressing the same thing.

To show the link, Table 8.04 shows some common fractions and their link to decimals and percentages.

Fraction	Meaning	Convert to decimal	Decimal	Percentage
$\frac{1}{2}$	The whole has divided into two parts and we've got one of them	key 1 ÷ 2	0.5	50%
$\frac{50}{100}$	The whole has divided into 100 parts and we've got 50 of them	key 50 ÷ 100	0.5	50%
$\frac{3}{4}$	The whole has divided into 4 parts and we've got 3 of them	key 3 ÷ 4	0.75	75%
$\frac{75}{100}$	The whole has divided into 100 parts and we've got 75 of them	key 75 ÷ 100	0.75	75%

Table 8.04 Common fractions

100% is 100 parts out of the 100 available and so the whole thing!

Here are some calculations with percentages:

Example 1

At the beginning of December, you notice that a shop has increased the price of a hi-fi system by 5% to make a bigger Christmas profit. However, at the beginning of January there is a sale in the store and you now see that all unsold items, including the hi-fi system, are now labelled 5% off! So would you now be paying the same, more or less than if you had bought the item in November?

As the question gives no actual values, choose an easy one and apply to the question accordingly.

So let's assume that the hi-fi cost £200 in November.
Then in December the store increases the price by 5%
5% is half of 10%, 10% of £200 would be £20 and therefore 5% must be £10.
So the December price must be **£210**.

In January the store then advertise 5% off the advertised price.
Remember that the advertised price is now £210.
As before, 5% is half of 10% and 10% of £210 is £21 and therefore 5% must be **£10.50**.

In January the sale price therefore works out as £210 − £10.50 = **£199.50**

So you would actually pay slightly less if you bought the hi-fi in the January sale.

Example 2

What is 35% of £18 000?

$\frac{35}{100} \times 18\,000$ gives us $0.35 \times 18\,000 = £6300$

Percentages assume that the whole is divided in to 100 parts, then saying that £18 000 (the whole thing in this example) has 100 parts and so dividing £18 000 by 100 gives 1%

$\frac{£18\,000}{100}$ comes to £180

Find 35% by multiplying £180 × 35 = £6300

Without a calculator we could do the calculation like this.

35% = 10% + 10% + 10% + 5%

10% of £18 000 is one tenth of £18 000 which is £1800.
5% is half of 10% which is £900
£1800 + £1800 + £1800 + £900 = £6300

We can also do a rough check by approximation.

You should know that 75% is three quarters, 50% is a half and we know that 33.333…% is one third.

A third of £18 000 (dividing 18 000 by 3) comes to £6000

As 33% is just under the 35% we are looking for, the real answer must be slightly **larger** than £6000 and so the answer £6300 is likely to be correct. Approximation is a very useful tool when checking those examination answers!

Example 3

I have a restaurant bill of £72 and want to leave the staff a 10% tip. How much will this be?

10% is $10 \div 100 = \frac{1}{10}$ or 0.1

0.1 × £72 is £7.20
We don't need a calculator as to find a tenth we divide by ten.
To divide by ten we move the digits one place to the right, giving us £7.20.

Algebra

People often think that algebra is very difficult, but it is actually just a way of writing down calculations without using specific numbers. Instead, we use letters or symbols to represent different quantities. Using algebra, we can write down relationships between different things, and then later we can replace the symbols with real numbers when we know them.

As a simple example, let's say we have 'x' girls in a class and 'y' boys in the same class. At the moment, it doesn't matter what numbers 'x' and 'y' represent. However, if someone asked you to find the total number of students in the class, which we can call 'z', then you know that you must add the total number of boys and girls together.

Written as an algebra equation this is $z = x + y$

If we are then told that $x = 11$ and $y = 15$, then by substituting the real numbers into the above equation, we can establish that z must $= 26$.

Rules for algebra

- In algebra, we don't use the multiplication sign. So D × E × F is written as DEF
- As in numerical calculations, in addition and multiplication, it doesn't matter which symbol comes first. So D + E = E + D, and DE = ED
- There are several ways of writing the same thing. So D(E + F) could be written as D × (E + F) or D(F + E). We could also write it as DE + DF

    As another example $\frac{D}{E} \times \frac{E}{F} = \frac{D}{F}$ (by cancelling)

Here's an example to demonstrate some of these rules.

> **Example**
>
> Work out the value of the expression 8DE − 2EF + DEF is when D = 1, E = 3 and F = 5.
>
> 8DE − 2EF + DEF becomes 8 × D × E − 2 × E × F + D × E × F
>
> Now replace the letters with the correct numbers:
>
> 8 × 1 × 3 − 2 × 3 × 5 + 1 × 3 × 5 becomes 24 − 30 +15 = **9**

Indices

If you remember, a few pages ago we talked about some little numbers that were called 'powers'.

A power, or an index to use its proper name, is used when we want to multiply a number by itself several times. The plural of index is indices. In other words, powers and indices are one and the same thing. However, it would be wise to cover the topic in a slightly different way to make sure we fully understand it and other related concepts.

If we multiply two identical numbers together, say 5 and 5, the answer is 25 and the process is usually expressed as: $5 \times 5 = 25$

However, we could express the same calculation as: $5^2 = 25$

The upper ^{2} is referred to as the index and the number it is applied to, in this case the number 5, is called the base. We can therefore say that the index is the number of times that the base is written down with multiplication signs between them.

Using the example above, sometimes this is referred to as five raised to the power of two, and using this logic we can therefore say that 5^3 means five written down 3 times with multiplication signs between them. In other words:

$5 \times 5 \times 5 = 125 = 5^3$

Be careful – Do not make the mistake of thinking $5^3 = 5 \times 3$. It is not!

Some other examples of using indices are:

> **Example**
>
> $3^3 = 3 \times 3 \times 3 = $ **27** $8^3 = 8 \times 8 \times 8 = $ **512** $6^2 = 6 \times 6 = $ **36**

How about multiplying two numbers together that both have indices? Consider $4^2 \times 4^2$

This could be shown as $(4 \times 4) \times (4 \times 4)$ or as, 4^4 meaning the indices ^{2} and ^{2} have been added together.

The answer? **256**

So the rule is: When multiplying numbers with indices, simply add the indices.

Here are two examples to illustrate both points:

> **Example**
>
> $4 \times 4^2 = 4^1 \times 4^2$ $5 \times 5^3 \times 5^2 = 5^1 \times 5^3 \times 5^2$
>
> Add the indices
>
> $ = 4^3$ $ = 5^6$
>
> $ = 4 \times 4 \times 4$ $ = 5 \times 5 \times 5 \times 5 \times 5 \times 5$
>
> $ = \mathbf{64}$ $ = \mathbf{15\,625}$

Let us now move on to the situation where we are dividing numbers that have indices.

> **Example**
>
> $\dfrac{2^5}{2^3} = \dfrac{2 \times 2 \times 2 \times 2 \times 2}{2 \times 2 \times 2}$ which if we check by multiplying out is $\dfrac{(32)}{(8)}$
>
> If we cancel out the twos:
>
> $\dfrac{2^5}{2^3} = \dfrac{\cancel{2} \times \cancel{2} \times \cancel{2} \times 2 \times 2}{\cancel{2} \times \cancel{2} \times \cancel{2}} = 2 \times 2$
>
> This leaves us with,
>
> $2 \times 2 = \mathbf{2^2}$ **or 4**   (32 ÷ 8 does indeed equal 4)

So the rule is: When dividing numbers with indices simply subtract the indices.

> **Example**
>
> $\dfrac{5^3 \times 1}{5^2}$ As anything multiplied by 1 remains the same, we could instead say $\dfrac{5^3}{5^2}$
>
> This means that the indices can now be subtracted. Therefore, subtracting the indices (i.e. 3 − 2) gives us:
>
> $\dfrac{5^3}{5^2} = 5^1$ and as we don't write the index of 1, our answer is **5**

In the above example we subtracted the indices (3 − 2). But 3 subtract 2 (3 − 2) can also be written as 3 add −2.

If you remember, the addition of indices belongs to multiplication of numbers with indices. So from this we can see that 5^3 divided by 5^2 will actually be the same as 5^3 multiplied by 5^{-2}

We can therefore say that: $\dfrac{1}{5^2}$ is the same as $1 \times 5^{-2} = 5^{-2}$

Further examples of this could be:

$$\frac{1}{4^3} = 4^{-3} \quad \text{and} \quad \frac{1}{2^6} = 2^{-6}$$

Where we can now see that indices can be moved above or below the line, but they will have an opposite value (i.e. 2^6 became 2^{-6}).

> **Example**
>
> $$\frac{5^6 \times 5^7 \times 5^{-3}}{5^4 \times 5^2} = \frac{5^{13} \times 5^{-3}}{5^6} = \frac{5^{10}}{5^6} = 5^{10} \times 5^{-6} = 5^4 = 5 \times 5 \times 5 \times 5 = \mathbf{625}$$

Transposition

Transposition is a method that uses the principles of mathematics to allow you to rearrange a formula or equation so that you can find an unknown quantity.

There is however one important rule that must always be followed … **without fail**!

What you do to one side of the equation, you must do to the other side.

Let's work through an example.

> **Example 1**
>
> Transpose (rearrange) the following formula to make Y the subject (the one we want):
>
> X = Y + Z
>
> First, think of the equation as being a pair of scales and remember that each side of the scales (each side of the equals sign) must be balanced.
>
> Second, when we want to remove something, we perform an opposite operation. In doing so, remember that if there is no + or − sign in front of a number, then the number is positive.
>
> We have to find Y; so to get Y by itself, we must remove Z.
> As Z has been added to Y we need to perform 'an opposite operation'.
> We need to subtract Z from both sides of the equation to keep it balanced.
>
> So: X = Y + Z   now becomes: X **− Z** = Y + Z **− Z**
>
> As **+ Z − Z** = 0, you are left with: **X − Z = Y or Y = X − Z**

Note that when something moves from one side of the equal sign to the other, it becomes the opposite value (+Z became –Z).

You can check your understanding of this new concept by using numbers instead of letters.

In our example   X = Y + Z   could become   8 = 6 + 2
When rearranged   X – Z = Y   would therefore be   8 – 2 = 6

Example 2

Transpose this equation A = R – LS to make R the subject of the equation.

To get R by itself, we need to move LS.
This is being subtracted from R, so we need to add it to both sides.
So: A **+ LS** = R – LS **+ LS**
as LS + LS = 0 then **A + LS = R** or **R = A + LS**

The previous examples showed how a formula made up of addition and subtraction can be changed around. Equally, we can view multiplication and division in much the same way. For example:

Example 3

Transpose the formula   V = I × R   to make I the subject.

We can see that I has been multiplied by R. Therefore, to leave I by itself, we must divide by R on both sides.

So: V = I × R   now becomes $\dfrac{V}{R} = \dfrac{I \times R}{R}$

Cancel $\dfrac{V}{R} = \dfrac{I \times \cancel{R}}{\cancel{R}}$ and we are left with an answer of: $I = \dfrac{V}{R}$

Again, note that when something moves from one side of the equal sign to the other, it becomes the opposite value (V = I × **R** ended up as I = V ÷ **R**).

Here is another example:

Example 4

This equation describes how resistance is related to length, area and type of material in a conductor (you will come across this equation later in this section).

$R = \dfrac{\rho L}{A}$ Transpose to find L.

First we need to get rid of the A.

As it is currently divided into ρL, we need to multiply both sides by A.

> **Example 4 continued**
>
> Therefore $R = \dfrac{\rho L}{A}$
>
> Becomes: $R \times A = \dfrac{\rho L \times A}{A}$
>
> Giving us $R \times A = \rho L$
>
> We now need to get rid of the symbol ρ
>
> This is currently multiplying L, so we must divide ρ on both sides.
>
> Therefore: $R \times A = \rho L$
>
> Becomes: $\dfrac{R \times A}{\rho} = \dfrac{\rho L}{\rho}$
>
> The two ρ on the right side cancel.
>
> Giving us: $\dfrac{R \times A}{\rho} = L$

Triangles and trigonometry

There are areas of electrical science and even practical installation work where knowledge of triangles and trigonometry is very useful. This section covers some basic principles.

Angles

An angle is the size of the opening between two lines.

Both lines start at point O, and the length of the lines makes no difference to the angle. Even if line B was twice as long as line A, the angle between them would be the same.

If we rotate line A (with one end still fixed to point O) anticlockwise, the angle will get bigger. Eventually, it will be on top of line B, and the angle will be zero. We divide this complete turn into 360 parts, called degrees, represented by the symbol '°' after the number (e.g. 360°).

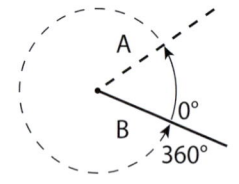

Figure 8.02 shows some common types of angle.

Acute angle

Right angle

The small square is often used in drawings to show a right angle

Obtuse angle

Straight line

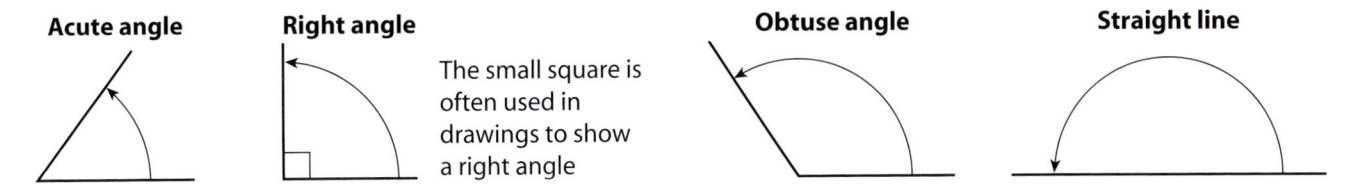

Figure 8.02 Types of angle

Angles on a straight line

In the diagrams so far, you can see that we can think of a straight line as an angle of 180° (which is not surprising as there are 360° in a full circle).

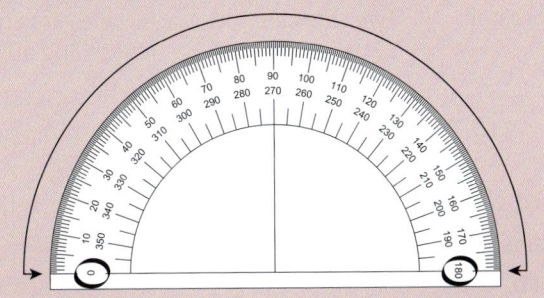

You can see this clearly on a protractor. Starting at one side and moving round, we end up at the 180° point.

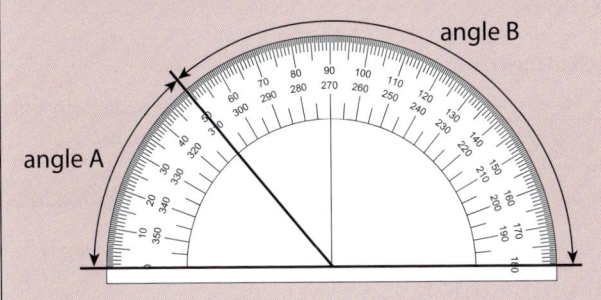

angle B

angle A

If we have two angles drawn on a straight line, then they must add up to 180°.
In our diagram, angle A is 50°, and angle B is 130°, which add up to 180°.

Angle sum of a triangle

In any triangle there are three internal angles at the corners and these must add up to 180°. This means that, if we are given two of the angles, we can easily find the third one. For example, if one angle is 40° and another is 60° (40 + 60 = 100), then the third must be 80° (100 + 80 = 180).

X

83°   45°

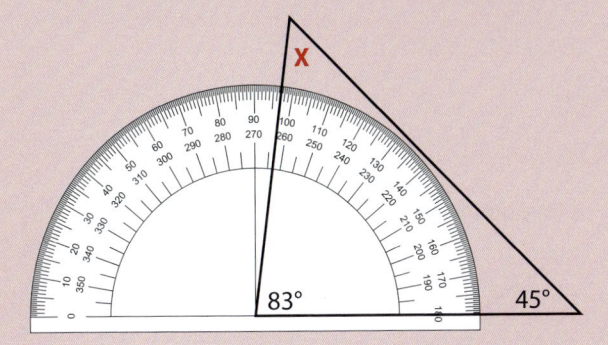

In this diagram, one angle is 83° the other is 45° and so the third, 'X', must be **52°**.

Triangles

There are five types of triangle as shown in Figure 8.03.

Scalene
Every angle is less than 90°

Right-angled
One angle is 90°

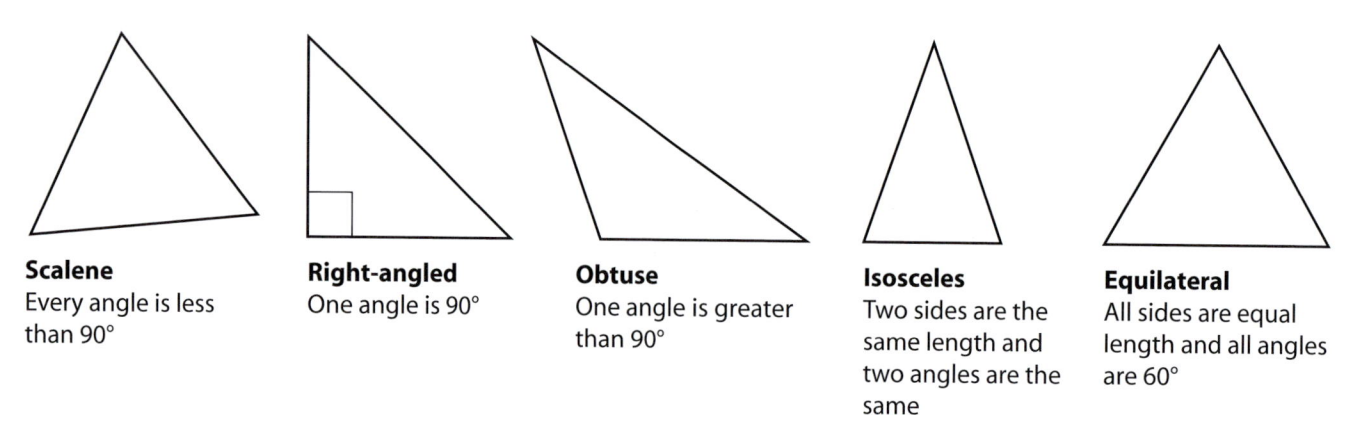

Obtuse
One angle is greater than 90°

Isosceles
Two sides are the same length and two angles are the same

Equilateral
All sides are equal length and all angles are 60°

Figure 8.03 Types of triangle

This brings us to one of the most useful formulas that you will ever be given: one that is seen in everyday use. Its name is **Pythagoras' Theorem**.

Pythagoras' Theorem

Pythagoras' Theorem states that:

> *For a right-angled triangle, the square of the hypotenuse is equal to the sum of the squares on the other two sides.*

This means that if you know the length of two of the sides, you can find out the length of the third.

Let's have a look at a right-angled triangle in Figure 8.04.

We can see that the right angle (represented with a square in the corner) is like an arrowhead and it always points at the longest side, which is called the hypotenuse. For ease, lets call the hypotenuse side (A) and the other sides (B) and (C). Remember the formula:

$A^2 = B^2 + C^2$

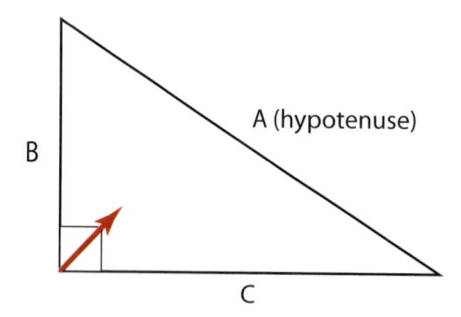

Figure 8.04 Right-angled triangle

We said that it could be used every day

Well, how do they get the sides of a building straight?

If you know that one side wall is 30 metres long and the other is 40 metres, then if you use Pythagoras' Theorem to work out the diagonal (hypotenuse) you can get both walls exactly straight by checking they form a right-angled triangle.

Using the formula:

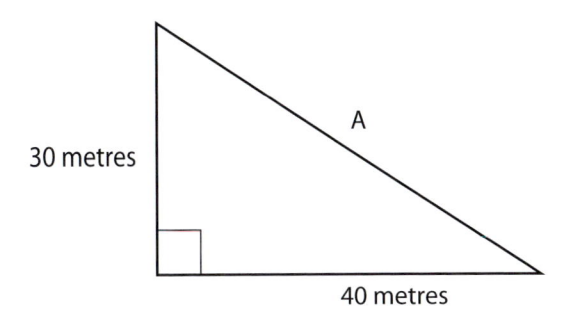

$A^2 = B^2 + C^2$
$A^2 = 30^2 + 40^2$
$A^2 = 900 + 1600$
Therefore $A^2 = 2500$
Therefore $A = \sqrt{2500}$
Therefore **A = 50 m**

Trigonometry

Trigonometry is about the relationship between the angles and sides of triangles, and on the previous page we discovered that in a right-angled triangle, we call the long side the hypotenuse and the right angle 'points' at it. The names of the other two sides will depend on the angle that we have to find, or intend to use!

The side which is opposite the angle being considered is called the **opposite**, and the side which is next to the angle under consideration and the right angle is called the **adjacent**.

Well, if placed on to a drawing we would have the following and in the drawing, (Ø) is the angle to be considered.

So, think of it as being like a torch and the beam is shining on to the opposite side.

We already know that the longest side is the hypotenuse, so the one that is left must be the adjacent.

These are shown in Figure 8.05.

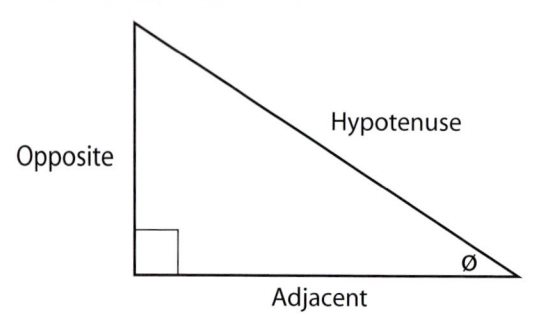

Figure 8.05 Sides of a triangle

The next step relates to the following terms: tangent, sine and cosine.

These are used to show the ratio between angles and sides and you choose which one you need depending upon the information that you have been given.

You need to know the following formulae:

$$\mathbf{S}\text{ine } \varnothing = \frac{\mathbf{O}\text{pposite}}{\mathbf{H}\text{ypotenuse}} \qquad \mathbf{C}\text{osine } \varnothing = \frac{\mathbf{A}\text{djacent}}{\mathbf{H}\text{ypotenuse}} \qquad \mathbf{T}\text{angent } \varnothing = \frac{\mathbf{O}\text{pposite}}{\mathbf{A}\text{djacent}}$$

To help you remember them, try to remember the following name: **SOHCAHTOA.**

It's another mnemonic and represents the letters that we have highlighted in each formula below

SOH (**S**ine = **O**pposite over **H**ypotenuse)
CAH (**C**osine = **A**djacent over **H**ypotenuse)
TOA (**T**angent = **O**pposite over **A**djacent)

Let us have a look at some examples. Remember that the formula you will use will depend upon the information that you have been given. Figure 8.06 shows the relevant keys on a calculator that you will need to use and which we refer to in the examples.

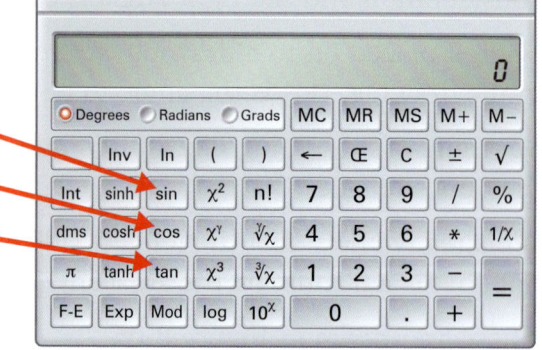

Figure 8.06 Using a calculator for trigonometry

Example

From the diagram, what is the value of angle Ø?

Because in this example we are given details about the opposite and adjacent sides, we will therefore use the tangent formula.

We have now found that the angle has a tangent of 1.4. Use your calculator to find the angle by entering the INV key and then the TAN key (on most calculators, this is labelled the TAN⁻¹ function) then 1.4 =.

The answer should be 54.5°.

$$\text{Tangent } \emptyset = \frac{\text{Opposite}}{\text{Adjacent}}$$

$$\text{Tangent } \emptyset = \frac{7}{5} = 1.4$$

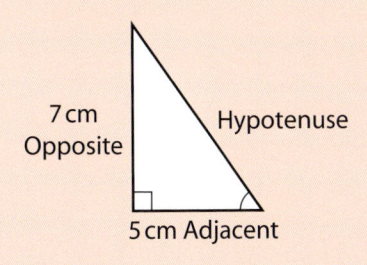

7 cm Opposite, Hypotenuse, 5 cm Adjacent

Remember

On some older calculators the buttons have to be used in a different order. Check to make sure you can get the correct answer.

Example

From the diagram, what is the value of angle Ø?

This time we have information about the opposite and hypotenuse and so we use the sine formula.

Now use your calculator to find the angle that has a sine of 0.375. Enter INV then the SIN key and then 0.375 =.

The answer should be 22°.

$$\text{Sine } \emptyset = \frac{\text{Opposite}}{\text{Hypotenuse}}$$

$$\text{Sine } \emptyset = \frac{3}{8} = 0.375$$

8 cm Hypotenuse, 3 cm Opposite

Example

From the diagram, what is the value of angle Ø?

This time we have information about the adjacent and hypotenuse and so we use the cosine formula.

Now find the angle that has a cosine of 0.81818. Use your calculator and enter the INV key then the COS key then enter 0.81818 =.

The answer should be 35.1°.

$$\text{Cosine } \emptyset = \frac{\text{Adjacent}}{\text{Hypotenuse}}$$

$$\text{Cosine } \emptyset = \frac{9}{11} = 0.81818$$

9 cm Adjacent, 11 cm Hypotenuse

In the previous examples we were using trigonometry to find an angle.

We can also use it to work out the length of a side of a triangle.

Example

In the following diagram, find the length of side **x**.

As in the previous examples, the formula that we use depends upon the information that we have been given.

In this question, we have been told to find the length of side **x** and this is opposite the angle that we have been given.

We are also told that the adjacent side is 5 cm long. Remember, if we are given details about the opposite and adjacent sides, then we use the Tangent formula.

We therefore now need to transpose the formula to make **x** the subject.

Doing so gives: **x** = Tangent 25° × 5 cm

Now using your calculator, press the TAN key enter 25 and then =. This will give you 0.4663.

Now multiply this by 5 and the answer should be 2.331 and so side **x** of the triangle is **2.33 cm** long.

$$\text{Tangent } 25° = \frac{\text{Opposite}}{\text{Adjacent}}$$

$$\text{Tangent } 25° = \frac{\text{Side x}}{5}$$

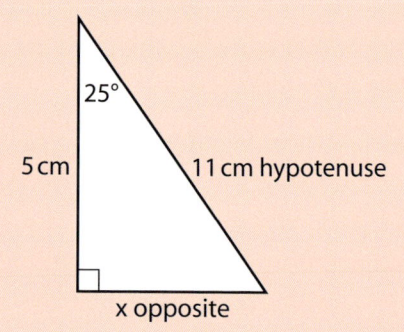

The area of a triangle

One last thing about triangles is how to calculate their area. Remember when you're looking at this, that a right-angled triangle will be half the area of a rectangle

The formula is: $\text{Area (A)} = \frac{1}{2} \text{base} \times \text{height}$ or $\frac{\text{base} \times \text{height}}{2}$

Here's an example using both formulae, noting that the large triangle we are trying to find the area of effectively contains two right-angled triangles. An example is shown below.

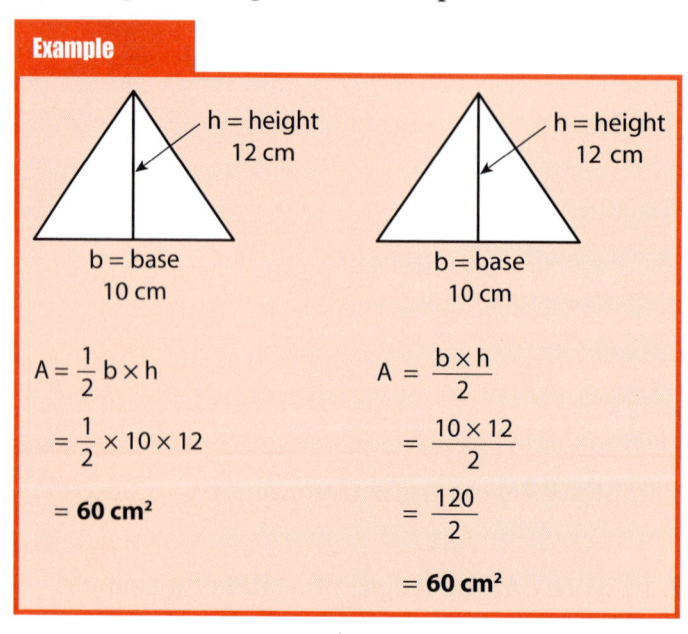

Example

h = height 12 cm, b = base 10 cm

$A = \frac{1}{2} b \times h$

$= \frac{1}{2} \times 10 \times 12$

$= 60 \text{ cm}^2$

$A = \frac{b \times h}{2}$

$= \frac{10 \times 12}{2}$

$= \frac{120}{2}$

$= 60 \text{ cm}^2$

Statistics

Charts

There are many ways to record and display data and in this section we will look at some of the basic techniques. We'll start with some **data** about an electrical company and then show how we can graphically represent the information.

Table 8.05 shows data that has been gathered about the workforce of an electrical contractor who employs a total of 40 people.

> **Key term**
>
> **Data** – factual information and statistics used as a basis for discussion, calculation or analysis

Type of staff	Ref Code	Number employed	Percentage of total workforce (%)
Electricians	A	14	35% (14 ÷ 40 × 100)
Apprentices	B	12	30% (12 ÷ 40 × 100)
Clerical staff	C	8	20% (8 ÷ 40 × 100)
Labourers	D	4	10% (4 ÷ 40 × 100)
Managers	E	2	5% (2 ÷ 40 × 100)

Table 8.05 Electrical contracting company staffing statistics

Having given each type of job a reference code, we can now represent this information pictorially and here are the two most common methods.

The pie chart

As the name suggests, this type of diagram shows the information as sections of a pie. To be able to do this we first have to convert information into angles. As we know that a full circle has 360°, this will represent our total number (100%).

In our data, this is a total of 40 employees. Therefore, we now need to find what angle represents each occupation.

We know that electricians (A) make up 14 out of the 40 and therefore for electricians the angle is:

$\dfrac{14}{40} \times 360 = 126$ In other words 126° out of 360° available belong to the electricians

Repeat the exercise for the other occupations and we have our complete chart, as shown in Figure 8.07.

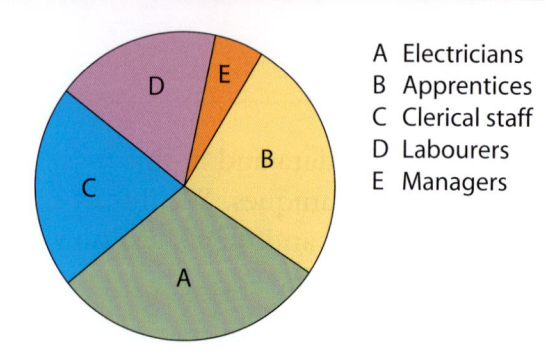

Figure 8.07 Pie chart of contracting company staffing statistics

A Electricians
B Apprentices
C Clerical staff
D Labourers
E Managers

The bar chart

A bar chart uses rectangular bars with lengths proportional to the values that they represent and it is the height of the bar that shows its magnitude.

Bar charts are used for comparing two or more values and the bars can be shown horizontally or vertically.

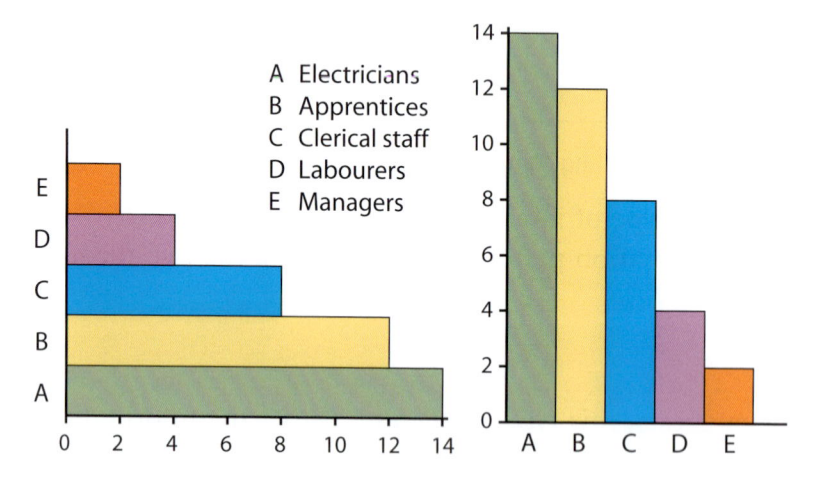

Figure 8.08 shows all our employer data using this method.

Figure 8.08 Bar chart of contracting company staffing statistics

Frequency distribution and tally charts

A frequency distribution is a representation of the number of counts of something (e.g. objects or responses), that is usually shown in the form of a table or graph.

Tally marks are a simple counting system that lets us record how many times something happens. The tally system works by drawing a vertical line every time something happens and every fifth occurrence a fifth line is drawn diagonally across the previous four lines to indicate a bunch of five.

If I was counting some bananas, here's how it would work:

1 I
2 II
3 III
4 IIII

And when I got to the fifth banana I would mark it down like this: ~~IIII~~

This makes working out an amount a lot easier, as it is easy to count in groups of five.

Here you will see both techniques in a worked example.

Example

Imagine that a company manufactures screws. A box that contains 100 screws is checked, the lengths of the screws are different and their lengths are shown in Figure 8.09.

A customer needs screws that can only be 15 mm long for some specialist equipment. and the manufacturer needs to know how many of each size there are. This is where the tally mark system can be used.

Using the tally system to count the various screws, gives Figure 8.10.

15.02	15.00	15.00	15.01	15.01
15.01	14.99	14.99	15.00	14.99
15.01	15.01	14.99	14.98	15.03
15.01	15.00	15.00	15.02	15.03
15.01	15.00	14.98	15.02	15.04
14.98	14.97	14.99	15.00	14.98
15.00	15.00	14.98	14.99	15.00
14.98	15.00	14.99	14.97	15.01
15.01	15.00	15.03	14.98	14.98
15.01	14.99	15.00	15.02	15.00
14.98	14.98	15.00	14.96	14.99
15.03	15.02	15.01	15.03	15.01
14.99	15.04	15.02	15.01	15.01
15.01	15.01	14.98	15.02	14.99
15.01	15.01	14.99	15.02	15.00
15.03	14.97	14.97	15.00	15.00
14.99	15.00	14.99	14.99	14.99
15.02	15.00	15.00	15.00	15.03
15.01	14.99	15.00	14.96	14.99
15.01	14.99	15.03	15.01	14.99

Figure 8.09 Lengths of screws

Length (mm)	Number of screws with this length	Frequency
14.96	II	2
14.97	IIII	4
14.98	~~IIII~~ I	11
14.99	~~IIII~~ ~~IIII~~ ~~IIII~~ ~~IIII~~	20
15.00	~~IIII~~ ~~IIII~~ ~~IIII~~ ~~IIII~~ III	23
15.01	~~IIII~~ ~~IIII~~ ~~IIII~~ ~~IIII~~ I	21
15.02	~~IIII~~ IIII	9
15.03	~~IIII~~ III	8
15.04	II	2

Figure 8.10 The tally system

From the tally chart we can see that there are twenty-three 15 mm screws and the customer can be supplied. It is easy to see on the tally chart how many we have of something (the frequency distribution), it can be easier and more useful if we portrayed the information pictorially.

The histogram

As we just said, frequency distribution becomes clearer if we draw a diagram and we would normally do this using a histogram.

A histogram is a type of bar chart where it is the area, rather than the height, of each bar that is representative of the frequency. The histogram provides a measure of spread and the bars always touch.

Figure 8.11 shows the information from the tally chart expressed as a histogram.

When you look at the histogram, the pattern of the variation in screw size is easy to understand, most of the values are grouped near the centre of the diagram with a few values more widely scattered.

This shows that the company generally make screws of the right length (15 mm) but that the accuracy could improve. There is possible waste of materials with longer ones and the shorter ones may be unusable.

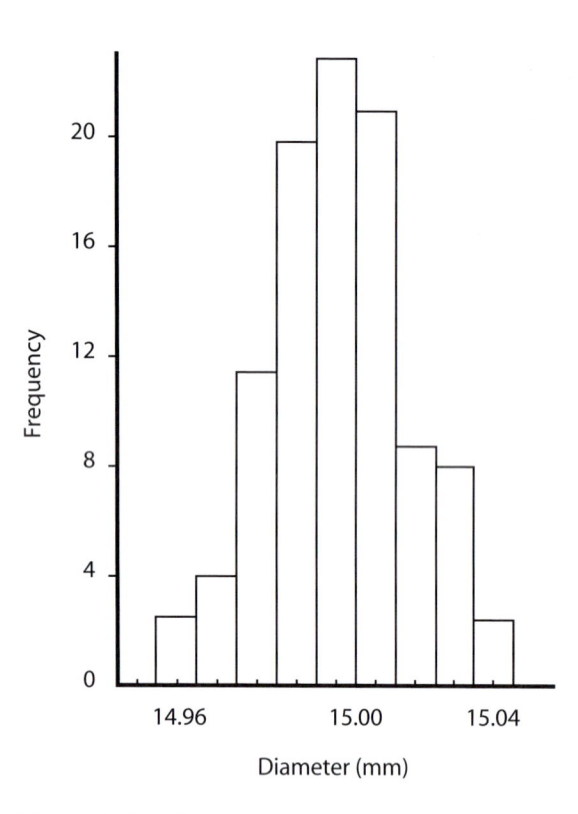

Figure 8.11 Histogram

Statistical averages

One last aspect of statistics involves the calculation of averages.

There are three common types of average:

Mean

This is the commonest type of average and it is determined by adding up all the items in the set and dividing the result by the number of items:

$$\text{Mean} = \frac{\text{The total of items}}{\text{The number of items}}$$

> **Example**
>
> The marks of an apprentice in four examinations were 86, 72, 68 and 78. Find the mean of his marks.
>
> $$\text{Mean} = \frac{86+72+68+78}{4}$$
>
> $$\text{Mean} = \frac{304}{4} = \mathbf{76}$$

Median

If a distribution is arranged so that all the items are in ascending (or descending) order of size, then the median is the value that is half-way along the series. Generally there will be an equal number of items below and above the median. If there is an even number of items the median is found by taking the average of the two middle items.

> **Example**
>
> The median of 3, 4, 4, 5, **6**, 8, 8, 9, 10 is 6
>
> The median of 3, 3, 5, **7**, **9**, 10, 13, 15 is $\left(\frac{7+9}{2}\right) = \mathbf{8}$

Mode

The observation or item which occurs most frequently in a distribution, is called the mode.

In the data set 3, 3, 5, 7, 9, 10, 13, 15 the number 3 occurs most often, so 3 is the mode.

> **Progress check**
>
> 1. Identify the SI units for the following:
>   - Length
>   - Mass
>   - Time
>   - Electric current
>   - Temperature
>   - Luminous intensity
>
> 2. Convert the following quantities:
>   - 6 km into metres
>   - 3500 mm into metres
>   - 1 MΩ into ohms
>
> 3. Your supervisor tells you to leave 10% extra cable at the end of a run to allow for terminations. If the cable run is 34.7m long what will the overall length be?

K3. Understand basic mechanics and the relationship between force, work, energy and power

The human race is very inventive. We have devised many means of overcoming simple problems, such as lifting a heavy object and moving it from one place to another.

Now that we understand the basics behind mathematical procedures, we need to start applying these to some of scientific and mechanical concepts you will be dealing with as an electrician.

The difference between mass and weight

We need to understand a very important concept:

The difference between weight (a force) and mass.

Mass

This is simply the amount of stuff or matter contained in an object. Assuming we do not cut or change the object, the mass of an object will stay the same wherever we are.

The unit of mass is the kilogram (kg)

Weight

This is a force and depends on how much gravity pulls on a mass. This can vary according to where we are (the higher above sea level you go, the gravitational pull of the Earth is less and so you weigh less).

This change in weight is tiny but can be measured with very sensitive and expensive scientific equipment.

The unit of weight (and all other forces) is the newton (N)

On Earth, if we disregard the effect of height above sea level, the weight acting on 1 kg of mass is equal to 9.81 newtons (N). So, **1 kg weighs 9.81 N**. In many situations this can be rounded up to 10 N.

> **Remember**
>
> It is vital you understand the difference between weight and mass.

Principles of basic mechanics

A simple machine is a device that helps us to perform our work more easily when a force is applied to it. A screw, wheel and axle and lever are all simple machines.

A machine also allows us to use a smaller force to overcome a larger force and can also help us change the direction of the force and work at a faster speed. The most common simple machines are shown as follows.

Levers

Levers let us use a small force to apply a larger force to an object. They are grouped into three classes, depending on the position of the fulcrum (the pivot).

Class 1

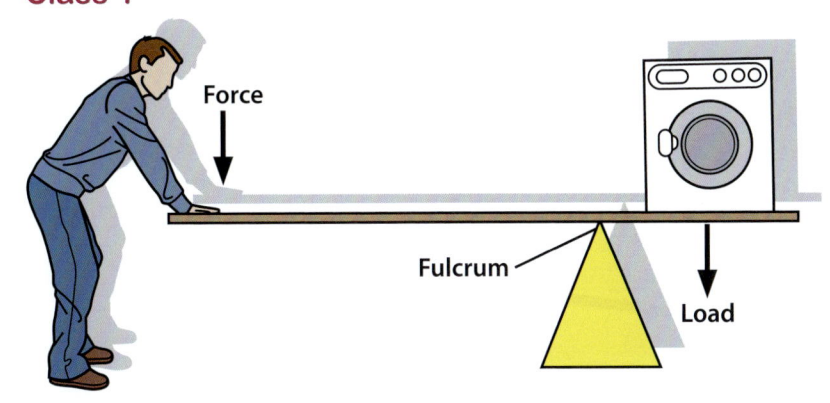

> **Remember**
> To make any simple machine work for us, we need to apply a force on it.

> **Did you know?**
> In medieval times, siege engines were used to hurl rocks at enemy castles. A siege engine is simply a class 1 lever.

The fulcrum is between the force and the load, like a seesaw.

Class 2

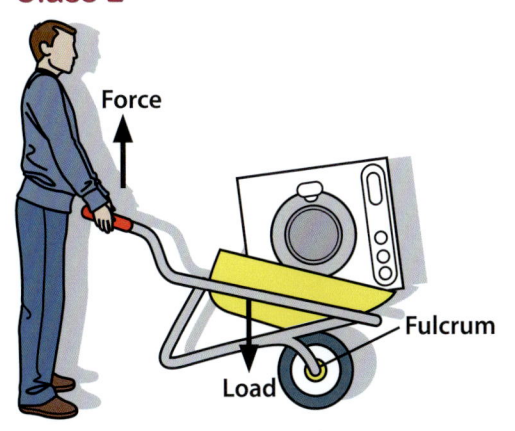

The fulcrum is at one end, the force at the other end, and the load is in the middle. A wheelbarrow is a good example.

Class 3

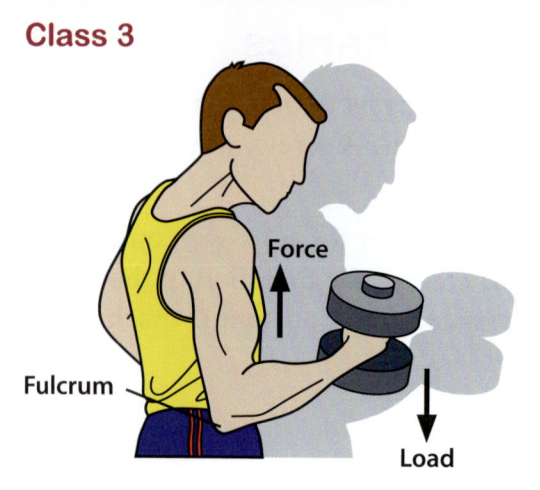

The fulcrum is at one end, the load at the other end and the force in the middle, like a human forearm.

A small force at a long distance will produce a larger force close to the pivot:

$$10 \times 2 = F \times 0.5$$

$$F = \frac{10 \times 2}{0.5} \qquad \frac{20}{0.5} = 40 \text{ N}$$

Figure 8.12 Large force close to pivot

Gears

Gears are wheels with teeth; the teeth of one gear fit snugly into those around it. You can use gears to slow things down or speed them up, to change direction, or to control several things at once. Each gear in a series changes the direction of rotation of the previous gear. A smaller gear will always turn faster than a larger gear and in doing so, turns more times.

Figure 8.13 Gears

The inclined plane

The inclined plane is the simplest machine of all, as it is basically a ramp or sloping surface. The shortest distance between two points is a straight line, but it is easier to move a heavy object to a higher point by using stairs or a ramp. If you think of the height of a mountain, the shortest distance is straight up from the bottom to the top. However, we usually build a road up a mountain as a slowly-winding inclined plane from bottom to top.

As an electrician, you will use the inclined plane in the form of a screw, which is simply an inclined plane wound around a central cylinder.

So the inclined plane works by saving effort, but to do this you must move things a greater distance.

Figure 8.14 The inclined plane

Pulleys

A pulley is made with a rope, belt or chain wrapped around a wheel and can be used to lift a heavy object (load). A pulley changes the direction of the force, making it easier to lift things. There are two main types of pulleys: the single fixed pulley and the moveable pulley.

A **single fixed pulley** is the only pulley that uses more effort than the load in order to lift the load from the ground. The fixed pulley, when attached to an unmoveable object, e.g. a ceiling or wall, acts as a class 1 lever with the fulcrum being located at the axis but with a minor change – the bar becomes a rope. The advantage of the fixed pulley is that you do not have to pull or push the pulley itself up and down. The disadvantage is that you have to apply more effort than the load.

A **moveable pulley** is one that moves with the load. The moveable pulley allows the effort to be less than the weight of the load. The moveable pulley also acts as a class 2 lever. The load is between the fulcrum and the effort.

There are many combinations of pulleys, the most common being the block and tackle, that use the two main types as their principle of operation. The example on page 251 shows how these work.

Mechanical advantage

The common theme behind all of these machines is that because of the machine we can increase our ability and gain an advantage. This is a relationship between the effort needed to lift something (input) and the load itself (output) and we call this ratio the **mechanical advantage**.

When a machine can put out more force than is put in, the machine is said to give a good mechanical advantage. Mechanical advantage can be calculated by dividing the load by the effort. There are no units for mechanical advantage, it is just a number.

$$\text{Mechanical Advantage (MA)} = \frac{\text{Load}}{\text{Effort}}$$

> **Remember**
> Mechanical advantage is just a number. It has no units.

Example

Using the final diagram from the example on page 251, what is the mechanical advantage of the pulley system?

$$MA = \frac{\text{Load}}{\text{Effort}} = \frac{2000 \text{ N}}{500 \text{ N}} = 4$$

In a lever, an effort of 10 N is used to move a load of 50 N. What is the mechanical advantage of the lever?

$$MA = \frac{\text{Load}}{\text{Effort}} = \frac{50}{10} = 5$$

This effectively means that for this lever, any effort will move a load that is five times larger. To summarise:

- **Where MA is greater than 1:** The machine is used to magnify the effort force (e.g. a class 1 lever).
- **Where MA is equal to 1:** The machine is normally used to change the direction of the effort force (e.g. a fixed pulley).
- **Where MA is less than 1:** The machine is used to increase the distance an object moves or the speed at which it moves (e.g. the siege machine).

Example

Let us look at examples of the two main types of pulleys to understand their operating principles. Imagine that you have the arrangement of a 2 kg weight suspended from a rope, but actually resting on the ground as shown in the diagram.

If we want to have the 2 kg load suspended in the air above the ground, then we have to apply an upward force of 20 N (1 kg will exert a force of approximately 10 N) to the rope in the direction of the arrow. If the rope was 3 m long and we wanted to lift the weight up 3 m above the ground, we would have to pull in 3 m of rope to do it.

Now imagine that we add a single fixed pulley to the scenario, as shown in the diagram. The only thing that has changed is the direction of the force we have to apply to lift the load. We would still have to apply 20 N of force to suspend the load above the ground, and would still have to reel in 3 m of rope in order to lift the weight 3 m above the ground. This type of system gives us the convenience of pulling downwards instead of lifting.

This diagram shows the arrangement if we add a second, moveable pulley. This new arrangement now changes things in our favour because effectively the load is now suspended by two ropes rather than one. That means the weight is split equally between the two ropes, so each one holds only half the weight, or 10 N. That means that if you want to hold the weight suspended in the air, you only have to apply 10 N of force (the ceiling exerting the other 10 N of force on the other end of the rope). However, if you want to lift the weight 3 m above the ground, then you have to reel in twice as much rope – i.e. 6 m of rope must be pulled in.

The more pulleys we have the easier it is to lift heavy objects. As rope is pulled from the top pulley wheel, the load and the bottom pulley wheel are lifted. If 2 m of rope are pulled through, the load will only rise 1 m (there are two ropes holding the load and both have to shorten by the same amount).

With pulley systems, to calculate the effort required to lift the load, we divide the load by the number of ropes (excluding the rope connected to the effort). The diagram shows a four pulley system, where the person lifting a 200 kg mass or 2000 N load has to exert a pull equal to only 500 N (i.e. 2000 N divided by 4 ropes).

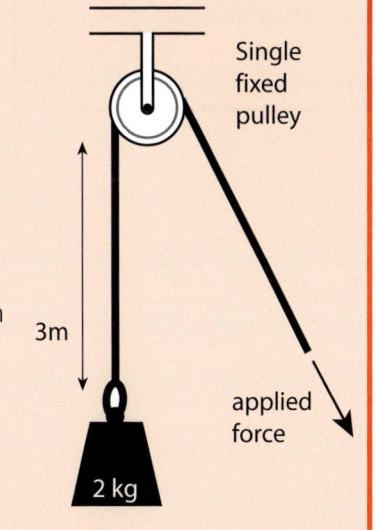

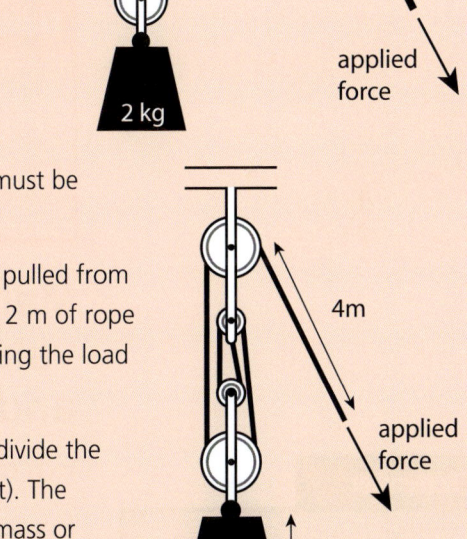

> **Remember**
> The VR of any pulley system is equal to the number of pulley wheels.

Velocity ratio

Sometimes machines translate a small amount of movement into a larger amount (or vice versa). For example, in Figure 8.15, a small movement of the piston causes the load to move a much greater distance. This property is known as the velocity ratio, and is found by dividing the distance moved by the effort by the distance moved by the load (in the same period of time). There are no units for velocity ratio, it is just a number.

$$\text{Velocity Ratio (VR)} = \frac{\text{Distance effort moves}}{\text{Distance load moves}}$$

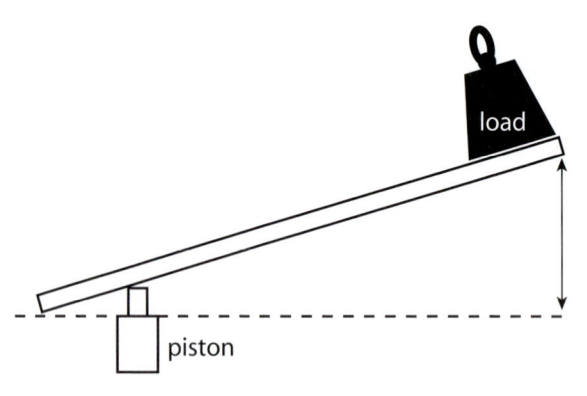

Figure 8.15 Velocity ratio

> **Example**
> In the diagram, the piston moves 1 m to move the load 5 m. The velocity ratio is:
>
> $$VR = \frac{\text{Distance effort moves}}{\text{Distance load moves}} = \frac{1}{5} = \mathbf{0.2}$$

Main principles and calculating values of force, work, energy, power and efficiency

Force

Force is a push or pull that acts on an object. If the force is greater than the opposing force, the object will change motion or shape. Obvious examples of forces are gravity and the wind. Force is measured in **newtons**.

The presence of a force is measured by its effect on a body, e.g. a heavy wind can cause a stationary football to start rolling; or a car colliding with a wall causes the front of the car to deform and the

> **Remember**
> Mass and weight are **not** the same. Mass is the amount of material in an object. Weight is a force – e.g. a person who weighs a certain amount on Earth would weigh less on the Moon due to the decreased gravitational force, but they would still have the same mass.

occupants of the car to be forced forwards towards the windscreen (hence the use of seat belts).

Equally, gravitational force will cause objects to fall towards the Earth. Therefore, a spring will extend if we attach a weight to it, because gravity is acting on the weight.

As the force of gravity acts on any mass, that mass tends to accelerate. This acceleration due to gravity is universally taken to be 9.81 m/s^2 at sea level and therefore a mass of 1 kg will exert a force of 9.81 N.

Expressed as a formula:

Force (N) = Mass × Acceleration

(Note: In calculations, the value of acceleration is often taken as 10 m/s^2 to simplify the calculations.)

> **Remember**
>
> Do not assume that the acceleration due to gravity is = 10m/s^2. It is 9.81m/s^2. Only take it as 10m/s^2 if you are told to do so in a question, or if you are doing a rough calculation for your own purposes.

Work

If an object is moved, then work is said to have been done. The unit of work done is the joule. (This is also the unit for energy.) Work done is a relationship between the effort (force) used to move an object and the distance that the object is moved. Expressed as a formula:

Work done (J) = Force (N) × Distance (m)

> **Example**
>
> A distribution board has a mass of 50 kg. How much work is done when it is moved 10 m?
>
> Work = Force × Distance
> = (50 × 9.81) × 10
> = 490.5 × 10
> = **4905 J**

Energy

Energy, measured in joules, is the ability to do work, or to cause something to move or the ability to cause change. Machines cannot work without energy and we are unable to get more work out of a machine than the energy we put into it.

Energy is wasted in a machine because of **friction**. Friction occurs when two substances rub together. Try rubbing your hands together. Did you feel them get warmer?

> **Key term**
>
> **Friction** – force that opposes motion

Work produced (output) is usually less than the energy used (input). Energy can be transferred from one form to another, but energy cannot be created or destroyed.

The energy lost by friction is converted into heat.

There are many forms of energy, but there are only two types:

- **potential energy** (energy of position or stored energy)
- **kinetic energy** (energy due to the motion of an object).

Some forms of energy are: solar, electrical, heat, light, chemical, mechanical, wind, water, muscles and nuclear.

Potential energy

Anything may have stored energy, giving it the potential to cause change if certain conditions are met. The amount of **potential** energy something has depends on its position or condition. A brick on the top of scaffolding has potential energy because it could fall under the influence of gravity. The bow used to propel an arrow has no energy in its resting position, but drawing the bow back requires energy and this is then stored as elastic potential energy. A change in its condition (releasing it) can cause change (propelling the arrow).

Potential energy due to height above the Earth's surface is called gravitational potential energy, and the greater the height, the greater the potential energy.

There is a direct relation between gravitational potential energy and the mass of an object; more massive objects have greater gravitational potential energy. There is also a direct relation between gravitational potential energy and the height of an object. The higher an object is above the Earth, the greater its gravitational potential energy. These relationships are expressed by the following equation:

PE_{grav} = mass of an object × gravitational acceleration × height

$$PE_{grav} = m \times g \times h = mgh$$

Another example of potential energy is the spring inside a clockwork watch. The wound spring transforms potential energy to kinetic energy of the wheels and cogs etc. as it unwinds.

Kinetic energy

Kinetic energy is energy in the form of motion and the greater the mass of a moving object, the more kinetic energy it has.

The formula is $KE = \frac{1}{2}mv^2$

Power

When we do work in a mechanical system, the energy we put into the system does not appear all at once. It takes a certain time to move an object, lift a weight etc. The power that we put into a system must depend not only on the amount of work we do but also how fast we carry out the work.

To try to understand this, think of a 100 metre runner and a marathon runner. The sprinter has a burst of energy for maybe 10 seconds or so whereas the marathon runner may use a similar amount of energy but at a much slower pace. It is clear that the sprinter had greater power because he used his energy very quickly.

We usually say that:

Power = the rate of doing work.

In terms of equations we can say that:

$$\text{Power (P)} = \frac{\text{Work done (W)}}{\text{Time taken to do that work (t)}}$$

or

$$\text{Power (P)} = \frac{\text{Energy used (E)}}{\text{Time taken to do that work (t)}}$$

Energy or work is measured in joules (J) and time is measured in seconds (s). Power is measured in joules per second or J/s, also known as watts (W).

1000 watts (W) = 1 kilowatt (kW).

The power of an electrical device is also measured in Watts or kW.

Returning to a previous example:

Remember

Be really careful, the shorthand for work is W and the units for power is W. Do not get them confused with each other.

> **Example**
>
> A distribution board has a mass of 50 kg and it is moved 10 m.
>
> Work = Force × Distance = (50 × 9.81) × 10 = 490.5 × 10 = 4905 J
>
> Now what if it took 20 s to move the distribution board by 10 m. How much power did we put into moving it? In this case:
>
> $$P = \frac{\text{Work done in moving the distribution board by 10 m (W)}}{\text{The time it took to move the distribution board (t)}}$$
>
> Now, W = 4905 J, and t = 20 s. So:
>
> $$P = \frac{4905}{20}$$
>
> Therefore: **P = 245.25 W**

Progress check

1. Calculate the amount of power required if a load requires 5000 joules of energy and takes 20 seconds to move to its final position.
2. Calculate the output power if 2 kW of input power is used and the efficiency is 85%.
3. Identify the units for the following:
   - Power
   - Force
   - Energy
   - Mass
   - Weight

Efficiency

We often think of machines as having an input and an output. A machine actually has two outputs, one that is wanted and one that is not and is therefore wasted (as frictional heat, noise etc.). The greater the unwanted component, the less efficient the machine is.

In all machines, the power at the input is greater than the power output, because of losses that occur in the machine such as friction, heat or vibration. This difference, expressed as a ratio of output power over input power, is called the **efficiency** of the machine. The symbol sometimes used for efficiency is the Greek letter η (eta).

That is:

$$\text{Efficiency} = \frac{\text{Output Power}}{\text{Input Power}}$$

To give the efficiency as a percentage, which is usually more convenient and understandable, we can say that

$$\% \text{ efficiency} = \frac{\text{Output Power}}{\text{Input Power}} \times 100$$

We will now follow a series of steps to end up with a final equation for efficiency.

It is quite complicated and the most important thing is to remember the final formula.

Now, for any machine:

Work done at the input = Effort × Distance the effort moves (force × distance)

and

Work done at the output = Load × Distance the effort moves (force × distance)

Dividing these two equations gives us:

$$\frac{\text{Work at Output}}{\text{Work at Input}} = \frac{\text{Load} \times \text{Distance moved by load}}{\text{Effort} \times \text{Distance moved by effort}} = \text{Efficiency}$$

Which can be rewritten as:

$$\frac{\text{Work at Output}}{\text{Work at Input}} = \frac{\text{Load}}{\text{Effort}} \times \frac{\text{Distance moved by load}}{\text{Distance moved by effort}} = \text{Efficiency}$$

Now, you already know that:

$$\text{Mechanical Advantage (MA)} = \frac{\text{Load}}{\text{Effort}}$$

And that:

$$\text{Velocity Ratio (VR)} = \frac{\text{Distance effort moves}}{\text{Distance load moves}}$$

So:

$$\frac{1}{\text{VR}} = \frac{\text{Distance moved by effort}}{\text{Distance moved by load}}$$

So:

$$\text{Efficiency} = \frac{\text{Work at Output}}{\text{Work at Input}} = \frac{\text{Load}}{\text{Effort}} \times \frac{\text{Distance moved by load}}{\text{Distance moved by effort}} = \frac{\text{MA} \times 1}{\text{VR}}$$

Therefore:

$$\text{Efficiency} = \frac{\text{Mechanical Advantage}}{\text{Velocity Ratio}} = \frac{\text{MA}}{\text{VR}} \quad \text{or} \quad \%\text{ Efficiency} = \frac{\text{MA}}{\text{VR}} \times 100$$

If a machine has low efficiency, this does not mean it is of limited use. A car jack, for example, has to overcome a great deal of friction and therefore has a low efficiency, but it is still a very useful tool as a small effort allows us to lift the whole weight of a car to change a tyre.

> **Did you know?**
>
> The unit of velocity (m/s) and acceleration (m/s²) are also written as ms⁻¹ and ms⁻².

Let us look at a couple of examples that will illustrate these concepts.

Example 1

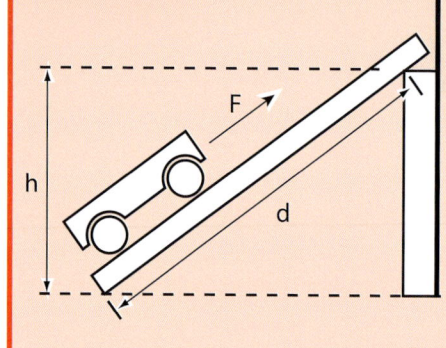

In the following diagram, a trolley containing lighting fittings is pulled at constant speed along an inclined plane to the height shown. Assume that the value of the acceleration due to gravity is 10 m/s².

If the mass of the loaded cart is 3.0 kg and the height shown is 0.45 m, then what is the potential energy of the loaded cart at the height shown?

$PE = m \times g \times h$

$PE = 3 \times 10 \times 0.45$

PE = 13.5 J

If a force of 15.0 N was used to drag the trolley along the incline for a distance of 0.90 m, then how much work was done on the loaded trolley?

$W = F \times d$

$W = 15 \times 0.9$

W = 13.5 J

Example 2

A motor control panel arrives on site. It is removed from the transporter's lorry using a block and tackle that has five pulley wheels. Establish the percentage efficiency of this system given that the effort required to lift the load was 200 N, the panel has a mass of 80 kg and acceleration due to gravity is 10 m/s².

The load = Mass × Acceleration due to gravity
= 80 × 10
= 800 N

Mechanical advantage $= \dfrac{\text{Load}}{\text{Effort}} = \dfrac{800 \text{ N}}{200 \text{ N}} = 4$

Remembering that velocity ratio is equal to the number of pulley wheels

Velocity Ratio = the number of pulley wheels = 5

Efficiency $= \dfrac{\text{Mechanical advantage}}{\text{Velocity ratio}} = \dfrac{4}{5} = 0.8$

So, % efficiency = 0.8 × 100 = 80%

Therefore the system is 80% efficient.

K4. Understand the relationship between resistance, resistivity, voltage, current and power

This section is where we really start to look at electricity and electrical circuits in detail. You cannot even consider becoming an electrician unless you have a sound knowledge of the principles involved and this starts with the atomic theory of matter and how this gives rise to an electric current. In this section we will therefore be looking at the following areas:

- states of matter
- molecules and atoms
- the electric circuit
- the causes of an electric current
- the effects of an electric current
- resistance.

Basic principles of electron theory

States of matter

It has in the past been thought that there were three states of matter, solid, liquid and gas. However, with advances in science and technology it is currently felt that there are five main states of matter: solid, liquid, gas, plasma and Bose-Einstein condensate.

We can think of each of these states as being a phase that matter can move from and to when affected by other things such as temperature. The effect of temperature is an easy one to see as if we apply enough heat to a block of metal it melts. In other words the metal has moved from a solid state to a liquid state.

Matter can change from one phase to another but it is still the same substance.

As an example consider a solid block of ice. If we apply gentle heat it will become a liquid pool of water. If we apply more heat to our pool of water it will evaporate into a gas.

However, through all the changes of phase, it is still water and always has the same chemical properties.

(A chemical change would be needed to change the water into something completely new.)

The three classic states of matter

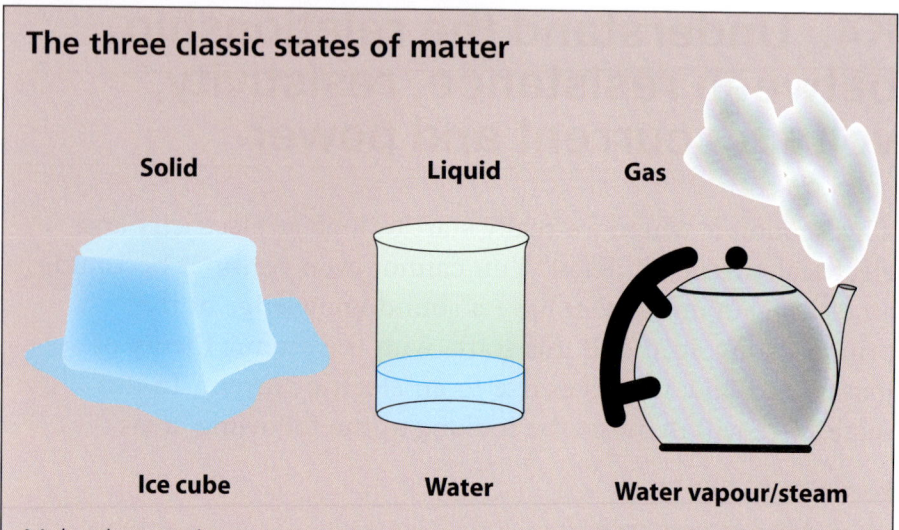

Solid — Ice cube
Liquid — Water
Gas — Water vapour/steam

Molecules are always in a state of rapid motion, but when they are densely packed together, this movement is restricted and the substance formed by these molecules is solid. When the molecules of a substance are less tightly bound, there is a great deal of free movement and the substance is a liquid. Finally, when the molecule movement is almost unrestricted, the substance can expand and contract in any direction and is a gas.

Molecules and atoms

As a starting point for understanding these, we can say that particles of matter are used to create atoms. Atoms are then used to create molecules and elements are used to create molecules.

Molecules are electrically neutral groups of at least two atoms held together by a chemical bond. A molecule may consist of atoms of a single chemical element such as oxygen (O), or of different elements such as water (two hydrogen atoms and one oxygen atom H_2O).

Atoms have a nucleus that is made of protons and neutrons, with very small electrons orbiting around it (rather like planets revolving round the Sun).

Protons and neutrons are now known to be made from even smaller particles called nucleons and quarks. You may know that nuclear chemists and physicists work together using particle accelerators such as the Large Hadron Collider, 17 miles in circumference and buried 150 metres underground in Switzerland, to study these sub-atomic particles.

Figure 8.16 Large Hadron Collider

Structure of the atom

Even though smaller atomic particles exist there are three basic particles within an atom: electrons, protons, and neutrons.

At the centre of each atom is the nucleus, which is made up from protons and neutrons. Protons are said to possess a positive charge (+), and neutrons are electrically neutral and act as a type of 'glue' that holds the nucleus together.

You probably already know that:

- Like charges repel each other (+ and + or – and –).
- Unlike charges attract each other (+ and –).

So without neutrons, the positively charged protons would repel each other and the nucleus would fly apart. The neutrons hold the nucleus together. However, as neutrons are electrically neutral, they play no part in the chemical or electrical properties of atoms.

The remaining particles in an atom are electrons and these circulate in orbits of varying radius around the nucleus and have a negative charge (–).

There are many different kinds of atoms, one for each type of element, and there are currently 118 different known elements (such as oxygen).

For any particular atom the number of electrons is usually the same as the number of protons. If the numbers are the same, the atom is seen as balanced, with the positive and negative charges being cancelled out thus leaving the atom electrically neutral.

The simplest atom is a hydrogen atom which has one proton and one neutron balancing each other.

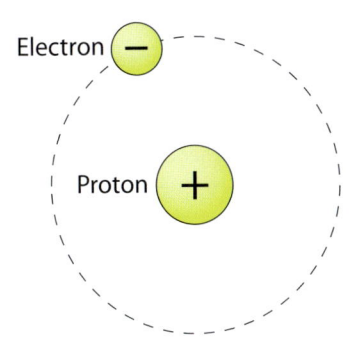

Figure 8.17 Hydrogen atom

Electrons in orbit nearest the nucleus are generally held tightly in place. However, those furthest away are more loosely attached and in some cases it is possible to remove or add an electron to a neutral atom. This leaves the atom with a net positive or negative charge. Such 'unbalanced' atoms are known as ions.

Since all atoms 'want' to be balanced, the atom that is unbalanced will attract a 'free' electron to fill the place of the missing one or lose an extra electron to return to its neutral state. These are wandering or 'free' electrons moving about the molecular structure of a material that give rise to what we refer to as electricity.

Electricity is the movement of free electrons along a suitable material (conductor). A material that does not allow the easy movement of free electrons is an insulator.

Identifying and differentiating between insulators and conductors

Insulators

We need to protect ourselves and contain the flow of electricity otherwise we would get an electric shock every time we used a piece of electrical equipment. The materials that we use to do this are called insulators. Insulators are poor conductors of electricity. They do not allow free passage of electrons through them. Surprisingly, one insulator that is used in cable manufacture is paper. Some others are shown Table 8.06.

> **Remember**
>
> A good insulator has high resistance.

> **Key term**
>
> **hygroscopic** – the ability to absorb water

Rubber/plastic	• Very flexible • Easily affected by temperature • Used in cable insulation
Impregnated paper	• Stiff and **hygroscopic** • Unaffected by moderate temperature • Used in large cables
Magnesium oxide	• Powder, therefore requires a containing sheath • Very hygroscopic • Resistant to high temperature • Used in cables for alarms and emergency lighting
Mica	• Unaffected by high temperature • Used for kettle and toaster elements
Porcelain	• Hard and brittle • Easily cleaned • Used for carriers and overhead line insulators
Rigid plastic	• Less brittle and less costly than porcelain • Used in manufacture of switches and sockets

Table 8.06 Common insulators

Conductors

Conductors have a molecular/atomic structure that allows electrons to move freely through them, meaning they have a low resistance to electron flow. Many are metals but graphite (carbon) and some liquids also conduct electricity.

Gold and silver are among the best conductors, but cost inhibits their use.

Table 8.07 is a guide to the most common conductors and their uses.

Aluminium (Al)	• Low cost and weight • Not very flexible • Used for large power cables
Brass (an **alloy** of copper and zinc)	• Easily machined • Corrosion resistant • Used for terminals and plug pins
Carbon (C)	• Hard • Low friction in contact with other materials • Used for machine brushes
Copper (Cu)	• Good conductor • Soft and ductile • Used in most cables and busbar systems
Iron/Steel (Fe)	• Good conductor • Corrodes • Used for conduit, trunking and equipment enclosure
Lead (Pb)	• Flexible • Corrosion resistant • Used as an earth and as sheath of a cable
Mercury (Hg)	• Liquid at room temperature • Quickly vaporises • Used for contacts • Vapour used for lighting lamps
Sodium (Na)	• Quickly vaporises • Vapour used in lighting lamps
Tungsten (W)	• Extremely ductile • Used for filaments in light bulbs

Table 8.06 Common conductors

> **Key term**
>
> **Alloy** – a mixture of two elements

Applying electron theory to electrical circuits

Measuring electricity

Electricity is invisible, so what exactly do we measure?

Electricity is simply the flow of free electrons along a conductor, so it would seem obvious to count the number of electrons moving along the conductor. However, the electron is far too small to be seen or counted. So we measure the number of larger groups of electrons moving along. These groups are coulombs, and contain an unimaginable 6 240 000 000 000 000 000 electrons or 6.24×10^{18} (give or take a couple).

A plumber will measure the amount of water flowing in gallons not drops, as drops are too small to measure. If the plumber wishes to know how much water is being used at any one time, in other words 'the rate of flow' of the water, this would be measured in gallons or litres per second.

This movement of the water can be thought of as its current.

Similarly, the electrician may wish to know the amount of electrons flowing at any one time (rate of flow of electrons). In electricity, just as with water, this rate of flow of electrons is called the current and is defined as being one coulomb (an imaginary 'bucket-full') of electrons passing by every second.

If one coulomb of electrons passes along the conductor every second, we say that the current flowing along is a current of one ampere.

We use the letter I to represent current.

The electric circuit

We now know that electricity is the movement of electron charges along a conductor and that the rate of flow is known as the current. But what makes the electron charges move?

> **Remember**
> Think of a coulomb as an imaginary bucket of electrons.

> **Remember**
> One ampere equals one coulomb of electrons passing by every second.

Battery

In this circuit (Figure 8.18), the battery cell has an internal chemical reaction that provides what is known as an electromotive force (e.m.f. for short), that will push the electrons along the conducting wire and into the lamp. The electrons will then pass through the lamp filament, causing it to heat up and glow and then leave via the second conductor, returning to the battery and thus completing the circuit.

Figure 8.18 A simple battery

In other words, a battery is a chemically fuelled electron pump and, like every other pump in the world, the battery does not supply the electrons that it pumps. When a battery runs down it is because its chemical 'fuel' is exhausted, not because any charges have been lost.

Electron flow and conventional current flow

An electromotive force (e.m.f.) is needed to cause the flow of electrons. This has the symbol E and the unit symbol V (volt). Any apparatus which produces an e.m.f. (such as a battery) is called a power source and as we saw in Figure 8.18, will require wires or cables to be attached to its terminals to form a basic circuit.

If we take two dissimilar metal plates and place them in a chemical solution (an electrolyte) a reaction will take place in which electrons from one plate travel through the electrolyte and collect on the other plate.

One plate now has an excess of electrons, which will make it more negative than positive. The other plate will now have an excess of protons, which makes it more positive than negative. This process is the basis of how a simple battery or cell works.

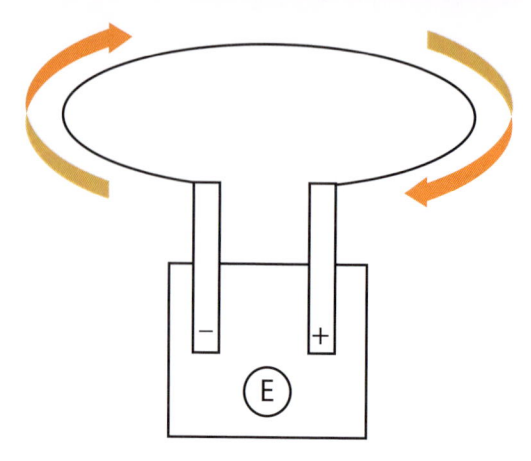

Figure 8.19 Electron flow through an electrolyte

Now select a piece of wire as a good conductor and connect this wire to the ends of the plate as shown in Figure 8.19.

Since unlike charges are attracted towards each other and like charges repel each other, you can see that the negative electrons will move from the negative plate, through the conductor towards the positive plate.

This drift of free electrons is what we know as electricity and this process will continue until the chemical action of the battery is exhausted and there is no longer a difference between the plates.

As we can see, the actual electron flow is from positive to negative inside the battery and then from negative to positive through the conductor.

Note that early science actually thought the opposite and that flow ran externally from positive to negative. This is called conventional current flow.

Potential difference

The chemical energy within a battery is used to do work on a charge in order to move it from the negative terminal, out through a conductor and then returning to the positive terminal.

To do this, the battery is raising the potential of the electrons.

If you hold a stone and raise your hand in the air, the stone has a potential energy. The higher up it is, the higher the potential and we measure that potential against a reference point.

If you now let go of the stone it will fall towards that reference point losing potential along the way.

> **Remember**
> Be careful, historically, circuit drawings show the current as flowing from the positive to the negative. The actual flow is of electrons in the opposite direction.

(It may be easier to understand if I said the reference point was your head and potential means the stone's ability to hurt you. The higher it is above your head the more likely it is to hurt you.)

In a battery, the electric charges at the negative terminal have more potential energy than they will have when they get to the positive terminal, in other words they can 'fall downhill' from the negative terminal to the positive terminal via the conductor that makes up the circuit. The electrons then pick up energy in the cell of the battery which pushes it out to the components of the circuit. Therefore at the point when they return to the cell, they have given up all the energy they gained.

We can therefore say that potential difference (p.d.) is the difference in electrical potential energy between any two points in a circuit, p.d. is therefore also measured in volts.

We measure an amount of charge in a unit called a coulomb so we could also say that potential difference is a measure of the amount of joules of work required to push one coulomb along the circuit between our two points. The units of this would be measured in joules/coulomb more commonly referred to as the volt.

Therefore **One volt = One joule/coulomb**

Our battery is therefore acting as an energy conversion system, converting chemical energy into electric potential energy. This work increases the potential energy of the charge and thus its electric potential. By chemical reaction the charge is moved from a 'low potential' terminal to a 'high potential' terminal inside the internal circuit of the battery. Once there it will then move through the external circuit (the conductor and equipment), before returning to the low potential terminal. The difference between our terminals is referred to as the potential difference and without it there can be no flow of charge.

As the charge moves through the external circuit, it can pass through different types of component each of which acts as an energy conversion system, for example, the lamp in Figure 8.18. Here the moving charge is doing work on the lamp filament to produce different forms of energy: heat and light. However, in doing so it is losing some of its electric potential energy and therefore on leaving the lamp, it is less energised.

Looking again at Figure 8.18, we can see that at a point just prior to entering the lamp (or any circuit component) we have a higher electric potential compared to a point just after leaving the lamp. This loss of potential across any circuit component is also called the volt drop and we will discuss this later on page 282.

> **Remember**
>
> Think of a marathon runner starting fresh and full of energy. As the race progresses, the runner uses up energy until at the end the runner has no energy left.

Controlling a circuit

In Figure 8.18 we have a complete circuit connected across a healthy battery and the lamp is therefore lit.

However, if either of the two wires becomes broken or disconnected the flow of electricity will be interrupted and the lamp will go out.

It is this principle that we use to control electricity in a circuit.

As we can see in Figure 8.20, by inserting a switch into one of the wires connected to the lamp, we can physically 'break the circuit' with the switch and thus switch the lamp off and on.

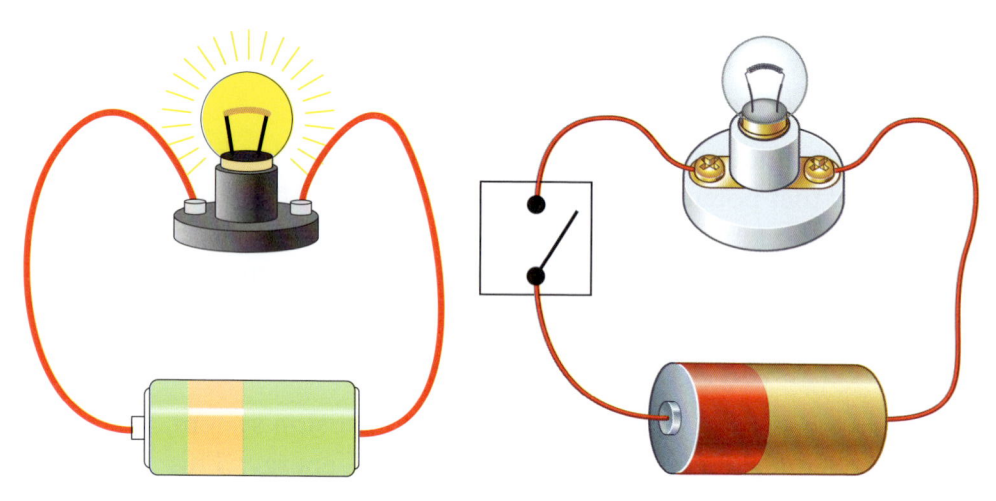

Figure 8.20 Simple battery circuit and broken circuit

To summarise, for practical purposes a working circuit should:

- have a source of supply (such as the battery)
- have a device (fuse/MCB) to protect the circuit
- contain conductors through which current can flow
- be a complete circuit
- have a load (such as a lamp) that needs current to make it work
- have a switch to control the supply to the equipment (load).

Chemical and thermal effects of electrical currents

The causes of an electric current

We need an electromotive force (e.m.f.) to drive electrons through a conductor. The principal sources of an e.m.f. that will cause current to flow can be classed as being:

Chemical

When we take two electrodes of dissimilar metal and immerse them in an electrolyte, we have effectively created a battery. So as we have seen earlier, the chemical reactions in the battery cause an electric current to flow.

Thermal

When a closed circuit consists of two junctions, each junction made between two different metals, a potential difference will occur if the two junctions are at different temperatures. This is known as the Seebeck effect, based upon Seebeck's discovery of this phenomenon in 1821.

If we now connect a voltmeter to one end (the cold end) and apply heat to the other, then our reading will depend upon the difference in temperature between the two ends. When we have two metals arranged like this, we have a thermocouple.

We can apply this to measure temperatures, with the 'hot end' being placed inside the equipment (such as an oven or hot water system) and the 'cold end' connected to a meter that has been located in a suitable remote position.

Magnetic

A magnetic field can be used to generate a flow of an electrons. We call this situation electromagnetic induction. If a conductor (wire) is moved through a magnetic field, then an e.m.f. will be induced in it. Provided that a closed circuit exists, this e.m.f. will then cause an electric current.

The effects of an electric current

The effects of an electric current are categorised in the exact same way as the causes, namely chemical, thermal and magnetic.

Looking at the circuit in Figure 8.21 we can see a d.c. supply enters a contactor. When we close the switch on the contactor the coil is energised and becomes an electromagnet, this attracts anything in the magnetic field that contains iron towards it. Once the circuit is made then the main supply flows through to the distribution centre.

From the distribution centre a supply is taken to a change-over switch. In its current position, this switch allows current to flow into the electrolyte (dilute sulphuric acid and water) via one of the two lead plates. The current returns to the distribution centre via the other lead plate.

Find out

Electromagnets are covered in detail on page 291.

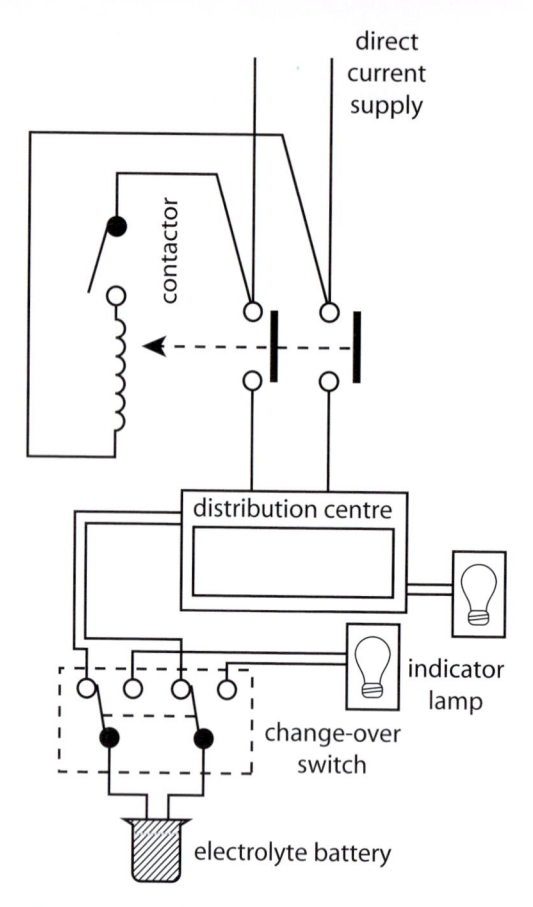

Figure 8.21 d.c. supply

Also fed from the distribution centre is a filament lamp. We could have equally used an electric fire. This is because when current flows through a conductor heat is generated. The amount of heat varies according to circumstances, such as the conductor size. If sized correctly, we can make the conductor glow white-hot (a lamp) or red-hot (a fire).

If we run the system like this for a few minutes and then switch off the contactor, obviously we would see our filament lamp go out. However, if we now move our change-over switch into its other position, we would see that our indicator lamp would glow for a short while.

If we were to look at the lead plates, we would see that one of the plates has become discoloured. This is because the current has caused a chemical reaction, changing the lead into an oxide of lead. In this respect the plates acted as a form of rechargeable battery, also known as a secondary cell.

Table 8.08 indicates how various pieces of equipment use these effects of current as their principle of operation.

Chemical effect	Heating effect	Magnetic effect
Cells	Filament lamp	Bell
Batteries	Heater	Relay
Electro-plating	Cooker	Contactor
	Iron	Motors
	Fuse	Transformers
	Circuit-breaker	Circuit-breaker
	Kettle	

Table 8.08 Principles of operation

Resistance and resistivity in relation to electrical circuits

So far we have considered the amount of electron charges flowing in a conductor every second and the force that pushes them along the conductor. But does anything interfere with this flow?

If we turn our minds back to our marathon runners again, would they rather be running on a brand new athletics track or through a field of sticky mud four feet deep?

Obviously the new track would be the easiest to run on as it is least likely to affect their ability to run.

Or, in other words, the new track will offer a lower resistance to their progress than the muddy field.

In electrical circuits, just like the muddy field, electrical conductors, connections and known resistors will offer a level of resistance to the electrons trying to flow through them.

You could also think of resistance as hurdles that electrons have to jump over on their way around the circuit, and the more hurdles there are, then the longer it will take to get around the circuit. Watch the Olympics if you need further proof!

There is a scientific law that we can apply to resistance.

Ohm's Law

So far we have established that **current** is the amount of electrons flowing by every second in a conductor and that a force known as the e.m.f. (or **voltage**) is pushing them. We now also know that the conductor will try to oppose the current, by offering a **resistance** to the flow of electrons.

Ohm's Law was named after the nineteenth-century German physicist G.S. Ohm who researched how current, potential difference and resistance are related to each other. It's probably the most important electrical concept you will need to understand and is stated as follows:

> The current flowing in a circuit is directly proportional to the voltage applied to the circuit, and indirectly proportional to the resistance of the circuit, provided that the temperature affecting the circuit remains constant.

In simple language we could re-write Ohm's Law as follows: The amount of electrons passing by every second will depend upon how hard we push them, and what obstacles are put in their way.

We can prove this is true, because if we increase the voltage (push harder), then we increase the number of electrons that we can get out at the other end.

Try flicking a coin along the desk. The harder you flick it, the further it travels along the desk. This is what we mean by **directly proportional**. If one thing goes up (voltage), then so will the other thing (current).

Equally we could prove that if we increase the resistance (put more obstacles in the way), then this will reduce the amount of electrons that we can get along the wire.

This time put an obstacle in front of the coin before you flick it. If flicked at the same strength, it will obviously not go as far as it did before. This is what we mean by indirectly proportional. If one thing goes up (resistance), then the other thing will go down (current).

Ohm's Law is therefore expressed by the following formula:

$$\text{Current (I)} = \frac{\text{Voltage (V)}}{\text{Resistance (R)}}$$

> **Remember**
>
> Remember Ohm's Law, $I = \dfrac{V}{R}$

Resistivity

Take away all resistors from a circuit and you still have some resistance there – caused by the conductor itself. Electrons find it easier to move along some materials than others and each material has its own resistance to the electron flow. This individual material resistance is called resistivity, represented by the Greek symbol rho (ρ) and measured in micro ohm millimetres ($\mu\Omega$mm).

But in considering a conductor there are also some other factors at work!

How long is it? Would you rather run for 100 metres or 25 miles? *The shorter distance of course.* So would the electron!

What is its cross-sectional area (CSA)?

Which is easier, to walk along a 3 metre high corridor, or to crawl along a 1-metre high pipe on your stomach?

Walking, of course.

To summarise: The amount of electrons that can flow along a conductor will be affected by how far they have to travel, what material they have to travel through and how big the object is that they are travelling along.

As an electrical formula, this is expressed as follows:

$$\text{Resistance} = \frac{\text{Resistivity} \times \text{Length}}{\text{Cross-Sectional Area}} \quad \text{or} \quad R = \frac{\rho \times \ell}{a}$$

We find the value of resistivity for each material, by measuring the resistance of a 1-metre cube of the material. Then, as cable dimensions are measured in square millimetres (e.g. 2.5 mm²), this figure is divided down to give the value of a 1 millimetre cube.

This resistivity, as we found out earlier, is given in $\mu\Omega$mm, or in other words we will encounter a resistance of so many millionths of an ohm for every millimetre forward that we travel through the conductor.

The accepted value for copper is 17.8 $\mu\Omega$mm and the accepted value for aluminium is 28.5 $\mu\Omega$mm.

Let us now look at a typical question involving resistivity.

> **Did you know?**
>
> Even if we removed all resistors from the circuit, we would still have some resistance caused by the actual conductors such as the connecting wires.

Example 1

Find the resistance of the field coil of a motor where the conductor cross-sectional area (CSA) is 2 mm², the length of wire is 4000 m and the material resistivity is 18 μΩmm.

$$R = \frac{\rho \times L}{A}$$

$$\overset{\text{Problem 1}}{\phantom{\frac{18}{1000000}}} \overset{\text{Problem 2}}{\phantom{\frac{4000000}{2}}}$$

$$R = \frac{18}{1\,000\,000} \times \frac{4\,000\,000}{2}$$

What has happened here?

Problem 1: The value of ρ is given in millionths of an ohm millimetre. If we have 18 μΩmm, then we have 18 millionths of an ohm and we therefore write it as 18 divided by one million, or:

$$\frac{18}{1\,000\,000}$$

Problem 2: Remember, when doing calculations, all units must be the same. Here the length is in metres, but everything else is in millimetres. Therefore, note that 4000 m has now become 4 000 000 mm.

So back to the calculation:

$$R = \frac{18}{1\,000\,000} \times \frac{4\,000\,000}{2} = \frac{72\,000\,000}{2\,000\,000} = 36\,\Omega$$

Another way of doing this calculation, **without the calculator**, would have been to cancel the zeros down (division):

$$R = \frac{18}{\cancel{1\,000\,000}} \times \frac{4\,\cancel{000\,000}}{2}$$

$$R = \frac{18 \times 4}{2} = \mathbf{36\,\Omega}$$

Here are two further examples:

Remember

All measurements should be of the same unit, i.e. all in metres or millimetres. In these examples, millimetres have been used. Therefore in Example 2, cable length is multiplied by 10^3 and resistivity by 10^{-6}.

Example 2

A copper conductor has a resistivity of 17.8 μΩmm and a CSA of 2.5 mm². What will be the resistance of a 30 m length of this conductor?

$$R = \frac{\rho \times L}{A} \quad \text{then} \quad R = \frac{17.8 \times 30\,000}{1\,000\,000 \times 2.5} = \frac{534\,000}{2\,500\,000} \quad \text{so} \quad \mathbf{R = 0.2136\,\Omega}$$

Example 3

A copper conductor has a resistivity of 17.8 $\mu\Omega$mm and is 1.785 mm in diameter. What will be the resistance of a 75 m length of this conductor?

We must first convert the diameter into the CSA. This is carried out by using one of the following formulas, which you may remember from school.

(a) $CSA = \dfrac{\pi d^2}{4}$ Where d = diameter

Or:

(b) $CSA = \pi r^2$ Where r = radius and π = **3.142**

Using the first formula:

(a) $CSA = \dfrac{\pi d^2}{4}$

Step 1: Put in the correct values: **Step 2:** Multiply the top line: **Step 3:** Divide by 4

$CSA = \dfrac{3.142 \times 1.785 \times 1.785}{4}$ $CSA = \dfrac{10.01}{4}$ CSA = **2.5 mm²**

Therefore, using this method, the CSA is 2.5 mm².

Using the second formula:

$CSA = \pi r^2$

Step 1: Put in the correct values: **Step 2:** Multiply out

$CSA = 3.142 \times 0.8925 \times 0.8925$ CSA = **2.5 mm²**

Using the second method, the CSA is still 2.5 mm². We can now proceed with the example:

$R = \dfrac{\rho \times L}{A}$

Step 1: Put in the correct values: **Step 2:** Calculate out the top line:

$R = \dfrac{17.8 \times 10^{-6} \times 75 \times 10^3}{2.5}$ $R = \dfrac{17.8 \times 10^{-3} \times 75}{2.5}$

Which is the same as:

$R = \dfrac{17.8 \times 75}{2.5 \times 10^3}$

So:

$R = \dfrac{1335}{2500}$

Therefore: **R = 0.534 Ω**

Current, voltage and resistance in parallel and series circuits

We have now started looking seriously at circuits in terms of what is in them and how current, resistance and potential difference are all related. However, a circuit can contain many resistors and they can be connected in many ways. In this section, we will be applying Ohm's Law and looking at:

- series circuits
- parallel circuits
- parallel-series circuits
- voltage drop.

Series circuits

If a number of resistors are connected together end to end and then connected to a supply, as shown in Figure 8.22, the current can only take one route through the circuit. We call this type of connection a series circuit.

Features of a series circuit

- The total circuit resistance (R_t) is the sum total of all the individual resistors. In Figure 8.22, this means:

$$R_t = R_1 + R_2 + R_3$$

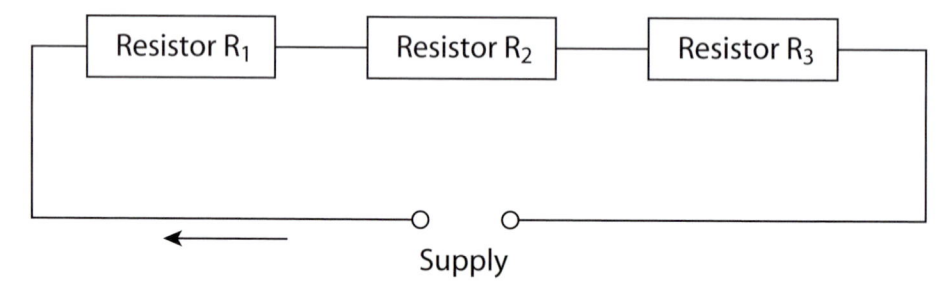

Figure 8.22 Series circuit

- The total circuit current (I) is the supply voltage divided by the total resistance. This is Ohm's Law:

$$I = \frac{V}{R}$$

- The current will have the same value at every point in the circuit.

- The potential difference across each resistor is proportional to its resistance. If we think back to Ohm's Law, we use voltage to push the electrons through a resistor. How much we use depends upon the size of the resistor. The bigger the resistor, the more we use. Therefore:

$$V_1 = I \times R_1 \quad V_2 = I \times R_2 \quad V_3 = I \times R_3$$

- The supply voltage (V) will be equal to the sum of the potential differences across each resistor. In other words, if we add up the p.d. across each resistor (the amount of volts 'dropped' across each resistor), it should come to the value of the supply voltage. We show this as:

$$V = V_1 + V_2 + V_3$$

- The total power in a series circuit is equal to the sum of the individual powers used by each resistor.

Calculation with a series circuit

Example

Two resistors of 6.2 Ω and 3.8 Ω are connected in series with a 12 V supply as shown in Figure 8.23.

We want to calculate:

(a) total resistance
(b) total current flowing
(c) the potential difference (p.d.) across each resistor.

(a) Total resistance

For series circuits, the total resistance is the sum of the individual resistors:

$R_t = R_1 + R_2 = 6.2 + 3.8 =$ **10 ohms**

(b) Total current

Using Ohm's Law:

$$I = \frac{\text{Voltage}}{\text{Resistance}} = \frac{12}{10} = \mathbf{1.2\ A}$$

(c) The p.d. across each resistor

$V = I \times R$, therefore:

Across R_1: $V_1 = I \times R_1 = 1.2 \times 6.2 =$ **7.44 V**

Across R_2: $V_2 = I \times R_2 = 1.2 \times 3.8 =$ **4.56 V**

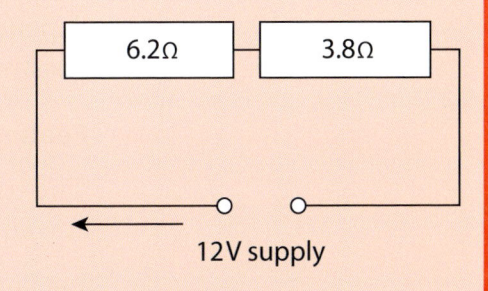

Figure 8.23 Series circuit

Parallel circuits

If a number of resistors are connected together as shown in Figure 8.24, so that there are two or more routes for the current to flow, then they are said to be connected in parallel.

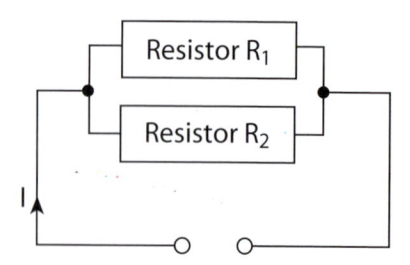

Figure 8.24 Parallel circuit

In this type of connection, the total current splits up and divides itself between the different branches of the circuit. However, note that the pressure pushing the electrons along (voltage), will be the same through each of the branches. Therefore any branch of a parallel circuit can be disconnected without affecting the other remaining branches.

Explanation

If we think about the definition of Ohm's Law, we know that the amount of electrons passing by (current) depends upon how hard we are pushing.

In a parallel circuit, the voltage is the same through each branch. Try to push two identical pencils in the same direction as the current flow towards a point on the circuit where the two branches split (shown as black circles in Figure 8.24). When they reach that point, one pencil will travel towards R1 and the other towards R2. But look how the force pushing the pencils has stayed the same.

However, how easily a pencil can then pass through a branch will depend upon the size of the obstacle in its way (the resistance of a resistor).

Features of a parallel circuit

The total circuit current (I) is found by adding together the current through each of the branches:

$I = I_1 + I_2 + I_3$

The same potential difference will occur across each branch of the circuit:

$V = V_1 = V_2 = V_3$

Where resistors are connected in parallel and, for the purpose of calculation, it is easier if the group of resistors is replaced by one total equivalent resistor (R_t). Therefore:

$$\frac{1}{R_t} = \frac{1}{R_1} + \frac{1}{R_2} + \frac{1}{R_3}$$

Calculation with a parallel circuit

Example

Three resistors of 16 Ω, 24 Ω and 48 Ω are connected across a 240 V supply. There are two ways to find out the total circuit current.

Method 1

Find the equivalent resistance, then use Ohm's Law:

$$\frac{1}{R_t} = \frac{1}{R_1} + \frac{1}{R_2} + \frac{1}{R_3}$$

Therefore:	And therefore:	Giving us:	Rearranging the equation:
$\frac{1}{R_t} = \frac{1}{16} + \frac{1}{24} + \frac{1}{48}$	$\frac{1}{R_t} = \frac{3+2+1}{48}$	$\frac{1}{R_t} = \frac{6}{48}$	$R_t = \frac{48}{6}$

and thus, $R_t = 8$ Ω

Now using the formula:	We can say that:	And therefore:
$I = \frac{V}{R}$	$I = \frac{2430}{8}$	$I = 30$ A

Method 2

Find the current through each resistor and then add them together.

Now, for R_1:	Gives:	So:
$I_1 = \frac{V}{R_1}$	$I_1 = \frac{240}{16}$	$I_1 = 15$ A
For R_2:	Gives:	So:
$I_1 = \frac{240}{16}$	$I_2 = \frac{240}{24}$	$I_2 = 10$ A
For R_3:	Gives:	So:
$I_3 = \frac{V}{R_3}$	$I_3 = \frac{243}{48}$	$I_3 = 5$ A
As:	Then:	
$I_t = I_1 + I_2 + I_3$	$I_t = 15 + 10 + 5 = $ **30 A**	

Series/parallel circuits

This type of circuit combines the series and parallel circuits as shown in the diagram in the example below. To calculate the total resistance in a combined circuit, we must first calculate the resistance of the parallel group. Then, having found the equivalent value for the parallel group, we simply treat the circuit as being made up of series connected resistors and now add this value to any series resistors in the circuit, thus giving us the total resistance for the whole of the network.

Here is a worked example.

Example 1

Calculate the total resistance of this circuit and the current flowing through the circuit, when the applied voltage is 110 V.

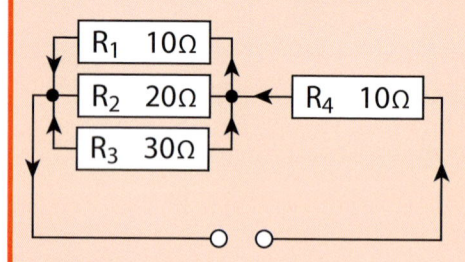

[Circuit diagram: R_1 10Ω, R_2 20Ω, R_3 30Ω in parallel, with R_4 10Ω in series]

Step 1: Find the equivalent resistance of the parallel group (R_p)

$$\frac{1}{R_p} = \frac{1}{R_1} + \frac{1}{R_2} + \frac{1}{R_3}$$

$$\frac{1}{R_p} = \frac{1}{10} + \frac{1}{20} + \frac{1}{30}$$

$$\frac{1}{R_p} = \frac{6+3+2}{60}$$

$$\frac{1}{R_p} = \frac{11}{60}$$

Therefore: $R_p = \frac{60}{11} =$ **5.45 Ω**

Step 2: Add the equivalent resistor to the series resistor R_4

$R_t = R_p + R_4$
$R_t = 5.45 + 10$
$R_t =$ **15.45 Ω**

Step 3: Calculate the current

$$I = \frac{V}{R_t} = \frac{110}{15.45} = 7.12 \text{ A}$$

Using the rules we have learned, what appear to be complicated diagrams of interconnected resistors can be resolved to a single value.

The following example shows the process step by step.

Example 2

Calculate the total resistance (R_t) of the resistor arrangement shown in the following diagram.

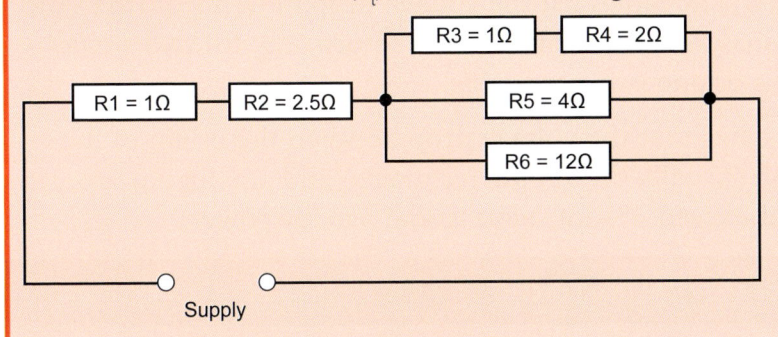

[Circuit diagram: Supply connected in series with R1 = 1Ω, R2 = 2.5Ω, then a parallel section containing three branches: (R3 = 1Ω in series with R4 = 2Ω), R5 = 4Ω, and R6 = 12Ω.]

Start by reducing the branch with resistors R_3 and R_4. As they are series connected we can add the resistances together meaning that we could now redraw the diagram as follows:

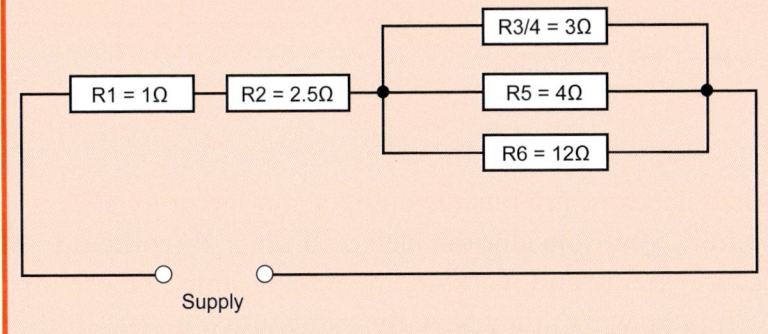

[Circuit diagram: Supply with R1 = 1Ω, R2 = 2.5Ω, and parallel branches R3/4 = 3Ω, R5 = 4Ω, R6 = 12Ω.]

Now we can reduce down the parallel group (R_p) of resistors $R_{3/4}$, R_5 and R_6.

Therefore, $\dfrac{1}{R_p} = \dfrac{1}{3} + \dfrac{1}{4} + \dfrac{1}{12}$ which gives us $\dfrac{4+3+1}{12} = \dfrac{8}{12}$

Therefore, $R_p = \dfrac{12}{8}$ so $\mathbf{R_p = 1.5\,\Omega}$

We can now redraw the circuit again to reflect this:

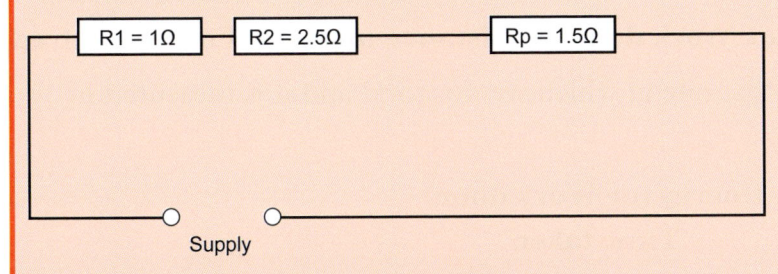

[Circuit diagram: Supply with R1 = 1Ω, R2 = 2.5Ω, Rp = 1.5Ω in series.]

We now have three series connected resistors. This means our final calculation will be:

$R_t = R_1 + R_2 + R_p = 1 + 2.5 + 1.5$

Therefore $\mathbf{R_t = 5\,\Omega}$

Voltage drop

Cables in a circuit are also resistors, in that the longer a conductor is, the higher its resistance becomes and thus the greater the voltage drop. We can calculate this by using Ohm's Law.

To determine voltage drop quickly in circuit cables, BS 7671 and cable manufacturer data include tables of voltage drop in cable conductors. The tables list the voltage drop in terms of (mV/A/m) and are listed as conductor feed and return, e.g. for two single-core cables or one two-core cable.

BS 7671 states that the voltage drop between the origin of the installation (usually the supply terminals) and any load point should not exceed 3% for lighting and 5% for power.

For a 230 V lighting circuit	For a 230 V power circuit
$\dfrac{3}{100} \times 230 = 6.9 \text{ V}$	$\dfrac{5}{100} \times 230 = 11.5 \text{ V}$

Table 8.09 Voltage drop in circuits

Power

We know that electrons are pushed along a conductor by a force called the e.m.f. Now consider the electrical units of work and power. Because energy and work are interchangeable (we use up energy to complete work) they have the same units, joules. Both can be measured in terms of force and distance.

If a force is required to move an object some distance, then work has been done and some energy has been used to do it. The greater the distance and the heavier the object, then the greater the amount of work done.

We already know that:

Energy (or Work done) = Distance moved × Force required

Power is, 'the rate at which we do work' and it is measured in watts. So:

$$\text{Power} = \frac{\text{Energy (or Work done)}}{\text{Time taken}}$$

$$= \frac{\text{Distance moved} \times \text{Force required}}{\text{Time taken}}$$

For example, we could drill two holes in a wall – one using a hand drill, and the other with an electric drill. When we have finished, the work done will be the same in both cases, there will be two identical holes in the wall, but the electric drill will do it more quickly because its power is greater.

If power is therefore considered to be the ratio of work done against the time taken to do the work, we may express this as follows:

$$\text{Power (P)} = \frac{\text{Work done (W)}}{\text{Time taken (t)}} = \frac{\text{Energy used}}{\text{Time}}$$

The units are:

$$\text{watts} = \frac{\text{joules}}{\text{seconds}}$$

Earlier in this unit, we considered the e.m.f. and defined it as being the amount of joules of work necessary to move one coulomb of electricity around the circuit, measured in joules per coulomb, also known as the volt.

Noting that 1 volt = 1 joule/coulomb and rearranging the formula, this could be expressed as:

joules = volt × coulombs

and since:

coulombs = amperes × seconds

we can substitute this to get:

joules = volts × amperes × seconds

and, since joules are the units of work:

Work = V × I × t (joules)

Taking this one step further, we can show how we arrive at some of our electrical formulae. It goes as follows.

If:

$$\text{Power} = \frac{\text{work}}{\text{second}} \quad \text{this means:} \quad P = \frac{V \times I \times t}{t}$$

So cancelling gives

P = V × I **or** P = I × V

In Ohm's Law:

V = I × R

Therefore by showing V as (I × R):

P = I × (I × R) **thus** P = I² × R

Also in Ohm's Law:

$$I = \frac{V}{R}$$

Therefore by showing I as $\left(\frac{V}{R}\right)$

$$P = \frac{V}{R} \times V \quad \textbf{thus} \quad P = \frac{V^2}{R}$$

Simple when you know how!

And when you have practised changing equations dozens of times!

Some examples of power calculations

Example 1

Two 100 Ω resistors are connected in series to a 100 V supply. What will be the total power dissipated?

Firstly find the total resistance, which for a series circuit is $R_t = R_1 + R_2$

Therefore R_t = 100 + 100 = 200 Ω

Power can be found by $\frac{V^2}{R}$ Therefore $P = \frac{100^2}{200} = \frac{10000}{200}$

Therefore P = **50 W**

Remember

Note: as all time measurements are given in seconds, we have to change hours or minutes into seconds. So, in the example:

1 hour = 60 minutes = 60 × 60 seconds = **3 600 s**

Example 2

How much energy is supplied to a 100 W resistor that is connected to a 150 V supply for one hour?

$P = \frac{V^2}{R}$ therefore $\frac{150 \times 150}{100}$ = **225 W**

Now:

E = P × t

Therefore:

E = energy supplied = 225 × 3600 joules

So:

E = **810,000 joules**

Power calculations for a parallel circuit are essentially the same as those used for the series circuit.

Since power dissipation in resistors consists of a heat loss, power dissipations will be additive irrespective of how the resistors are connected in the circuit and the total power is equal to the sum of the power dissipated by each individual resistor.

Example 3

Three resistors are connected in parallel as shown.

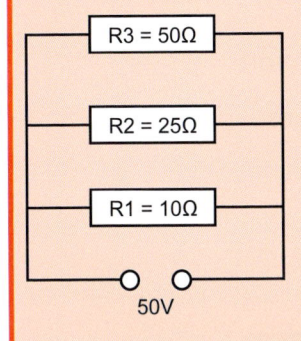

$R3 = 50\Omega$
$R2 = 25\Omega$
$R1 = 10\Omega$
50V

Calculate the power dissipated by each resistor and in total.

Power for R_1 $\quad P = \dfrac{V^2}{R} = \dfrac{50^2}{10} = \dfrac{2500}{10} =$ **250 W**

Power for R_2 $\quad P = \dfrac{V^2}{R} = \dfrac{50^2}{25} = \dfrac{2500}{25} =$ **100 W**

Power for R_3 $\quad P = \dfrac{V^2}{R} = \dfrac{50^2}{50} = \dfrac{2500}{50} =$ **50 W**

As power dissipation is additive, total power = 250 + 100 + 50 = **400 W**

An alternative method to find the total power would be to collapse the parallel network to an equivalent resistance.

In which case our calculation would be:

$\dfrac{1}{R_t} = \dfrac{1}{R_1} + \dfrac{1}{R_2} + \dfrac{1}{R_3}$ therefore $\dfrac{1}{R_t} = \dfrac{1}{10} + \dfrac{1}{25} + \dfrac{1}{50}$ which gives us $\dfrac{1}{R_t} = \dfrac{5+2+1}{50} = \dfrac{8}{50}$

If $\dfrac{1}{R_t} = \dfrac{8}{50}$ then $R_t = 50 \div 8 = 6.25\ \Omega$

If $P = \dfrac{V^2}{R}$ then gives us $\dfrac{50^2}{6.25} = \dfrac{2500}{6.25}$ therefore again, P = **400 W**

Kilowatt hour

It should be noted that the joule is far too small a unit for sensible energy measurement. For most applications, we use something called the kilowatt hour.

The kilowatt hour is defined as the amount of energy used when one kilowatt (1000 watts) of power has been used for a time of one hour (3600 seconds).

From this we can see that:

1 joule (J) = 1 watt (W) for one second (s)

1000 joules (J) = 1 kilowatt (kW) for one second

In one hour there are 3600 seconds. Therefore:

3600 s × 1000 J = 1 kW for one hour (kWh)

So:

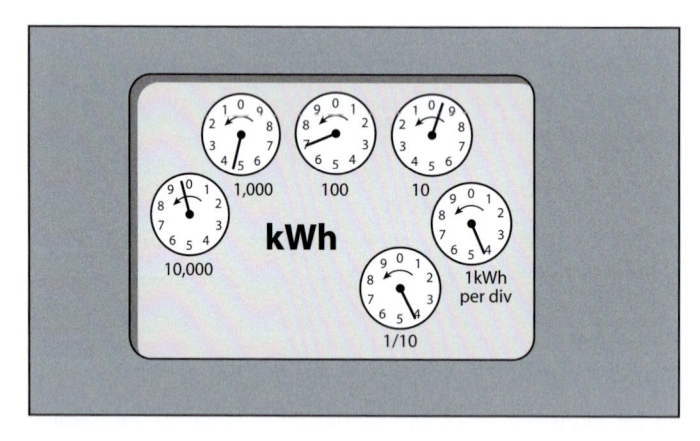

1 kWh = 3.6 × 10^6 J

Figure 8.25 Typical electric meter dials

The kilowatt hour is the unit used by the electrical supply companies to charge their customers for the supply of electrical energy. Have a look in your house. You will see that the electric meter is measuring in kWh. However, these are more often referred to as Units by the time they appear on your bill!

Example

If a small house has the following items connected to the supply each day, calculate how much energy would be consumed over a seven day period.

- Four 100 W light fittings each used for 2 hours.
- One 3 kW electric fire used for 2 hours.
- One 3 kW kettle for a total of 1 hour.

Remembering that E = P × t and applying this to each load:

Lights   = (4 × 100) × 2 = 800 = 0.8 kWh

Fire   = 3 × 2 = 6 kWh

Kettle   = 3 × 1 = 3 kWh

This gives a total daily consumption of 9.8 kWh

Therefore over seven days the consumption will be 9.8 × 7 = **68.6 kWh**

Efficiency

We have looked at efficiency in the mechanics section. The calculations and theory for efficiency applied to electrical circuits are very similar.

You already know that:

$$\textbf{Percentage efficiency} = \frac{\textbf{Output}}{\textbf{Input}} \times 100$$

Let us have a look at two examples.

Example 1

Calculate the efficiency of a water heater if the output in kilowatt hours is 25 kWh and the input energy is 30 kWh.

$$\text{Efficiency (\%)} = \frac{\text{Output}}{\text{Input}} \times 100 = \frac{25}{30} \times 100 = \textbf{83.33\%}$$

Example 2

The power output from a generator is 2700 W and the power required to drive it is 3500 W. Calculate the percentage efficiency of the generator.

$$\text{Efficiency (\%)} = \frac{\text{Output}}{\text{Input}} \times 100 = \frac{2700}{3500} \times 100 = \textbf{77.1\%}$$

Progress check

1. In terms of electron movement what is the major difference between a conductor and an insulator?
2. What are the essential components of a practical circuit?
3. One of the most important formulas for electricians, named after German physicist G.S. Ohm, is Ohm's law. State Ohm's law and identify the quantities involved.
4. Calculate the resistance of 100 m of cable where the cross-sectional area is 2.5 mm² and the resistivity 7.41 mΩ/m.
5. Calculate the total current in a circuit and the power dissipated, where three resistors of 20 Ω, 40 Ω and 50 Ω are connected in parallel with a 200 V supply.

K5. Understand the fundamental principles which underpin the relationship between magnetism and electricity

We can really go no further with circuit theory until we have looked more closely at the magnetic behaviour of materials and the way this affects the interaction between electrical currents and magnetic fields.

Magnetic effects of electrical currents

The word magnetic originated with the ancient Greeks, who found natural rocks possessing this characteristic.

Magnetic rocks such as magnetite, an iron ore, occur naturally. The Chinese observed the effects of magnetism as early as 2600 BC when they saw that stones like magnetite, when freely suspended, had a tendency to assume a north and south direction. Because magnetic stones aligned themselves north–south, they were referred to as lodestones or leading stones.

Magnetism is hard to define – we all know what its effects are: the attraction or repulsion of a material by another material, but why does this happen? And why do we only see it in some materials, notably metals and particularly iron? The physics behind this is too complex to cover here, but it is useful to remember that magnetism is a fundamental force (like gravity) and it arises due to the movement of electrical charge. Magnetism is seen whenever electrically charged particles are in motion.

Materials that are attracted by a magnet, such as iron, steel, nickel and cobalt, have the ability to become magnetised. These are called magnetic materials.

For the purpose of this book, we are only interested in two types of magnet:

- the permanent magnet
- the electromagnet (temporary magnet).

The permanent magnet

A permanent magnet is a material that when inserted into a strong magnetic field will exhibit a magnetic field of its own, and continue to exhibit a magnetic field once it has been removed from the original field.

> **Remember**
>
> For the electrician, we say that a magnet is any device that produces an external magnetic field.

This remaining field would allow the magnet to exert force (the ability to attract or repel) on other magnetic materials. This magnetic field is continuous without losing strength, as long as the material is not subjected to a change in environment (temperature, de-magnetising field, etc.).

The ability to continue exhibiting a field while withstanding different environments helps to define the capabilities and types of applications in which a magnet can be successfully used.

Magnetic fields in permanent magnets come from two atomic causes: the spin and orbital motions of electrons. Therefore, the magnetic characteristics of a material can change when alloyed with other elements.

For example, a non-magnetic material such as aluminium can become magnetic in materials such as alnico or manganese-aluminium-carbon.

When a ferromagnetic material (a material containing iron) is magnetised in one direction, it will not relax back to zero magnetisation when the imposed magnetising field is removed. The amount of magnetisation it retains is called its remanence and it can only be driven back to zero by a field in the opposite direction; the amount of reverse driving field required to de-magnetise it is called its coercivity.

We have probably all experienced at some time the effect of a permanent magnet (although we cannot see the magnetic field with the naked eye), even if it is just to leave a message on the fridge door.

The magnetic field looks like a series of closed loops that start at one end (pole) of the magnet, arrive at the other and then pass through the magnet to the original start point.

At school you probably did the famous experiment where you took a magnet, placed it on a piece of paper and then sprinkled iron filings over it?

If you did, you would see that it looks like Figure 8.28 on page 290, with the attraction of the magnet causing the filings to line up on the lines of magnetic flux (the direction of the field).

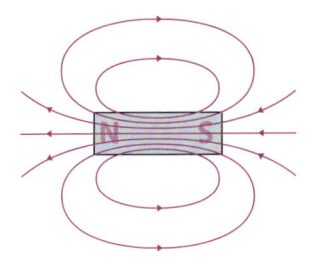

Figure 8.26 Bar magnet

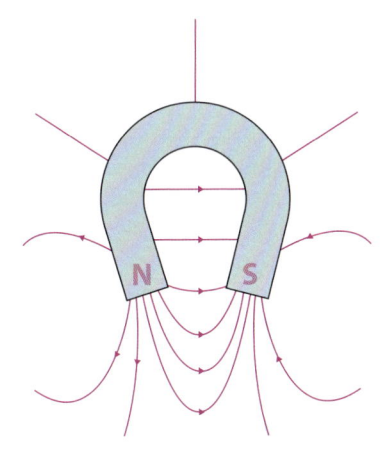

Figure 8.27 Horseshoe magnet

> **Remember**
>
> Like poles repel each other and unlike poles attract.

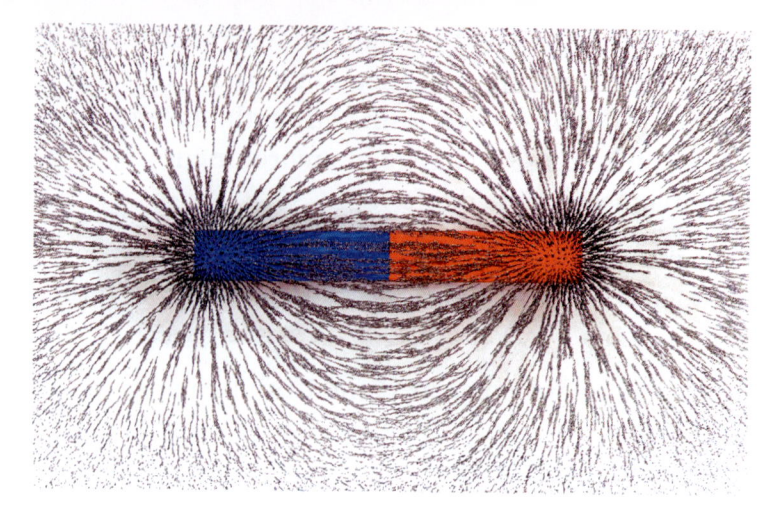

Figure 8.28 Iron filings around a bar magnet

If we were to add a small compass to the experiment, we would find that the lines run externally from the North pole of the magnet to the South and that they have the following properties:

- They will never cross, but may become distorted.
- They will always try to return to their original shape.
- They will always form a closed loop.
- Outside the magnet they run north to south.
- The higher the number of lines of magnetic flux, the stronger the magnet.

If we could count the lines, we could establish the magnetic flux (which we measure in webers), and we would find that the more lines that there were, the stronger the magnet would be. In other words the bigger the magnet the bigger the flux produced.

The strength of the magnetic field at any point is calculated by counting the number of lines that we have at that point and this is then called the flux density (measuring webers/square metre, which are given the unit title of a tesla).

We define the tesla as follows: If one weber of magnetic flux was spread evenly over a cross-sectional area of one square metre, then we have a flux density of one tesla. In other words the flux density depends upon the amount of magnetic flux lines and the area to which they are applied.

We use the following formula to express this:

$$\text{Flux density B (tesla)} = \frac{\text{magnetic flux}}{\text{CSA}} = \frac{\Phi}{A} \ (\text{webers}/\text{m}^2)$$

Here is an example to help you understand what is going on.

> **Example**
>
> The field pole of a motor has an area of 5 cm² and carries a flux of 80 µWb. What will be the flux density?
>
> Using the formula:
>
> $$\text{Flux density B (tesla)} = \frac{\text{(magnetic flux)}}{\text{(CSA)}} = \frac{\Phi}{A} \text{ (webers/m}^2\text{)}$$
>
> We need to make allowance for the area being given in cm² and the flux in mWb.
>
> Therefore:
>
> $$\text{Flux density B (tesla)} = \frac{\text{(magnetic flux)}}{\text{(CSA)}} = \frac{80 \times 10^{-6} \text{ Wb}}{5 \times 10^{-4} \text{ m}^2} = \mathbf{0.16 \text{ T}}$$

The electromagnet

An electromagnet is produced where there is an electric current flowing through a conductor, as a magnetic field is produced around the conductor. This magnetic field is proportional to the current being carried, since the larger the current, the greater the magnetic field.

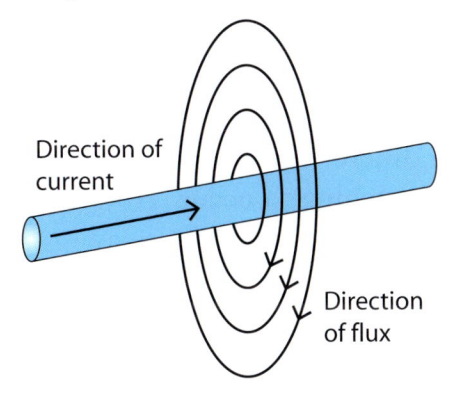

Figure 8.29 Lines of magnetic force set up around a conductor

An electromagnet is defined as being a temporary magnet because the magnetic field can only exist while there is a current flowing. If we have a typically shaped conductor such as a wire, then, as shown in Figure 8.29, the magnetic field looks like concentric circles and these are along the whole length of the conductor. However, the direction of the field depends on the direction of the current.

The screw rule

The direction of the magnetic field is traditionally determined using the 'screw rule'.

Rotation of screw = Rotation of magnetic field

Direction of screw = Direction of current

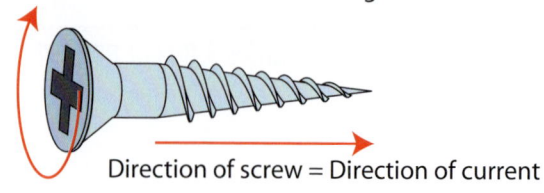

Figure 8.30 The screw rule

In the screw rule we think of a normal right-hand threaded screw, where the movement of the tip represents the direction of the current through a straight conductor and the direction of rotation of the screw represents the corresponding direction of rotation of the magnetic field.

Rotation of screw = rotation of magnetic field
Direction of screw = direction of current

Let's have a quick look at how we can use the magnetic effect in our industry.

The relay

A relay is an electromechanical switch used in many types of electronic device to switch voltages and electronic signals.

The most common electromechanical switch is the simple one-way wall switch used to control the lights in your home, as this type of switch requires a human to perform the 'switching' between on and off. Relays operate differently as they require no human interaction in order for the switching to occur. However, considering the one-way switch helps to explain how a relay works.

Figure 8.31 One-way switch off position

In the one-way circuit, if we want to put the light on in a room, the switch is operated by your finger. When we do this, we are closing the internal switch contact and the contact is mechanically held in place across the terminals. Consequently when we take our finger off the switch, it remains in position and the light stays on.

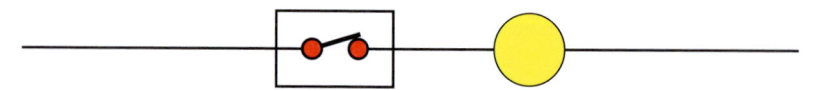

Figure 8.32 One-way switch on position

However let's say we don't want that sort of switch, but want to control the light using a relay. The concept is similar if you think of a relay as being an assembly that contains a one-way switch and a coil. We'll draw the switch contact in a slightly different way this time, but the idea is exactly the same, i.e. electricity will pass from one terminal to the other when the contact is closed.

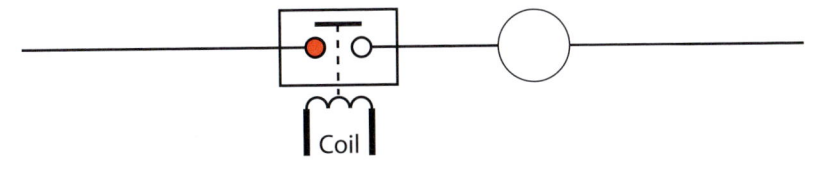

Figure 8.33 One-way switch and coil – off position

If we were now to energise the coil, the resulting magnetic field would pull the contact across the two terminals, thus closing the circuit and the light would come on. This time, instead of the switch contact being held in place mechanically, it is being held in place by the magnetic field produced by the coil in the relay. It will only remain this way while the coil is energised.

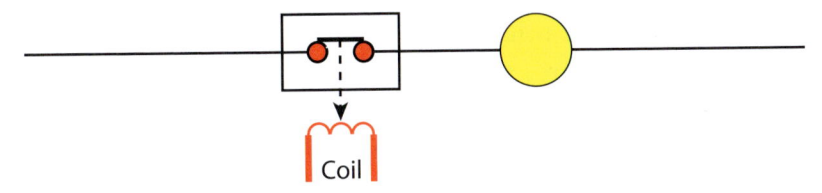

Figure 8.34 When the coil is energised, the switch is on

We describe this type of relay as having 'normally open' contacts, this means when the coil is de-energised, the contact opens and no electricity can pass through the relay.

It is possible to have a relay where the exact opposite function takes place, i.e. when the coil is energised the contact is pulled away from the terminals. In such a relay the supply would normally be passing through the closed contact and operating the coil will therefore break the circuit. We say that such a relay has 'normally closed' contacts.

A relay is therefore an electromechanical switch that uses an electromagnet to create a magnetic field which opens or closes one or many sets of contacts.

Figure 8.35 A relay

Applications of the relay

Relays can be used to:

- control a high voltage circuit with a low voltage signal, as in some types of modem
- control a high current circuit with a low current signal, as in all the lights in the hall of a leisure centre being controlled from a 5 A switch in reception
- control a mains powered device from a low voltage switch.

When choosing a relay there are several things to consider:

- **Coil voltage** – this indicates how much voltage (230 V, 24 V) and what kind (a.c. or d.c.) must be applied to energise the coil. Therefore, make sure that the coil voltage matches the supply fed into it.
- **Contact ratings** – this indicates how heavy a load the relay can control (e.g. 0.5 A or 10 A).
- **Contact arrangement** – there are many kinds of switches, so there are many kinds of relays. The contact geometry indicates how many poles there are, and how they open and close.

For example, a changeover relay has one moving contact and two fixed contacts. One of these is normally closed when the relay is switched off, and the other is normally open. Energising the coil will cause the normally open contact to close and the normally closed contact to open.

So far we have looked at the force on an object in a magnetic field produced by a permanent magnet or electromagnet. But what happens when we place a current-carrying conductor (a wire that has a current flowing through it) inside a magnetic field or next to another current-carrying conductor?

Force between current-carrying conductors

We already know that a magnetic field, in the form of concentric circles, will be produced by a current-carrying conductor. If we took two such conductors and placed them side by side, we would see a force exists between them due to the magnetic flux and the direction of this force will be dictated by the direction of the current.

> **Remember**
>
> Relays are useful as they can be used to switch current between circuits or turn a circuit on and off.

> **Did you know?**
>
> In a relay we can operate a high voltage circuit with a very low, safe, switching circuit, because the coil-energising circuit is completely separate to the contact circuit(s).

For example, in Figure 8.36 the direction of current in both conductors is different and therefore the direction of the flux in between the conductors is the same. This means that the largest concentration of flux is between the two conductors and a lesser amount on the outside of them and therefore the resultant force will try to push them apart.

The reverse is true, in that if the direction of current in both conductors was the same, then the direction of the flux in between the conductors will be in opposite directions. This cancels flux levels out and leaves more flux on the outside of the two conductors and therefore the resultant force will try to push them together.

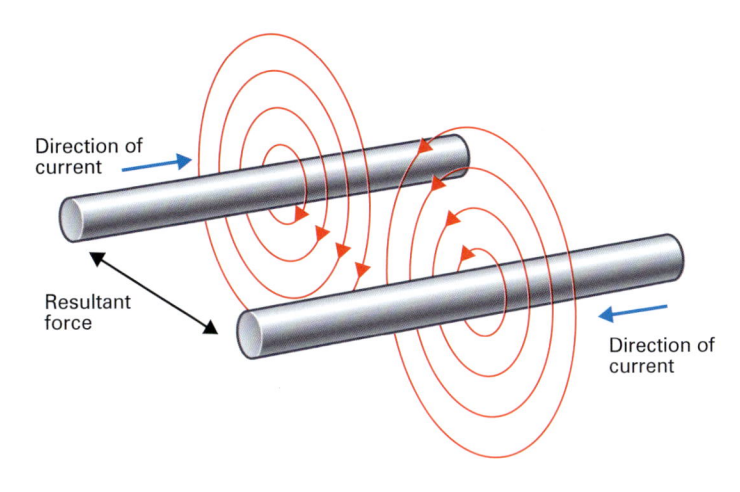

Figure 8.36 Force between current-carrying conductors

Force on a current-carrying conductor in a magnetic field

In the section on the permanent magnet, we said that lines of flux can be distorted but will never cross, and it is this principle that we will consider in this section.

If we place a current-carrying conductor between two magnetic poles we can look at the field caused only by the conductor. As shown in Figure 8.37, you will see that the current is going away from you, therefore the field is clockwise. This is the screw rule and can be remembered by imagining a corkscrew being twisted into a cork in a bottle, as you turn the corkscrew to the right (this symbolises the magnetic field) and the corkscrew is moves away from you into the cork, this symbolises the current flowing away from you.

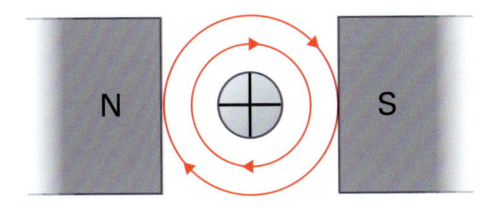

Figure 8.37 Current-carrying conductor between two magnetic poles

This time in Figure 8.38 we will look at the same arrangement, but only looking at the field caused by the two magnetic poles.

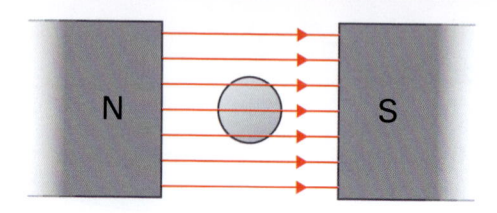

Figure 8.38 Field caused by two magnetic poles

Now bearing in mind that lines of flux cannot cross, Figure 8.39 is what the result would be if we actually had a current-carrying conductor in a magnetic field.

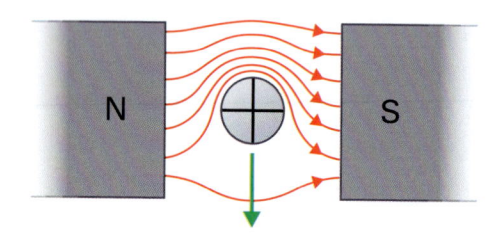

Figure 8.39 Current-carrying conductor in a magnetic field

In Figure 8.39 we can see that the main field now becomes distorted and, that as the two fields above the conductor are in the same direction, the amount of flux is high and therefore the force will move the conductor downwards. The reverse would be true if the two fields were in opposition.

The direction in which a current-carrying conductor tends to move when it is placed in a magnetic field can be determined by Fleming's left-hand (motor) rule.

This rule states that if the first finger, the second finger and the thumb of the left hand are held at right angles to each other as shown below, then with the first finger pointing in the direction of the Field (N to S), and the second finger pointing in the direction of the current in the conductor, then the thumb will indicate the direction in which the conductor tends to move.

First finger pointing in the direction of the **F**ield (N to S)

Se**C**ond finger pointing in the direction of the **C**urrent in the conductor

Thu**M**b points in the direction in which the conductor tends to **M**ove

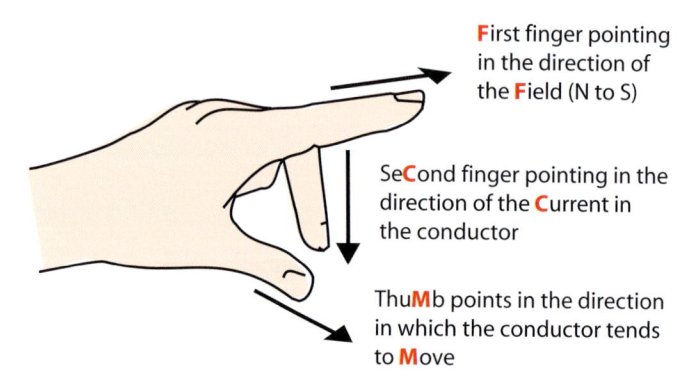

Figure 8.40 Fleming's left-hand rule

Calculating the force on a conductor

The force that moves the current-carrying conductor placed in a magnetic field depends on the strength of the magnetic flux density (B), the magnitude of the current flowing in the conductor (I), and the length of the conductor in the magnetic field (ℓ).

The following equation expresses this relationship:

Force (F) = B × I × ℓ

Where B is in tesla, ℓ is in metres, I is in amperes and F is in newtons.

Here are two examples.

Example 1

A conductor 15 m in length lies at right angles to a magnetic field of 5 tesla. Calculate the force on the conductor when:

(a) 15 A flows in the coil
(b) 25 A flows in the coil
(c) 50 A flows in the coil.

Answer: Using the formula F= B × I × ℓ:

F = 5 × 15 × 15 = 1125 N
F = 5 × 25 × 15 = 1875 N
F = 5 × 50 × 15 = 3750 N.

Example 2

A conductor 0.25 m long situated in, and at right angles to, a magnetic field experiences a force of 5 N when a current through it is 50 A. Calculate the flux density.

Answer: Transpose the formula F = B × I × ℓ for (B):

$$B = \frac{F}{I \times \ell}$$

Substitute the known values into the equation:

$$B = \frac{5}{0.25 \times 50} = \textbf{0.4 T}$$

The solenoid

A solenoid is a long hollow cylinder around which we wind a uniform coil of wire. When a current is sent through the wire, a magnetic field is created inside the cylinder.

A solenoid (Figure 8.41) usually has a length that is several times its diameter and the wire is closely wound around the outside of a long cylinder in the form of a helix with a small pitch. The magnetic field created inside the cylinder is quite uniform, especially far from the ends of the solenoid and the flux density is increased by winding on to an iron core instead of a hollow cylinder.

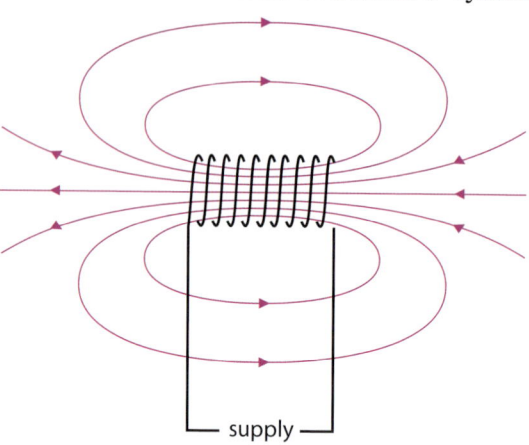

Figure 8.41 Construction of a solenoid

Essentially, the magnetic field produced by a solenoid is similar to that of a bar magnet. If an iron rod were then placed partly inside a solenoid and the current turned on, the rod will be drawn into the solenoid by the resulting magnetic field.

We can use this motion to move a lever or operate a latch to open a door and is most commonly seen in use inside a doorbell.

As shown in Figure 8.42, we can use a switch to energise the solenoid, the magnetic field will draw the iron rod in and therefore produce a mechanical action at a remote location, e.g. the doorbell. When the supply is not present the iron rod is returned to its original position under spring pressure.

However, instead of having a current-carrying conductor placed in a magnetic field which causes it to move, what if we took a conductor with **no** current flowing in it and instead moved the conductor through the magnetic field?

Instead of having the current causing a motion, we will now have the motion causing a current and in this case the magnetic field is responsible for the flow of an electric current. We call this electromagnetic induction.

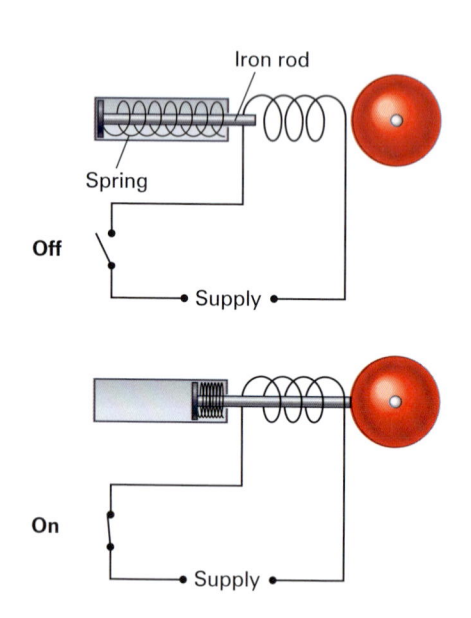

Figure 8.42 Operation of a switch in the off and on positions on a solenoid in a doorbell

Electromagnetic induction

Stated simply, if a conductor is moved through a magnetic field, then, provided there is a closed circuit, a current will flow through it.

We know that we need a 'force' to drive electrons along a conductor, and we can say that an e.m.f. must be producing the current. In this situation we are causing an e.m.f. This is known as the induced e.m.f. and it will have the same direction as the flowing current.

If we were to pass an electric current through a conductor this would generate a uniform magnetic field around the conductor and at right angles to the conductor. The strength of this magnetic field is directly proportional to the current flowing in the conductor and the strength of this magnetic field can be further increased by coiling the conductor to form a solenoid.

If the coil were connected to a d.c. supply the only resistance to the current flow would be the resistance of the conductor itself. However, if the coil is connected to an a.c. supply the situation must be looked at differently. Any change in the magnetic environment of a coil of wire will cause a voltage (e.m.f.) to be 'induced' in the coil. No matter how the change is produced, the voltage will be generated.

The change could be produced by changing the magnetic field strength, moving a magnet toward or away from the coil, moving the coil into or out of the magnetic field, rotating the coil relative to the magnet, etc.

Alternating current creates the effect of a continuously changing magnetic field inside the coil, which induces an e.m.f. in the coil that acts in opposition to the supply voltage and is therefore referred to as the back e.m.f.

The behaviour of a coil of wire in resisting any change of electric current (electron flow) through the coil is typically referred to as inductance, where the SI unit of inductance is known as the henry (H) and the symbol for inductance is L.

The unit of inductance is the rate of change of current in a circuit of 1 amp per second, which produces an induced electromotive force of 1 volt.

Values of inductors range from about 0.1 microhenry, written as $0.1\,\mu H$, to 10 henries (H). The inductance of a coil can be altered by:

- changing the number of turns of wire on the coil
- changing the material composition of the core (air, iron or steel)
- changing the diameter of the coil
- changing the material composition of the coil.

> ΔI means the change in I and so $\frac{\Delta I}{\Delta t}$ is the rate of change of I with time t

Thinking of the solenoid we have just been talking about, you realise that an e.m.f. is induced only when we have a changing situation. What could change in the solenoid set-up? Well, the following could change:

- the number of turns in the coil (N)
- the rate of change of current flowing in the coil $\left(\frac{\Delta I}{\Delta t}\right)$ – how quickly the current alternates in the coil
- the rate of change of magnetic flux $\left(\frac{\Delta \Phi}{\Delta t}\right)$ – how quickly the magnetic flux changes.

In the 19th century a scientist named Michael Faraday spent a lot of time looking at magnetic induction. He devised a law that tells how much e.m.f. is induced when a conductor is moving in a magnetic field. We will have a look at this and try to make it as simple as possible.

Faraday found that, for a conductor, the induced e.m.f. is given by:

$$\text{e.m.f.} = -\left(\frac{\Delta \Phi}{\Delta t}\right)$$

So we can find the induced e.m.f. by knowing the rate of change of flux. This is simply the same as how quickly the conductor cuts the lines of flux.

So, for a coil of N turns:

$$\text{e.m.f.} = -N\left(\frac{\Delta \Phi}{\Delta t}\right)$$

So we can find the induced e.m.f. by knowing the number of turns and rate of change of flux and

$$\text{e.m.f.} = -L\left(\frac{\Delta I}{\Delta t}\right)$$

So we can find the induced e.m.f. by knowing the inductance and rate of change of current.

These equations are true for both self and mutual inductance.

Note: In all these equations there is a negative (−) sign. This is because any induced e.m.f. will always be in opposition to the changes that created it.

When a number of inductors need to be connected together to form an equivalent inductance they follow the same rules as for resistors, therefore:

- to increase inductance, connect inductors in series.
- to decrease inductance and increase the current rating, connect inductors in parallel.

It takes time to build up to maximum current; however, this is important when connected to an a.c. supply because the rate of change of current with time can be calculated and adjusted so that a smoothing effect can be produced in the a.c. If the coil is suddenly switched off, the magnetic field collapses and a high voltage is induced across the circuit. This effect is used for starting fluorescent tube circuits.

The d.c. generator

We know that when a current is present in a conductor, a magnetic field is set up around that conductor that is always in a clockwise direction in relation to the direction of the current flow.

We also know that when a conductor is moved at right angles through a magnetic field, a current is induced into the conductor, the direction of the induced current being dependent on the direction of movement of the conductor. The strength of the induced current is determined by the speed at which the conductor moves.

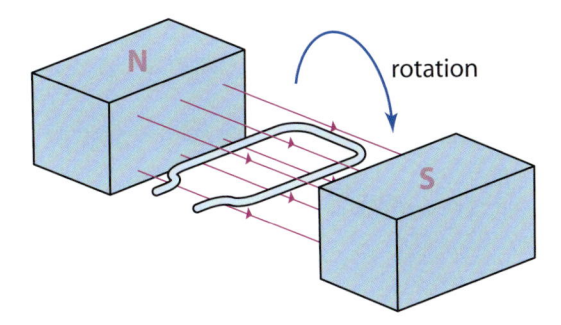

Figure 8.43 A d.c. generator

If we were to take this arrangement and form the conductor into a loop and then connect it to some device that would spin the wire loop within the permanent magnetic field, then it would look something like Figure 8.44.

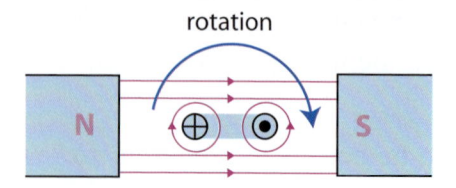

Figure 8.44 Wire loop within a permanent magnetic field

In this position, it could be said that the loop is lying in between the lines of magnetism (magnetic flux) and therefore we say that it is not 'cutting' any lines of flux (as shown in Figure 8.44).

However, as we slowly start to rotate the loop, it will start to pass through the lines of flux. When this happens, we say that we are cutting through the lines of flux and it is by this action that we start to induce an e.m.f.

The maximum number of lines that are cut through will occur when the loop has moved through 90° and the maximum induced e.m.f. in this direction will therefore occur at this point. Keep rotating the loop and the number will once again reduce to zero as we are again lying between the lines of flux.

Figure 8.45 shows this and how if we connect a load via the use of a commutator and fixed carbon brushes, a current will flow around the circuit and the load would work. In other words we have created a d.c. generator, sometimes called a dynamo.

> **Did you know?**
>
> In reality we don't have just one loop, we have many. This arrangement of loops is called the armature.

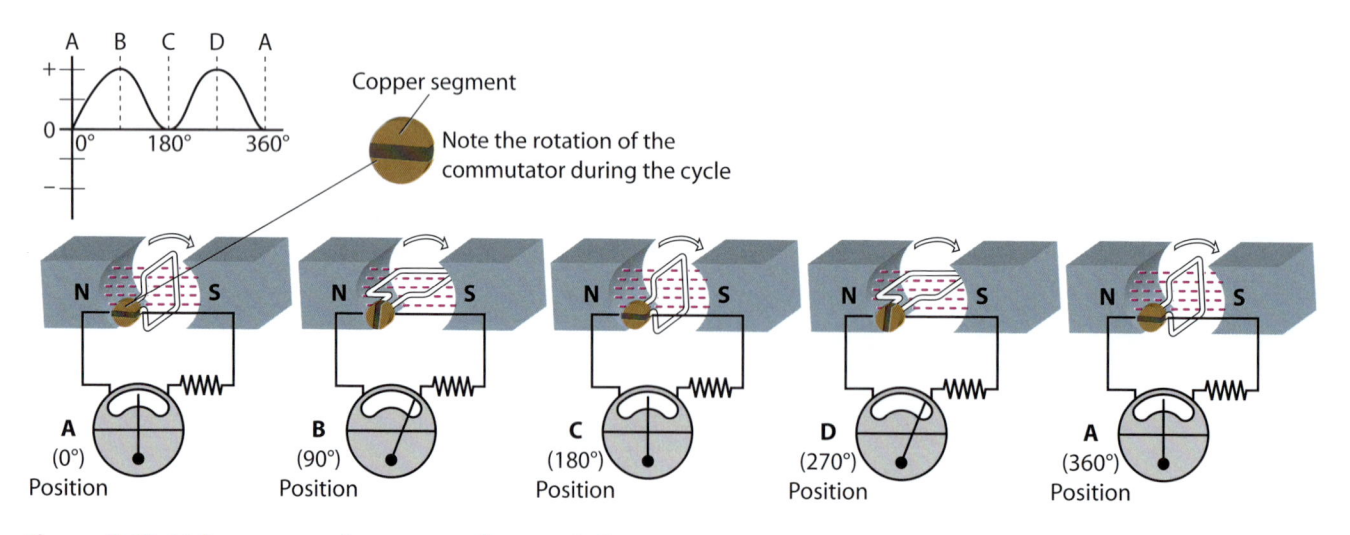

Figure 8.45 Voltage output for one complete revolution

This type of generator produces a voltage/current that alternates in magnitude but flows in one direction only: in other words it has no negative parts in its cycle and we have a direct current (d.c.), but because the generator only has one loop, it provides a pulsating d.c. output as shown by the wave form above. In general use a number of coils are used to produce a more stable output.

In the loop assembly, the armature, revolves between two stationary field poles, the current in the armature moves in one direction during one half of each revolution and in the other direction during the other half. To produce a steady flow of direct current from such a device, it is necessary to provide a means of reversing the current flow outside the generator once during each revolution.

As shown in Figure 8.46, in older machines this reversal is accomplished by means of the previously mentioned commutator. The commutator is a split metal ring mounted on the shaft of the armature, where the two halves of the ring are insulated from each other and serve as the terminals of the armature coil.

Fixed carbon brushes are then held against the commutator as it revolves, connecting the coil electrically to external wires and devices.

As the armature (our loop) turns, each brush is in contact alternately with the halves of the commutator, changing position at the moment when the current in the armature coil reverses its direction, because when the coil turns past the 'dead spot' where the brushes meet the gap in the ring, the connections between the ends of the coil and external terminals are reversed. Consequently there is a flow of d.c. in the outside circuit to which the generator is connected.

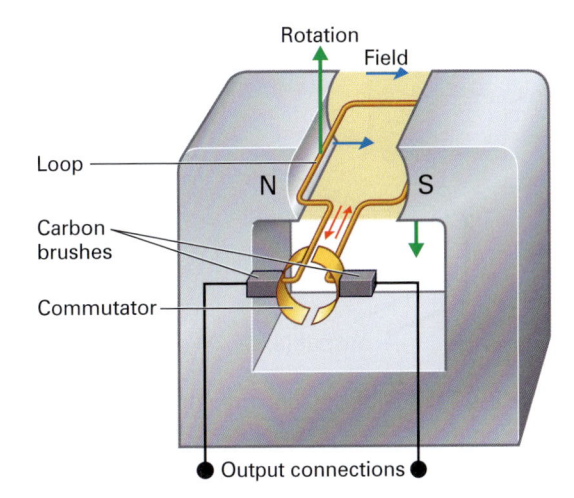

Figure 8.46 Use of a commutator

Modern d.c. generators use drum armatures that usually consist of a large number of windings set in longitudinal slits in the armature core and connected to appropriate segments of a multiple commutator. In an armature having only one loop of wire, the current produced will rise and fall depending on the part of the magnetic field through which the loop is moving.

A commutator of many segments used with a drum armature always connects the external circuit to one loop of wire moving through the high-intensity area of the field, and as a result the current delivered by the armature windings is virtually constant.

What is alternating current?

Alternating current (**a.c.**) is a flow of electrons, which rises to a maximum value in one direction and then falls back to zero before repeating the process in the opposite direction. In other words, the electrons within the conductor do not drift (flow) in one direction, but actually move backwards and forwards.

The journey taken, i.e. starting at zero, flowing in both directions and then returning to zero, is called a cycle. The number of cycles that occur every second is said to be the frequency and this is measured in **hertz** (Hz).

The a.c. generator

Sometimes referred to as an alternator, the operating principle of the a.c. generator is much the same as that of the d.c. version described in pages 301–303. However, instead of the ends of our rotating loop terminating via carbon brushes at the commutator, they are instead terminated via brushes at slip rings as shown in Figure 8.47.

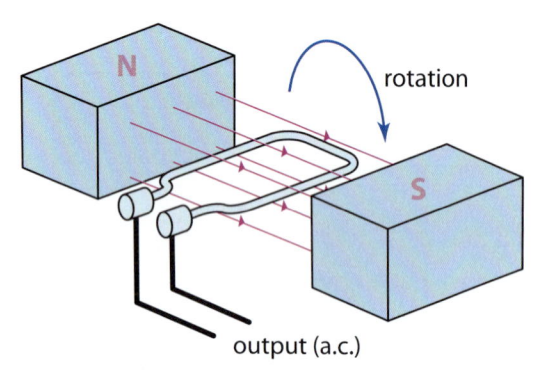

Figure 8.47 Ends of the loop are connected via slip rings

With our loop revolving between two stationary field poles, the maximum number of lines that are cut through will occur when the loop has moved through 90° and the maximum induced e.m.f. in this direction will therefore occur at this point.

Keep rotating the loop and the e.m.f. will reduce to zero as we are again lying between the lines of flux. The loop has now completed what is known as the positive half cycle.

Repeat the process and an e.m.f. will be induced in the opposite direction (the negative half cycle) until the loop returns to its original starting position.

If we were to plot this full 360° revolution (cycle) of the loop as a graph, we would see the e.m.f. induced in the loop as what is known as a sine wave (Figure 8.48).

As we can see, the sine wave shows the e.m.f. rising from zero as we start to cut through more lines of flux, to its maximum after 90° of rotation. This is known as the peak value. However, after completing 180° (half a rotation or cycle), the e.m.f. passes through zero and then changes direction.

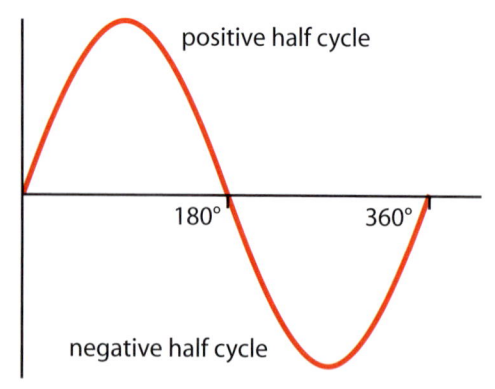

Figure 8.48 Sine wave

The opposite directions of the induced e.m.f. will still drive a current through a conductor, but that current will alternate as the loop rotates through the magnetic field. The current will be flowing in the same direction as the induced e.m.f. and consequently the current will rise and fall in the same way as the induced e.m.f. When this happens, we say that they are in phase with each other.

To access this a.c. output, the ends of the loop are connected via slip rings as shown above, and because the two brushes contact two continuous rings, the two external terminals are always connected to the same ends of the coil, hence out sinusoidal output.

We say that such a device is called an a.c. generator and is producing an alternating current (a.c.), where the number of complete revolutions (cycles) that occur each second is known as the frequency, measured in hertz (Hz) and given the symbol f. The frequency of the supply in this country is 50 Hz.

Sine waves and a.c. motors

This will be covered in depth later in this unit on pages 312–335.

Alternating current or direct current?

In the late 1880s there was conflict over the use of alternating current as advocated by George Westinghouse and Nikola Tesla, versus the Thomas Edison promotion of direct current for electric power distribution.

Originally Edison's d.c. was used as the main method of transmitting electricity, but as technology developed, a.c. became the preferred choice for two main reasons.

Reason 1
Operating at 100 volts and therefore using smaller cables, the voltage drop in Edison's system was so high that generating plants had to be located within about a mile of the user. Higher voltages cannot easily be used with the d.c. system as transformers don't work with d.c. Consequently, adjusting a d.c. voltage means having to converting the d.c. to a.c., adjust the resulting a.c. voltage with a transformer and then convert the adjusted a.c. voltage back to a corresponding d.c. voltage. Clearly, adjusting d.c. voltage is more complicated, and not surprisingly more expensive, than adjusting a.c. voltages.

In Tesla's alternating current system, a transformer could be used between a high voltage distribution system and the customer loads. Large loads, such as industrial motors could therefore be served by the same distribution network that fed lighting, by using a transformer with a suitable secondary voltage and the transformers made it easy to adjust a.c. to a higher or lower voltage very efficiently. This was useful, because to transmit at high voltage reduces current and power loss and therefore allows smaller cable sizes and a reduction in costs.

Reason 2

Good a.c. motors (and generators) are easier and cheaper to build than good d.c. ones, as although motors are available for either a.c. or d.c., the structure and characteristics of a.c. and d.c. motors are quite different.

With a.c. it is easy to produce a magnetic field which rotates rapidly in space and any electric conductor placed within the rotating magnetic field rotates in the same direction as the field. Consequently, a metal armature rotates with the rotating magnetic field with little slippage and, through a shaft attached to the armature, can deliver mechanical power to a mechanical load such as a fan or a water pump. Called an a.c. induction motor, it is a reasonably simple means of converting electric power to mechanical power.

However, d.c. motors rely on a complex mechanical system of brushes and commutator switches. The mechanical complexity of d.c. motors, consequently, not only makes them more expensive to manufacture than a.c. motors, but also more expensive to maintain.

Progress check

1 Describe briefly the difference between a permanent magnet and an electromagnet.
2 Describe briefly what is meant by an alternating current.

K6/K7. Understand electrical supply and distribution systems and how different electrical properties can affect electrical circuits, systems and equipment

The basic method of generating electricity by turning a loop of wire between the poles of a magnet is still the basis of electricity generation in our power stations today.

How electricity is generated and transmitted for domestic and industrial/commercial consumption

There are several methods of generating electricity. A Van De Graaf generator for example can be used to generate static electricity and in a small number of power stations, nuclear reaction can be converted directly into electricity. However, in the main, electricity is generated using electromagnetic induction, in which mechanical energy is used to rotate the shaft of an a.c. generator.

But how do we get the generator to rotate?

In general, the shaft of a three-phase a.c. generator is turned by using steam turbines, and most electricity in the UK is produced by this method.

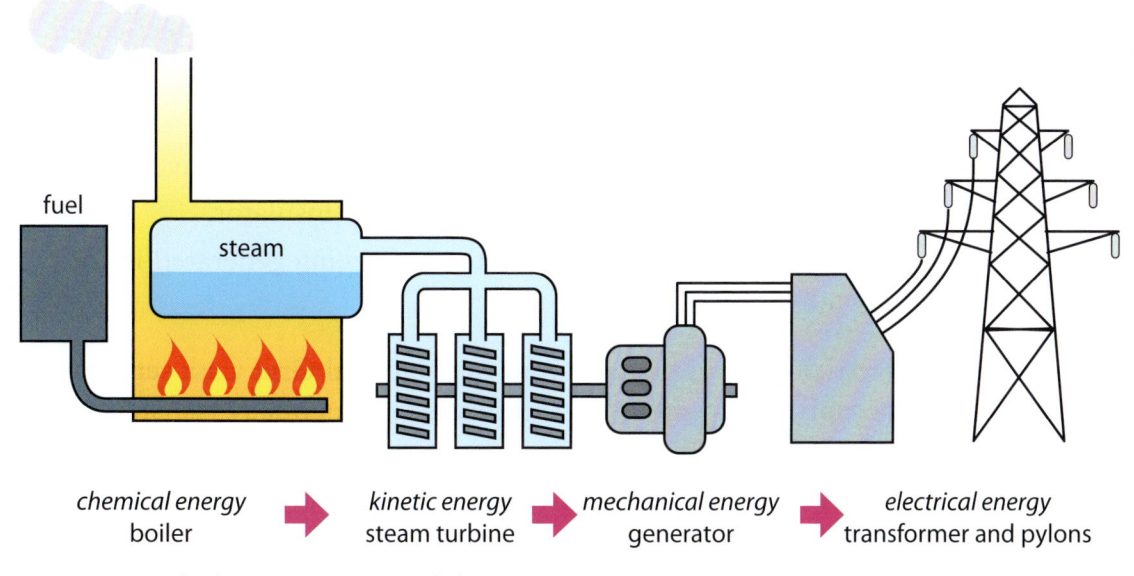

chemical energy → *kinetic energy* → *mechanical energy* → *electrical energy*
boiler → steam turbine → generator → transformer and pylons

Figure 8.49 The basic components of electricity generation systems

As we can see from Figure 8.49, water is heated by a fuel until it becomes high pressure steam. At this point the steam is forced onto the vanes of a steam turbine, which in turn rotates the generator. A variety of energy sources can be used to heat the water in the first place and the more common ones are coal, gas, oil and nuclear power.

We do have other methods of getting the generator to rotate, some being more eco-friendly than others. Such other options include hydro (where running water turns the generator), wind and tidal power.

Features and characteristics of transmission and distribution

Power station generator output is transformed upwards before transmission. Electricity is then transmitted at very high voltage (400 kV or 275 kV for the super grid) in order to reduce the power losses that occur in the power lines, as transmission at low voltage would necessitate the installation of very large cables and switchgear indeed.

At this point, the electricity is fed into the National Grid system (132 kV for the national grid), which is a network of nearly 5000 miles of overhead and underground power lines that link power stations together and are interconnected throughout the country.

The concept of the National Grid is that, should a fault develop in any one of the contributing power stations or transmission lines, then electricity can be requested from another station on the system to maintain supplies.

Electricity is transmitted around the grid, mainly via steel-cored aluminium conductors, which are suspended from steel pylons. We do this for three main reasons.

- The cost of installing cables underground is excessive.
- Air is a very cheap and readily available insulator.
- Air also acts as a coolant for the heat being generated in the conductors.

Electricity is then 'taken' from the National Grid via a series of appropriately located sub-stations that sequentially transform the grid supply down as follows:

- 66 kV and 33 kV for secondary transmission to heavy industry
- 11 kV for high-voltage distribution to lighter industry

- 415/400 V to commercial consumer supplies
- 240/230 V to domestic consumer supplies.

At the 11 kV stage we distribute electricity to a series of local sub-stations. It is their job to then take the 11 kV supply, transform it down to 400 V and then distribute this via a network of underground radial circuits to customers. In rural areas this distribution is usually with overhead lines.

It is also at this point that we see the introduction of the neutral conductor, which is normally done by connecting the secondary winding of the transformer in star and then connecting the star point to earth via an earth electrode beneath the sub-station.

Final distribution to the customer

The connection is then from the local sub-station to the customer. The connection within the customer's premises is called the main intake position.

There are many different sizes of installation, but generally speaking we will find certain items at every main intake position. These items, which belong to the supply company, are:

- a sealed overcurrent device that protects the supply company's cable
- an energy metering system to determine the customer's electricity usage.

It is after this point that we say we have reached the consumer's installation. The consumer's installation must be controlled by a main switch located as close as possible to the supply company equipment and be capable of isolating all live conductors. In the average domestic installation this device is combined with the means of distributing and protecting the final circuits in what we know as the consumer unit.

Generating electricity from other sources

In the ever-increasing search for 'green' technology and the need for energy conservation, a variety of alternative sources of electricity have developed. Methods such as solar photo-voltaic, wind energy, micro-hydro and CHP (combined heat and power) are fully covered in Book A Unit ELTK 02. However, there are three other sources that you should be familiar with.

> **Did you know?**
>
> Sub-stations are dotted throughout our cities. These small brick buildings are normally connected together on a ring circuit basis.

Cells and batteries

A cell is a device that produces electricity from a chemical reaction. A battery is generally perceived as two or more connected cells. However, we in normal day-to-day conversation, we tend to call a single cell device a battery as well. A cell and a battery can be the same thing.

It is generally accepted that the world's first 'battery' was invented by Alessandro Volta around about 1800. The principle he used remains the same today. Volta used two dissimilar metal plates separated by an electrolyte.

Such a cell consists of a negative electrode called the cathode, a positive electrode called the anode and an electrolyte that causes a chemical reaction, which in turn produces an e.m.f.

When the cell is connected to any external load, in this case a lamp, the negative electrode supplies a current of electrons that flow through the lamp before returning to the cathode. When the external load is removed, the chemical reaction stops.

A primary cell (battery) is a cell whose chemicals produce an e.m.f. until they run out. A secondary cell (battery) is a cell that can be recharged and can therefore be used many times.

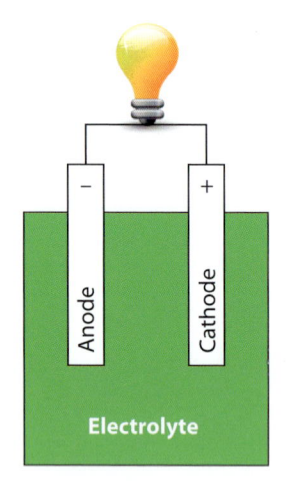

Figure 8.50 The basic structure of a battery

There are many different types of battery, mostly categorised by the chemicals used in them. Perhaps the most commonly seen are the cylindrical batteries, such as the AA or AAA that we use in portable equipment.

The construction of such a battery is shown in Figure 8.51, but you can see that the basic concept has not changed.

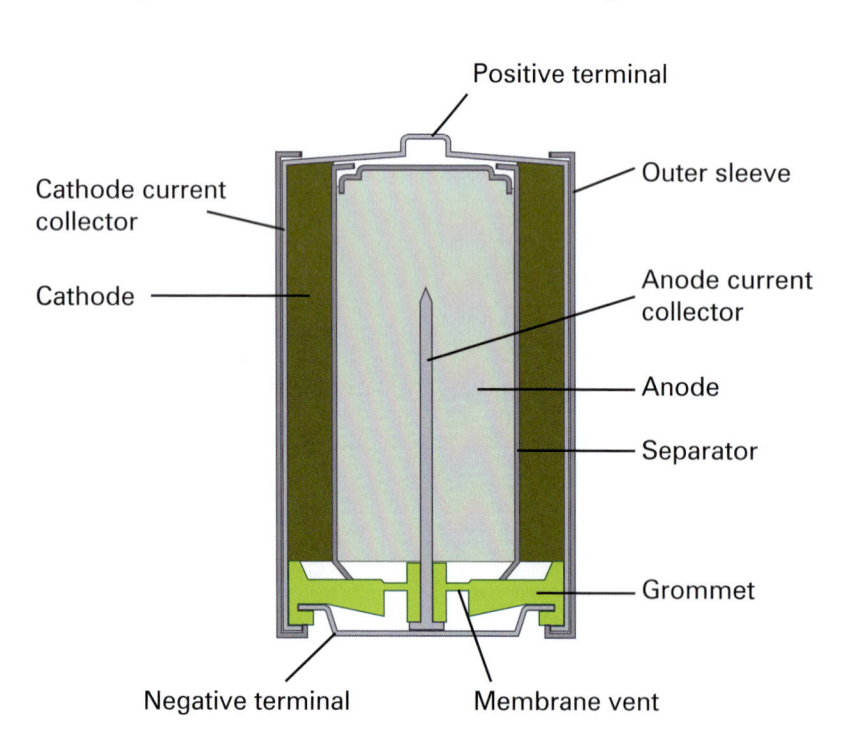

Figure 8.51 Construction of a cylindrical battery

Geothermal generation

Geothermally generated electricity was first produced in Italy in 1904 and geothermal power plants have three different methods of operation:

1. Steam is taken directly from fractures in the ground and used to drive a turbine.
2. Hot water (in excess of 200°C) is taken from the ground, allowed to boil as it rises to the surface and then steam is separated and used to turn the turbine.
3. Extracted hot water flows through heat exchangers, where it boils a fluid that in turn spins the turbine.

Geothermal power is considered to be sustainable because the heat extraction is small compared with the Earth's heat content. Also, much like the ground source heat pump, the condensed steam and remaining geothermal fluid from all three types of plants are injected back into the hot rock to collect more heat.

Countries such as Iceland and the Philippines generate about 20% of their electricity in this way.

Wave generated electricity

Waves are generated by wind passing over the surface of the sea. The height and frequency of the waves is then governed by wind speed, the time the wind has been blowing and the pattern of the sea floor.

Still very much an emerging technology, wave power devices are generally categorised by the method used to capture and convert the energy of the waves.

Some do this by using the vertical motion of buoys on the surface to create hydraulic pressure that in turn spins a generator. Others such as a 'Wave Dragon', use the principle of water from a wave being directed into the device and the falling water then turns the turbine.

Alternating current theory

As we have seen alternating current is the main method for supplying electricity. Alternating current (a.c.) uses a flow of electrons, which rises to a maximum value in one direction and then falls back to zero before repeating the process in the opposite direction. In other words, the electrons move backwards and forwards. This journey, from starting at zero, flowing in both direction and returning to zero is called a cycle. The number of cycles per second is called the frequency and measured in hertz (hz). A sine wave is used to show this.

If we look at the graph of a sine wave (Figure 8.48 on page 304), there are several values that can be measured from such an alternating waveform.

Instantaneous value

If we took a reading of induced electromotive force (e.m.f.) from the sine wave at any point in time during its cycle, this would be classed as an instantaneous value.

Average value

Using equally spaced intervals in our cycle (say every 30°) we could take a measurement of current as an instantaneous value. To find the average we would add together all the instantaneous values and then divide by the number of values used. As with the average of anything, the more values used, the greater the accuracy. *For a sine wave only*, we say that the average value is equal to the maximum value multiplied by 0.637. As a formula:

Average current = Maximum (peak) current × 0.637

Peak value

You will remember when the loop in an a.c. generator has rotated for 90° it is cutting the maximum lines of magnetic flux and therefore the greatest value of induced e.m.f. is experienced at this point. This is known as the peak value and both the positive and negative half cycles have a **peak value**.

Peak to peak value

We have said that the maximum value of induced e.m.f., irrespective of direction, is called the maximum or peak value. The voltage measured between the positive and negative peaks is known as the **peak to peak value**. The graph in Figure 8.52 shows this more clearly.

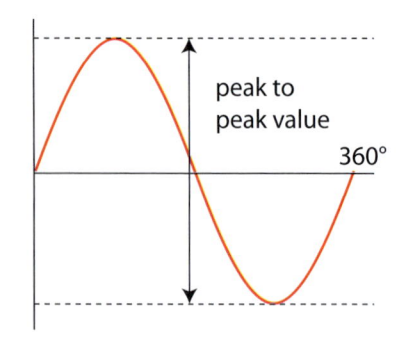

Figure 8.52 Peak to peak value

Root mean square (r.m.s.) or effective value of a waveform (voltage and current)

We have seen that in direct current (d.c.) circuits, the power delivered to a resistor is given by the product of the voltage across it and the current through it. However, in an a.c. circuit this is only true of the instantaneous power to a resistor as the current is constantly changing.

In most cases the instantaneous power is of little interest, and it is the average power delivered over time that is of most use. In order to have an easy way of measuring power, the r.m.s. method of measuring voltage and current was developed.

The r.m.s. or effective value is defined as being the a.c. value of an equivalent d.c. quantity that would deliver the same average power to the same resistor.

When current flows in a resistor, heat is produced. When it is direct current flowing in a resistor, the amount of electrical power converted into heat is expressed by the formulae.

$P = I^2 \times R$ or $P = V \times I$

However, an alternating current having a maximum (peak) value of 1 A does not maintain a constant value (see Figure 8.52). The alternating current will not produce as much heat in the resistance as will a direct current of 1 A. Consider the circuits in Figure 8.53.

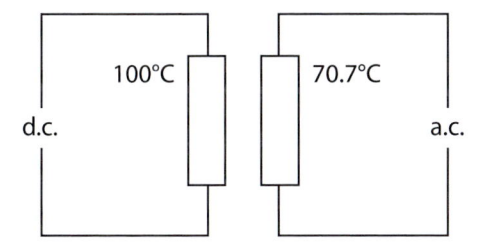

Figure 8.53 d.c. and a.c. circuits

In both the circuits in Figure 8.53, the supplies provide a maximum (peak) value of current of 1 A to a known resistor. However, the heat produced by 1 ampere of alternating current is only 70.7°C compared to the 100°C of heat that is produced by 1 ampere of direct current. We can express this using the following formula:

$$\frac{\text{Heating effect of 1A maximum a.c.}}{\text{Heating effect of 1A maximum d.c.}} = \frac{70.7}{100} = 0.707$$

Therefore, **the effective or r.m.s. value of an a.c. = 0.707 × I_{max}**

where I_{max} = the peak value of the alternating current.

We can also establish the maximum (peak) value from the r.m.s. value with the following formula:

$$I_{max} = I_{r.m.s.} \times 1.414$$

The rate at which heat is produced in a resistor is a convenient way of establishing an effective value of alternating current, and is known as the 'heating effect' method.

An alternating current is said to have an effective value of one ampere when it produces heat in a given resistance at the same rate as one ampere of direct current.

To see where the r.m.s. value sits alongside the peak value, look at Figure 8.54.

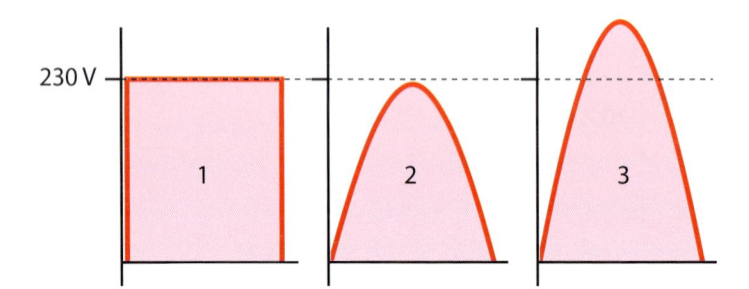

Figure 8.54 Peak value diagrams

The wave in Figure 8.54(1) represents a 230 V d.c. supply, running for a set period of time, with the heating effect produced shown as the shaded area.

The wave in Figure 8.54(2) represents an a.c. 50 Hz supply in which the voltage peaks at 230 V during one half-cycle, with the heating effect produced shown as the shaded area. As the wave only reaches 230 V for a small period of time, less heat is produced overall than in the d.c. wave.

The wave in in Figure 8.54 (3) represents an increased peak value of voltage to give the same amount of shaded area as in the d.c. example. In this diagram 230 V has become our r.m.s. value.

Important Note: Unless stated otherwise, all values of a.c. voltage and current are given as r.m.s. values.

Frequency and period

Remember that the number of cycles that occur each second is referred to as the frequency of the waveform and this is measured in hertz (Hz). The frequency of the UK supply system is 50 Hz.

In the basic arrangement of the a.c. generator loop, if one cycle of e.m.f. was generated with one complete revolution of the loop over a period of one second, then we would say the frequency was 1 Hz. If we increased the speed of loop rotation so that it was producing five cycles every second, then we would have a frequency of 5 Hz.

We can therefore say that the frequency of the waveform is the same as the speed of the loop's rotation, measured in revolutions per second. We can express this using the following equation:

Frequency (f) = Number of revolutions (n) × Number of pole pairs

If we apply this to the simple a.c. generator and rotate the loop at 50 revolutions per second, then:

Frequency = 50 × 1 (there is 1 × pole pair) = 50 Hz

The amount of time taken for the waveform to complete just one full cycle is known as the periodic time (T) or period. Therefore, if 50 cycles are produced in one second, one cycle must be produced in a fiftieth of one second. This relationship is expressed using the following equations:

$$\text{Frequency (f)} = \frac{1}{\text{Periodic time}} = \frac{1}{T}$$

$$\text{Periodic time (T)} = \frac{1}{\text{Frequency}} = \frac{1}{f}$$

> **Key term**
>
> **Pole pair** – any system consisting of a north and south pole

Power factor

When we are dealing with a.c. circuits, we often look at the way power is used in particular types of component within the circuit.

Generally, the **power factor** is a number less than 1.0, which is used to represent the relationship between the apparent power of a circuit and the true power of that circuit. In other words:

$$\text{Power Factor (PF)} = \frac{\text{True Power (PT)}}{\text{Apparent Power (PA)}} \quad \text{or}: PF = \frac{PT}{PA}$$

The units of power factor are $\frac{\text{Watts}}{\text{Watts}}$, for power factor has no units, it is a number. It is also determined by the phase angle, which we will cover shortly.

We will be returning to this over the next few pages, so by the end of the section you will have a good understanding of power factor.

Explain the relationship between, and calculate, resistance, inductance, capacitance and impedance

Resistance (R) and phasor representation

Although the sine wave is useful, it is also difficult and time-consuming to draw. We can therefore also represent a.c. by the use of phasors. A **phasor** is a straight line where the length is a scaled representation of the size of the a.c. quantity and the direction represents the relationship between the voltage and current, this relationship being known as the phase angle.

To see briefly how we use phasors, let us look at Figure 8.55, where a tungsten filament lamp has been included as the load.

Circuits like this are said to be **resistive**, and in this type of circuit the values of e.m.f. (voltage) and current actually pass through the same instants in time together. In other words, as voltage reaches its maximum value, so does the current (see Figure 8.55).

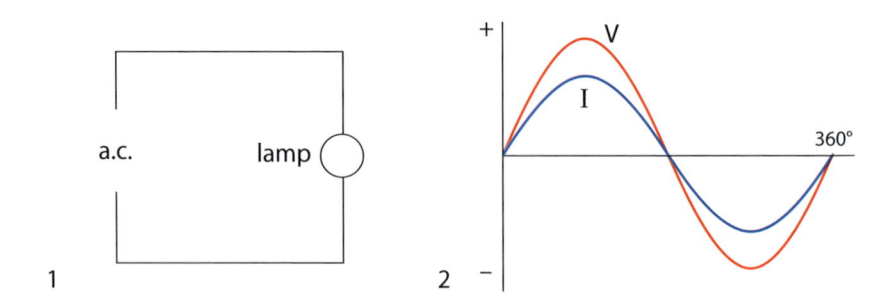

Figure 8.55 Circuit and sine wave diagrams

This happens with all resistive components connected to an a.c. supply and as such the voltage and current are said to be '**in phase**' with each other, or possess a zero phase angle.

The graph in Figure 8.55 shows this when represented by a sine wave. However, we could also show this by using a phasor diagram as shown in Figure 8.56.

Figure 8.56 Phasor diagram for zero phase angle

We can therefore say that a resistive component will consume power and we would carry out calculations as we would for a d.c. circuit (i.e. using $P = V \times I$).

We can also say that resistive equipment (filament lamps, fires, water heaters) uses this power to create heat, but such a feature in long cable runs, windings etc. would be seen as unsuitable power loss in the circuit (i.e. using $P = I^2R$).

Inductance (L)

If the load in our circuit were not a filament lamp, but a motor or transformer (something possessing windings), then we would say that the load is **inductive**.

With an inductive load the voltage and current become '**out of phase**' with each other. This is because the windings of the equipment set up their own induced e.m.f., which opposes the direction of the applied voltage, and thus forces the flow of electrons (current) to fall behind the force pushing them (voltage). However, over one full cycle, we would see that no power is consumed. When this happens, it is known as possessing a lagging phase angle or power factor.

As voltage and current are no longer perfectly linked, this type of circuit would be given a power factor of less than 1.0 (perfection), for example 0.8.

As we can see from Figure 8.57, the current is lagging the applied voltage by 90°. To make things easier in this exercise, we assumed that the above circuit is purely inductive. However, in reality this is not possible, as every coil is made of wire and that wire will have a resistance. The opposition to current flow in a resistive circuit is resistance.

The sine wave used to represent this inductive circuit would look like this:

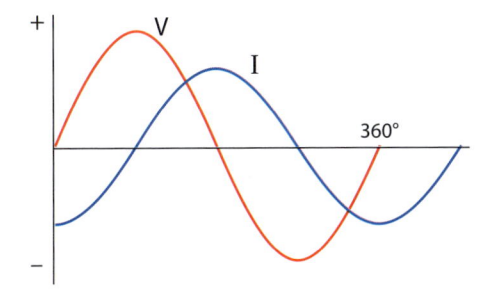

If we represented this as a phasor diagram we end up with:

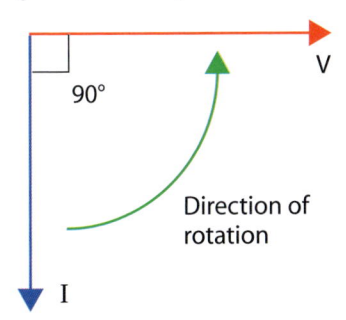

Figure 8.57 Sine wave and phasor diagrams for inductive circuit

The limiting effect to the current flow in an inductor is called the **inductive reactance**, which we are able to calculate with the following formula:

$X_L = 2\pi fL \; (\Omega)$

where:

X_L = inductive reactance (ohms – Ω)
f = supply frequency (hertz – Hz)
L = circuit inductance (henrys – H)

Let us now look at inductive and resistive circuits and see if the current is affected by a lagging phase angle.

To recap, we said that power factor is the relationship between voltage and current and that the ideal situation would seem to be the resistive circuit, where both these quantities are perfectly linked.

In the resistive circuit we know that the power in the circuit could only be the result of the voltage and the current (P = V × I). This is known as the apparent power, and possesses what we call unity power factor, to which we give the value one (1.0).

However, we now know that depending upon the equipment, the true power (actual) in the circuit must take into account the phase angle and will often be less than the apparent power … but never greater.

True power (in watts) is calculated using the cosine of the phase angle (cos Ø). The formula is:

P = VI cos Ø (remember we do not have to use the '×' sign)

When there is no phase lag, Ø = 0 and cos Ø = 1, a purely resistive circuit. To prove our previous points, let us consider the following example.

> **Example**
>
> If we have an inductive load, consuming 3 kW of power from a 230 V supply, with a power factor of 0.7 lagging, then the current (amount of electrons flowing) required to supply the load is:
>
> P = V × I × cos Ø
>
> cos Ø = Power Factor. Therefore by transposition:
>
> $$I = \frac{P}{V \times \cos \emptyset}$$
>
> Or in other words:
>
> $$I = \frac{P}{V \times PF}$$
>
> Therefore:
>
> $$I = \frac{3000}{230 \times 0.7}$$
>
> And so:
>
> I = 18.6 A
>
> However, if the same size of load was purely resistive, then cos Ø = 0, thus the power factor would be 1.0, and thus:
>
> P = V × I × PF
>
> Therefore by transposition:
>
> $$I = \frac{P}{V \times PF}$$
>
> Therefore:
>
> $$I = \frac{3000}{230 \times 1}$$
>
> And consequently:
>
> **I = 13 A**

In other words, the lower the power factor of a circuit, then the higher the current will need to be to supply the load's power requirement.

It therefore follows that if the power factor is low, then it will be necessary to install larger cables, switchgear etc. to be capable of handling the larger currents. There will also be the possibility of higher voltage drop due to the increased current in the supply cables.

Consequently, local electricity suppliers will often impose a financial fine on premises operating with a low power factor. Fortunately, we have a component that can help. It is called the capacitor.

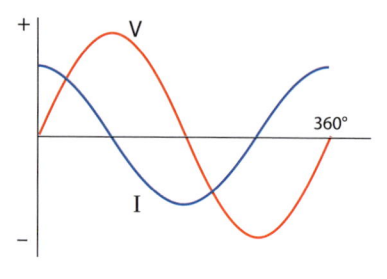

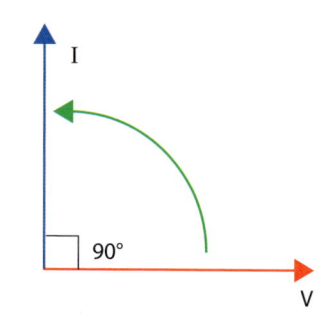

Figure 8.58 Sine wave and phasor diagrams for capacitive circuit

Capacitance (C)

Simply put, a **capacitor** is a component that stores an electric charge if a potential difference is applied across it.

The capacitor's use is then normally based on its ability to return that energy back to the circuit. When a capacitor is connected to an a.c. supply, it is continuously storing the charge and then discharging as the supply moves through its positive and negative cycles. But, as with the inductor, no power is consumed.

This means that in a capacitive circuit, we have a leading phase angle or power factor. The sine wave and phasors used to represent this would look as in Figure 8.58.

As we can see from Figure 8.58, the current leads the voltage by 90°. Consequently, the capacitor is able to help because it provides a leading power factor, and therefore if we connect it in parallel across the load, it can help neutralise the effect of a lagging power factor.

The opposition to the flow of a.c. to a capacitor is termed capacitive reactance, which, like inductive reactance, is measured in ohms and calculated using the following formula:

$$X_C = \frac{1}{2\pi f C} (\Omega)$$

where:

X_C = capacitive reactance (ohms – Ω)
f = supply frequency (hertz – Hz)
C = circuit **capacitance** (farads – F)

Since, in this type of circuit, we have voltage and current but no real power (in watts), the formula of $P = V \times I$ is no longer accurate. Instead, we say that the result of the voltage and current is reactive power, which is measured in reactive volt amperes (VAr).

The current to the capacitor, which does not contain resistance or consume power, is called reactive current.

Well, so far in our attempt to explain power factor, we have looked at a range of different subjects including resistance, inductance, capacitance and also talked about phasor diagrams. As if that wasn't bad enough, some circuits contain combinations of these components.

In order to work out these calculations, you will need to use trigonometry and also the following small section regarding the addition of phasors.

Phasors

When sine waves for voltage and current are drawn, the nature of the wave diagram can be based upon any chosen alternating quantity within the circuit. In other words, we can start from zero on the wave diagram with either the voltage or the current.

In electrical science we often need to add together alternating values. If they were 'in phase' with each other, then we would simply add the values together. However, when they are not in phase we cannot do this, hence the need for phasor diagrams.

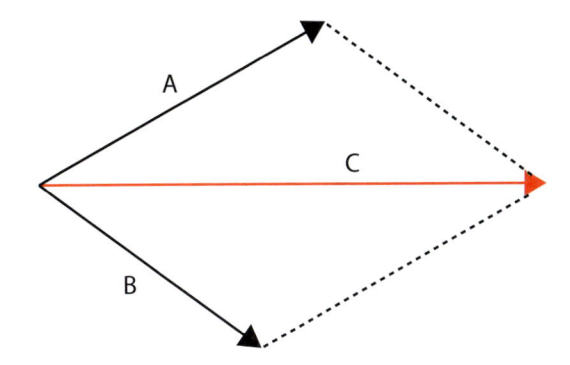

Figure 8.59 Phasor diagram

When we use phasor diagrams the chosen alternating quantity is drawn horizontally and is known as the reference.

When choosing the reference phasor, it makes sense to use a quantity that has the same value at all parts of the circuit. For example, in a series circuit the same current flows in each part of the circuit, therefore use current as the reference phasor. In a parallel circuit, the voltage is the same through each branch of the circuit and therefore we use voltage as the reference phasor.

Using the knowledge gained in the previous section, we can now measure all phase angles from this reference phasor.

Our answer (the resultant) is then found by completing a parallelogram. If we use Figure 8.59 as an example, we have been given the values of phasor A and phasor B. Therefore, the result of adding A and B together will be phasor C.

Impedance

Previously, we have been discussing components with an a.c. circuit. In fact, what those components are actually offering is opposition to the flow of current. By way of a summary, we could say that we now know that:

- The opposition to current in a resistive circuit is called resistance (R), is measured in ohms and the voltage and current are in phase with each other.
- The opposition to current in an inductive circuit is called inductive reactance (X_L), is measured in ohms and the current lags the voltage by 90°.
- The opposition to current in a capacitive circuit is called capacitive reactance (X_C), is measured in ohms and the current leads the voltage by 90°.

> **Key terms**
>
> **Impedance** – total opposition to current in a circuit
>
> **True** or **active power** – the rate at which energy is used
>
> **Apparent power** – in an a.c. circuit the sum of the true or active power and the reactive power

However, we also know that circuits will contain a combination of these components. When this happens we say that the total opposition to current is called the **impedance** (Z) of that circuit.

In summary:

- The power consumed by a resistor is dissipated in heat and not returned to the source. This is called the **true power**.
- The energy stored in the magnetic field of an **inductor** or the plates of a capacitor is returned to the source when the current changes direction.
- The power in an a.c. circuit is the sum of true power and reactive power. This is called the **apparent power**.
- **True power is equal to apparent power in a purely resistive circuit** because the voltage and current are in phase. Voltage and current are also in phase in a circuit containing equal values of inductive reactance and capacitive reactance. If the voltage and current are 90° out of phase, as would be the case in a purely capacitive or purely inductive circuit, the average value of true power is equal to zero. There are high positive and negative peak values of power, but when added together the result is zero.
- Apparent power is measured in volt-amps (VA) and has the formula: **P = VI**
- True power is measured in watts and has the formula: **P = VI cos Ø**
- **In a purely resistive circuit** where current and voltage are in phase, there is no angle of displacement between current and voltage. The cosine of a zero° angle is one, and so, the power factor is one. This means that all the energy that is delivered by the source is consumed by the circuit and dissipated in the form of heat.
- **In a purely reactive circuit**, voltage and current are 90° apart. The cosine of a 90° angle is zero so the power factor is zero. This means that the circuit returns all the energy it receives from the source, back to the source.
- **In a circuit where reactance and resistance are equal**, voltage and current are displaced by 45°. The cosine of a 45° angle is 0.7071, and so the power factor is 0.7071. This means that such a circuit uses approximately 70% of the energy supplied by the source and returns approximately 30% back to the source.

Resistance and inductance in series (RL)

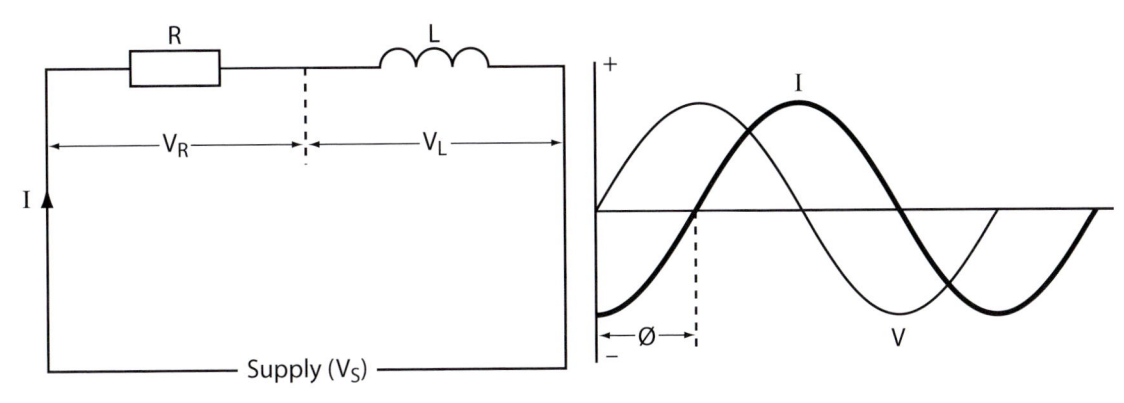

Consider Figure 8.60.

Figure 8.60 Resistor and inductor in series

Here we have a resistor connected in series with an inductor and fed from an a.c. supply. In a series circuit, the current (I) will be common to both the resistor and the inductor, causing **voltage drop** V_R across the resistor and V_L across the inductor.

The sum of these voltages must equal the supply voltage. Here's how to construct a phasor diagram for this circuit.

In a series circuit we know that current will be common to both the resistor and the inductor. It therefore makes sense to use current as our reference phasor. We also know that voltage and current will be in phase for a resistor. Therefore, the volt drop (**p.d.**) V_R across the resistor must be in phase with the current. Also, in an inductive circuit, the current lags the voltage by 90°.

If the current is lagging voltage, then we must be right in saying that voltage is leading the current.

This means in this case that the volt drop across the inductor (V_L) will lead the current by 90°. We can then find the value of the supply voltage (V_S), by completing the parallelogram that was discussed on page 321, Figure 8.59. When we draw phasors, we always assume that they rotate anticlockwise and the symbol Ø represents the phase angle.

There are two ways of doing the drawing (Figure 8.61).

We can see that in the second example, the phasors produce a right-angled triangle. We can therefore use Pythagoras' Theorem to give us the formula:

$$V_S^2 = V_R^2 + V_L^2$$

Remember

Because this is an a.c. circuit, you cannot just add the voltages together as you would have done for a d.c. circuit, you need to construct a phasor diagram to work it out.

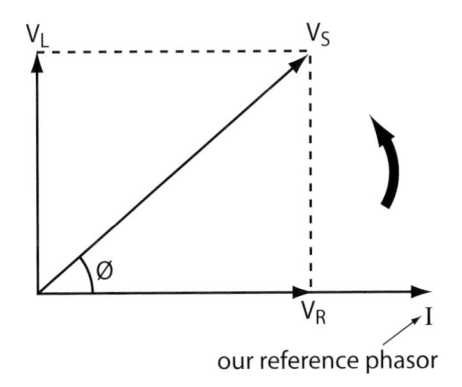

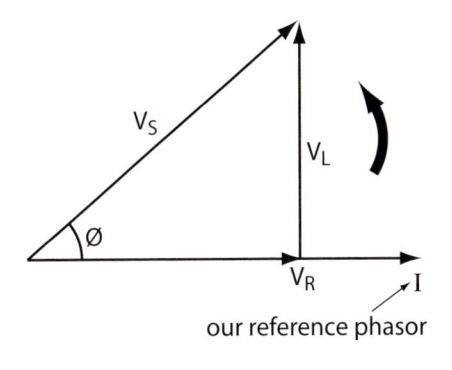

Figure 8.61 Two ways of drawing a phasor diagram

We can then use trigonometry to give us the different formulae, dependent on the values that we have been given:

$$\cos\emptyset = \frac{V_R}{V_S} \qquad \sin\emptyset = \frac{V_L}{V_S} \qquad \tan\emptyset = \frac{V_L}{V_R}$$

Example 1

A coil of 0.15 H is connected in series with a 50 Ω resistor across a 100 V 50 Hz supply. Calculate the following:

a. The inductive reactance of the coil
b. The impedance of the circuit
c. The circuit current

a. Inductive reactance (X_L)

For inductive reactance, we use the formula

$X_L = 2\pi fL$ (Ω)

Inserting the values, this would give us

$X_L = 2 \times 3.142 \times 50 \times 0.15$ therefore X_L = **47.13 Ω**

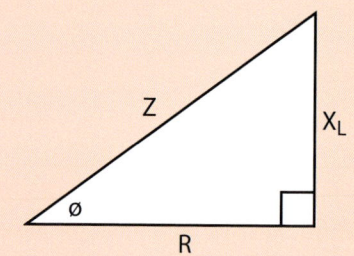

Figure 8.62 Impedance triangle

b. Circuit impedance (Z)

When we have resistance and inductance in series, we calculate the impedance using the following formula:

$Z^2 = R^2 + X_L^2$

which becomes

$Z = \sqrt{R^2 + X_L^2}$

In the case of the first formula, isn't this the same as Pythagoras' Theorem for a right-angled triangle ($A^2 = B^2 + C^2$)? We therefore sometimes refer to this as the impedance triangle and it can be drawn for this type of circuit, as shown. Here, the angle (Ø) between sides R and Z is the same as the phase angle between sides R and Z, is the same as the phase angle between current and voltage.

If we therefore apply some trigonometry, the following applies:

$$\cos\emptyset = \frac{R}{Z} \qquad \sin\emptyset = \frac{X_L}{Z} \qquad \tan\emptyset = \frac{X_L}{R}$$

However, using our formula

$Z = \sqrt{R^2 + X_L^2}$

then $Z = \sqrt{50^2 + 47.1^2}$ therefore Z = **68.69 Ω**

c. Circuit current (I)

As we are referring to the total opposition to current, we use the formula

$I = \dfrac{V}{R} = \dfrac{100}{68.69}$ = **1.46 A**

Example 2

A coil of 0.159 H is connected in series with a 100 Ω resistor across a 230 V 50 Hz supply. Calculate the following:

a. The inductive reactance of the coil
b. The circuit impedance
c. The circuit current
d. The p.d. across each component
e. The circuit phase angle

a. Inductive reactance (X_L)

$X_L = 2\pi fL$ therefore $X_L = 2 \times 3.142 \times 50 \times 0.159 =$ **50 Ω**

b. Circuit impedance (Z)

$Z = \sqrt{R^2 + X_L^2}$ therefore $Z = \sqrt{100^2 + 50^2} =$ **111.8 Ω**

c. Circuit current (I)

$I = \dfrac{V}{Z}$ therefore $I = \dfrac{230}{111.8} =$ **2.06 A**

d. The p.d. across each component (V)

$V_R = I \times R$ therefore $V = 2.06 \times 100 =$ **206 V**
$V_L = I \times X_L$ therefore $V = 2.06 \times 50 =$ **103 V**

e. Circuit phase angle (Ø)

Using our right-angled triangle:

$\tan\varnothing = \dfrac{V_L}{V_R}$ therefore $\tan\varnothing = \dfrac{103}{206} =$ **0.5**

If you then key [INV] [TAN] [0] [.] [5] [=] into your calculator you should get the number 26.6.

Therefore, the current is lagging voltage by **26.6°**

Resistance and capacitance in series (RC)

Figure 8.63 shows a resistor connected in series with a capacitor and fed from an a.c. supply. Once again, in a series circuit the current (I) will be common to both the resistor and the capacitor, causing voltage to drop (p.d.) V_R across the resistor and V_C across the capacitor.

As with the resistance/inductance (RL) circuit

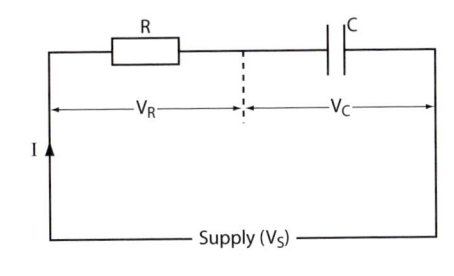

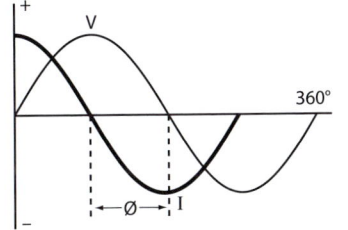

Figure 8.63 A resistor connected in series with a capacitor, fed from an a.c. supply

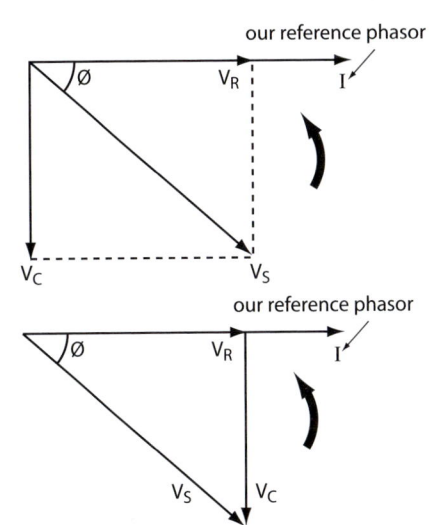

Figure 8.64 Phasor diagrams

previously, we can take current as the reference phasor. The voltage across the resistor will be in phase with that current. In a capacitive circuit the current leads the voltage by 90° so the voltage across the capacitor will be lagging the current. Now calculate the supply voltage (V_S) by completion of the parallelograms (see Figure 8.64). As with the inductor, we can apply Pythagoras' Theorem and trigonometry to give us the following formulae:

$$V_S^2 = V_R^2 + V_C^2$$

$$\cos\emptyset = \frac{V_R}{V_S} \quad \sin\emptyset = \frac{V_C}{V_S} \quad \tan\emptyset = \frac{V_C}{V_R}$$

Example

A capacitor of 15.9 μF and a 100 Ω resistor are connected in series across a 230 V 50 Hz supply. Calculate:

a. The circuit impedance
b. The circuit current
c. The p.d. across each component
d. The circuit phase angle

a. Circuit impedance (Z)

To be able to find the impedance we must first find the capacitive reactance.

$$X_C = \frac{1}{2\pi fC}$$

However, as the capacitor value is given in μF, we use $X_C = \frac{10^6}{2\pi fC}$

This gives us: $X_C = \frac{10^6}{2 \times 3.142 \times 50 \times 15.9} = \frac{10^6}{4995.76} = \mathbf{200\ \Omega}$

When we have resistance and capacitance in series, we use the following formula:

$Z^2 = R^2 + X_C^2$ which becomes $Z = \sqrt{R^2 + X_C^2}$

therefore $Z = \sqrt{100^2 + 200^2} = \sqrt{50\ 000} = \mathbf{224\ \Omega}$

b. Circuit current (I)

$I = \frac{V}{Z}$ therefore $I = \frac{230}{224} = \mathbf{1.03\ A}$

c. The p.d. across each component (V)

$V_R = I \times R$ therefore $V_R = 1.03 \times 100 = \mathbf{103\ V}$
$V_C = I \times X_L$ therefore $V_C = 1.03 \times 200 = \mathbf{206\ V}$

d. Circuit phase angle (Ø)

Using our right-angled triangle: $\tan\emptyset = \frac{V_C}{V_R}$ therefore $\tan\emptyset = \frac{206}{103} = 2$

If you then key [INV] [TAN] [2] [=] into your calculator you should get the number 63.4. Therefore, the current is leading voltage by **63.4°**

Resistance, inductance and capacitance in series (RLC)

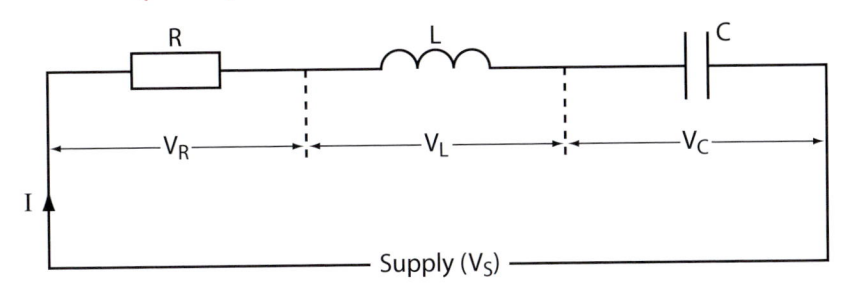

Figure 8.65 Resistor connected in series with an inductor and capacitor, fed from an a.c. supply

Figure 8.65 shows a resistor connected in series with an inductor and a capacitor then fed from an a.c. supply. This is often referred to as an RLC circuit or a **general series circuit**. Again, as we have a series circuit, the current (I) will be common to all three components, causing a voltage drop (p.d.) V_R across the resistor, V_L across the inductor and V_C across the capacitor.

Here V_R will be in phase with the current, V_L will lead the current by 90° (because the current lags the voltage) and V_C will lag the current by 90° (because current leads the voltage in a capacitive circuit). Because V_L and V_C are in opposition to each other (one leads and one lags), the actual effect will be the difference between their values, subtracting the smaller from the larger. We can once again calculate V_S by completing a parallelogram as shown in Figure 8.66.

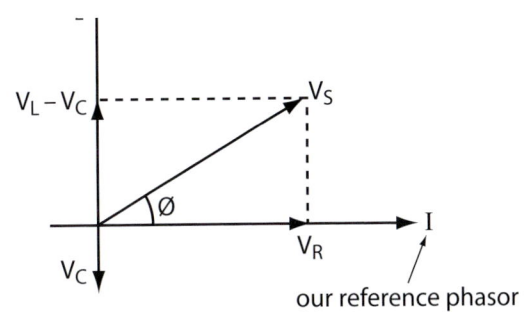

Figure 8.66 Phasor parallelogram

As before, we can now apply Pythagoras' Theorem and trigonometry to give us the following formulae depending on whether V_L or V_C is the larger:

$$V_S^2 = V_R^2 + (V_L - V_C)^2 \quad \text{or} \quad V_S^2 = V_R^2 + (V_C - V_L)^2$$

and because I is the same for each component we get:

$$Z = \sqrt{R^2 + (X_L - X_C)^2} \quad \text{or} \quad Z = \sqrt{R^2 + (X_C - X_L)^2}$$

and finally: $\cos\varnothing = \dfrac{V_R}{V_S} \quad \sin\varnothing = \dfrac{V_L - V_C}{V_S} \quad \tan\varnothing = \dfrac{V_L - V_C}{V_R}$

Example

A resistor of 5 Ω is connected in series with an inductor of 0.02 H and a capacitor of 150 µF across a 250 V 50 Hz supply. Calculate the following.

a. The impedance
b. The supply current
c. The power factor

a. Impedance (Z)

In order to find the impedance, we must first find out the relevant values of reactance.

Therefore: $X_L = 2\pi fL = 2 \times 3.142 \times 50 \times 0.02 =$ **6.28 Ω**

$X_C = \dfrac{1}{2\pi fC}$ allowing for microfarads

$= \dfrac{10^6}{2 \times 3.142 \times 50 \times 150} =$ **21.2 Ω**

If you remember, we said that the effect of inductance and capacitance together in series would be the difference between their values. Consequently, this means that the resulting reactance (X) will be found as follows:

$X = X_L - X_C$ or, in this case becasue X_C is the larger: $X = X_C - X_L$

and therefore: $X = 21.2 - 6.28 =$ **14.92 Ω**

We can now use the impedance formula as follows:

$Z = \sqrt{R^2 + (X_L - X_C)^2}$ or $Z = \sqrt{R^2 + (X_C - X_L)^2}$

which gives us

$Z = \sqrt{5^2 + 14.92^2} =$ **15.74 Ω**

Note: X_C is greater than X_L. Therefore we subtract X_L from X_C. Had X_L been the higher, then the reverse would be true. Also, as capacitive reactance is highest, the circuit current will lead the voltage. Had the inductive reactance been the higher, then the current would lag the voltage.

b. Supply current (I)

$I = \dfrac{V}{Z} = \dfrac{230}{15.74} =$ **14.6 A**

c. Power factor (Ø)

$\cos\varnothing = \dfrac{R}{Z} = \dfrac{5}{15.74} = $ **0.32**

Therefore PF = **0.32 leading**

Resistance, inductance and capacitance in parallel

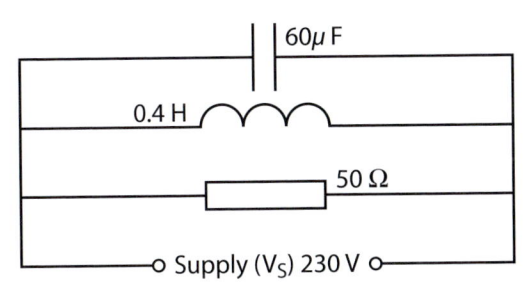

Consider Figure 8.67.

Figure 8.67 Resistor, capacitor and inductor in parallel connected to an a.c. supply

There can obviously be any combination of the above components in parallel. However, to demonstrate the principles involved, we will look at all three connected across an a.c. supply.

As we have a parallel circuit, the voltage (V_S) will be common to all branches of the circuit. Consequently, when we draw our parallelogram we will use voltage as the reference phasor.

In this type of circuit, the current through the resistor will be in phase with the voltage, the current through the inductor will lag the voltage by 90° and the current through the capacitor will lead the voltage by 90°.

Normally, when we carry out calculations for parallel circuits, it is easier to treat each branch as being a separate series circuit. We then draw to scale each of the respective currents and their relationships to our reference phasor, which is voltage.

As with voltage V_L and V_S in the RLC series circuit, the current through the inductor (I_L) and the current through the capacitor (I_C) are in complete opposition to each other. Therefore, the actual effect will be the difference between their two values. We calculate this value by the completion of our parallelogram. But, bear in mind that the bigger value (I_C or I_L) will determine whether the current ends up leading or lagging.

If we use the diagram as the basis for our example, calculate the circuit current and its phase angle relative to the voltage.

To calculate the circuit current and its phase angle relative to the voltage, we must first find the current through each branch:

$$I_R = \frac{V}{R} = \frac{230}{50} = \mathbf{4.6\ A}$$

$$X_L = 2\pi fl = 2 \times 3.142 \times 50 \times 0.4 = \mathbf{126\ \Omega}$$

Therefore $I_L = \dfrac{V}{X_L} = \dfrac{230}{126} = \mathbf{1.8\ A}$

$X_C = \dfrac{10^6}{2\pi fC} = \dfrac{10^6}{2 \times 3.142 \times 50 \times 60} = \mathbf{53\ \Omega}$

Therefore $I_C = \dfrac{V}{X_C} = \dfrac{230}{53} = \mathbf{4.3\ A}$

The actual effect will be $I_C - I_L$ which gives $4.3 - 1.8 = \mathbf{2.5\ A}$

Now add this to I_R by completing the scale drawing in Figure 8.67. This gives a current of 5.2 A that is leading voltage by an angle of 28°.

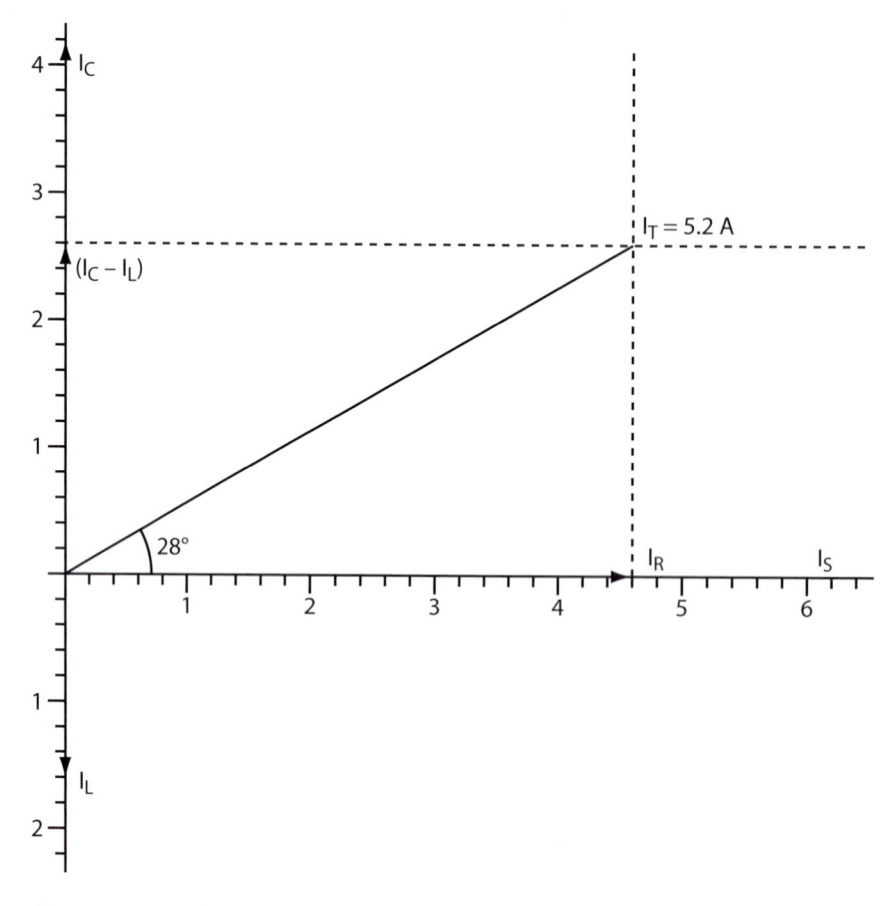

Figure 8.68 Scale drawing

Power in an a.c. circuit

If we were to try to push something against a resistance, we would get hot and bothered as we use up energy in completing the task. When current flows through a resistor, a similar thing happens in that power is the rate of using up energy – in other words, the amount of energy that was used in a certain time.

If a resistor (R) has a current (I) flowing through it for a certain time (t), then the power (energy being used per second) given in watts, can be calculated by the formula:

$P = I^2 \times R$

This power will be dissipated in the resistor as heat and reflects the average power in terms of the r.m.s. values of voltage and current.

What we are effectively saying here is that power is only dissipated by the resistance of a circuit and that the average power in a resistive circuit (one which is non-reactive, i.e. doesn't possess inductance or capacitance) can be found by the product of the readings of an ammeter and a voltmeter.

In other words, in the resistive circuit, the power (energy used per second) is associated with that energy being transferred from the medium of electricity into another medium, such as light (filament lamp) or heat (electric fire/kettle). We call this type of power the active power. When we look at the capacitive circuit, we find that current flows to the capacitor, but we have no power.

Look at this wave diagram for voltage in Figure 8.69.

During the first period of the cycle, the voltage is increasing and this provides the energy to charge the capacitor. However, in the second period of the cycle, the voltage is decreasing and therefore the capacitor discharges, returning its energy back to the circuit as it does so. The same is also true of the third and fourth periods.

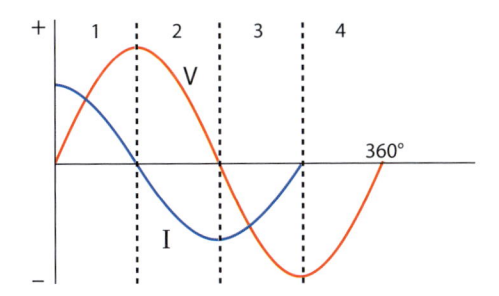

Figure 8.69 Sine wave diagram

This exchange of energy means that we have voltage and current, but no average power and therefore no heating effect. This means that our previous formula ($P = I^2 \times R$) is no longer useful.

We therefore say that the result of voltage and current in this type of circuit is called **reactive power** and we express this in reactive voltamperes (VAr). Equally, we say that the current in a capacitive circuit, where there is no resistance and no dissipation of energy, is called **reactive current**.

When we come to the inductive circuit we have a similar position. This time, as voltage increases during the first period of the cycle, the energy is stored as a magnetic field in the inductor. This energy will then be fed back into the circuit during the second period of the cycle as the voltage decreases and the magnetic field collapses. In other words, once again the exchange of energy produces no average power (energy used per second).

Circuits are likely to comprise combinations of resistance, inductance and capacitance. In a circuit, which has resistance and reactance, there will be a phase angle between the voltage and current. This relationship has relevance, as power will only be expended in the resistive part of the circuit.

Let's now look at this relationship via a phasor diagram (see Figure 8.70), for a circuit containing resistance and capacitance, remembering that for this type of circuit the current will lead the voltage.

This circuit has two components (resistance and capacitance), so we've drawn two current phasors. In reality, as we already know, these are not currents that will actually flow, but their phasor sum will be the actual current in the circuit I_p(actual).

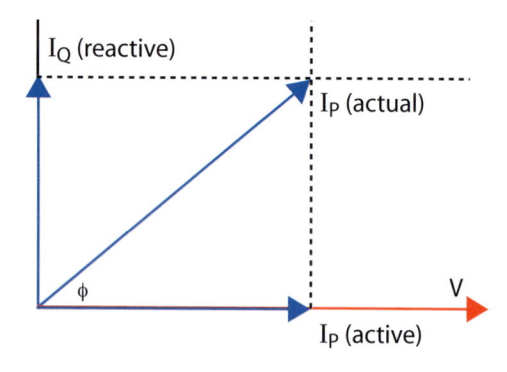

Figure 8.70 Phasor diagram

We also said that in a resistive circuit the voltage and current are in phase, and therefore this section of the current has been represented by the phasor I_p(active). This part of the current is in the active section and we therefore refer to this as the active current.

We then show that part of the current in the reactive section (capacitor) as leading the voltage by 90° and this has been represented by the phasor I_Q(reactive). As stated previously, we refer to this as the reactive current.

We also know that for a resistive circuit, we calculate the power by multiplying together the r.m.s. values of voltage and current (V × I).

Logically, as we know that no power is consumed in the reactive section of the circuit, we can therefore calculate the power in the circuit by multiplying together the r.m.s. value of voltage and the value of current, which is in phase with it (I_p(active)). This would give us the formula:

$P = V \times I_p$

But as the actual current will be affected by the reactive current and therefore the phase angle (cos Ø), our formula becomes:

$P = V \times I_p \text{(active)} \times \cos Ø$

If I_p (active) is zero then $\emptyset = 90°$ and $\cos \emptyset = 0$. Therefore $P = 0$.

We have now established that it is possible in an a.c. circuit for current to flow, but no power to exist.

We also say that the product of voltage and current is power given in watts. However, it would be fair to say that this is not the actual power of the circuit.

The actual (true) power of the circuit has to take on board the effect of the phase angle ($\cos \emptyset$), the ratio of these two statements being the power factor. In other words:

$$\text{Power Factor} (\cos\emptyset) = \frac{\text{True power (P)}}{\text{Apparent power (S)}}$$

$$= \frac{V \times I_p(\text{actual}) \times \cos\emptyset}{V \times I_p(\text{actual})} = \frac{\text{watts}}{\text{volts-amperes}}$$

To summarise, what we perceive to be the power of a circuit (the apparent power) can also be the true power, as long as we have a unity power factor (1.0). However, as long as we have a phase angle, then we have a difference between apparent power and reality (true power). This difference is the power factor (a value less than unity).

In reality we will use a wattmeter to measure the true power and a voltmeter and ammeter to measure the apparent power.

The power triangle

We can use Pythagoras' Theorem to help us calculate the different power components within a circuit.

> **Remember**
>
> **Pythagoras' Theorem:** The square of the hypotenuse is equal to the sum of the square on the other two sides.

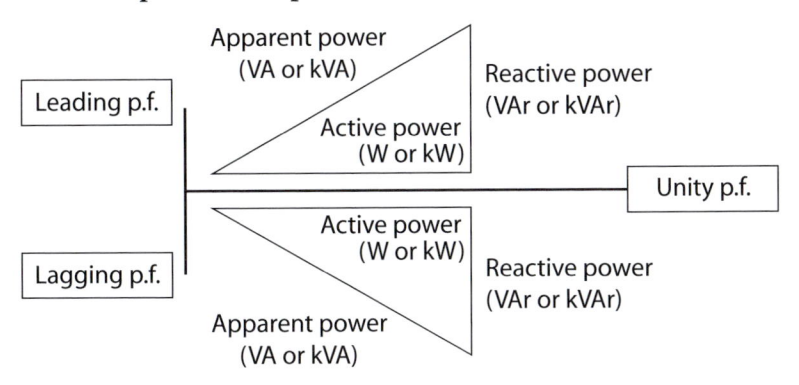

Figure 8.71 Power triangle

We do so by using Pythagoras' formula as follows:

$(VA)^2 = (W)^2 + (VAr)^2$

This is then applied in Figure 8.71, where we have shown both inductive and capacitive conditions.

Example

A resistor of 15 Ω has been connected in series with a capacitor of reactance 30 Ω. If they are connected across a 230 V supply, establish both by calculation and by drawing a scaled power triangle, the following:

(a) the apparent power
(b) the true power
(c) the reactive power
(d) the power factor.

By calculation

In order to establish the elements of power, we must first find the current. To do this we need to find the impedance of the circuit.

Therefore:

$$Z = \sqrt{R^2 + X_C^2} = \sqrt{15^2 + 30^2} = \sqrt{225 + 900} = \sqrt{1125}$$

Therefore Z = 33.5 Ω

$$I = \frac{V}{Z} = \frac{230}{35.5} = 6.9 \text{ A}$$

$$Pf(\cos \emptyset) = \frac{R}{Z} = \frac{15}{35.5} = 0.45$$

True power   = I × V × Pf
                         = 6.9 × 230 × 0.45
                         = 714.15 W   or   0.714 kW

Apparent power   = I × V
                             = 6.9 × 230
                             = 1587 VA   or   1.587 kVA

Reactive power   = sin Ø × apparent power

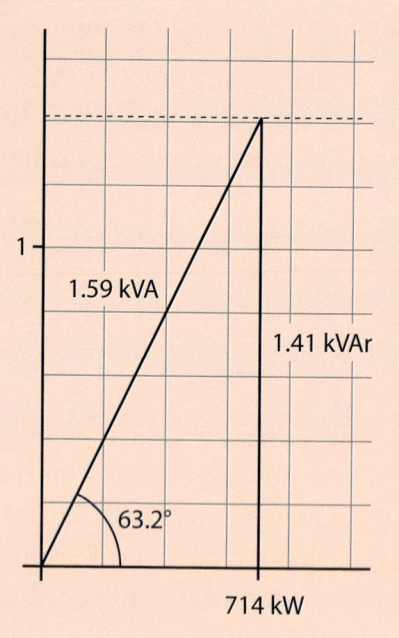

However, we do not know the value of sin Ø, we must therefore convert cos Ø to an angle and then find the sine of that angle.

cos Ø = 0.45   INV cos of 0.45 = 63.2
sine of 63.2° = 0.89

Therefore Reactive power   = sin Ø × apparent power
                                       = 0.89 × 1587
                                       = **1412.43 VAr**   or   **1.412 kVAr**

Power triangle: 1.59 kVA (hypotenuse), 1.41 kVAr (vertical), 714 kW (horizontal), angle 63.2°.

Characteristics of electrical supplies

We have already looked at the characteristics of single-phase electrical supplies earlier in this unit. In such a circuit, we normally use two conductors, where one delivers current and one returns it. In this section we will look at some of the characteristics of different electrical supplies.

Three-phase supplies

You might assume that we would need six conductors for a three-phase system, with two being used per phase. However, in reality we only use three or four conductors, depending upon the type of connection that is being used.

We call these connections either **star** or **delta**.

Remember that current flows along one conductor and returns along another called the neutral. But what if there were no neutral? And what exactly is the neutral conductor for?

Keeping things simple, when we generate an e.m.f., we do so by spinning a loop of wire inside a magnetic field. To get three phases, we just spin three loops inside the magnetic field. Each loop will be mounted on the same rotating shaft, but they'll be 120° apart.

> **Remember**
>
> Electricity is the movement of electrons along a conductor. The rate of electron flow, the current, is maintained by a force known as the e.m.f.

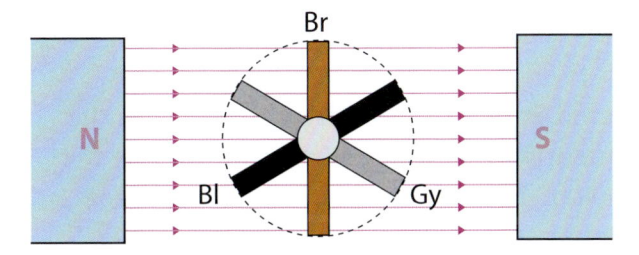

Figure 8.72 Three-phase generation

In Figure 8.72, each loop will create an identical sinusoidal waveform, or in other words, three identical voltages, each 120° apart.

Whether or not we need a neutral will now depend upon the load. If we accept that the e.m.f. being generated by each loop pushes current down the conductors (lines), then we would find that where we have a balanced three-phase system, i.e. one where the current in each of the phases (lines) is the same, then by phasor addition, we would find that the resultant current is zero.

If the current is zero then we do not need a neutral, because the neutral is used to carry the current in an out-of-balance system. Let us now look at the different types of connection.

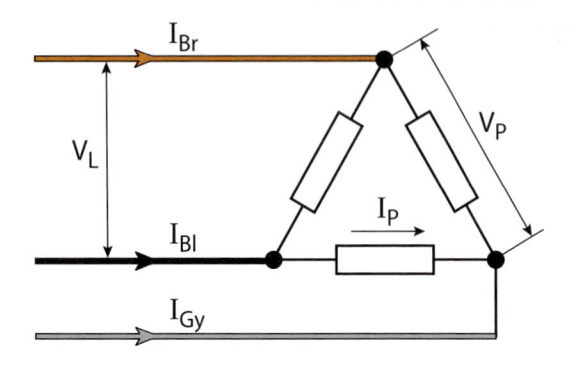

Figure 8.73 Delta connection

Delta connection

We tend to use the delta connection when we have a balanced load. This is because there is no need for a neutral connection and therefore only three wires are needed. We tend to find that this configuration is used for power transmission from power stations or to connect the windings of a three-phase motor.

In Figure 8.73 we have shown a three-phase load, which has been delta connected. You can see that each leg of the load is connected across two of the lines, e.g. Br–Gy, Gy–Bl and Bl–Br. We refer to the connection between phases as being the line voltage and have shown this on the drawing as V_L.

Equally, if each line voltage is pushing current along, we refer to these currents as being line currents, which are represented on the drawing as I_{Br}, I_{Gy} or I_{Bl}. These line currents are calculated as being the phasor sum of two phase currents, which are shown on the drawing as I_P and represent the current in each leg of the load. Similarly, the voltage across each leg of the load is referred to as the phase voltage (V_P).

In a delta connected balanced three-phase load, we are then able to state the following formulae:

$$V_L = V_P \qquad I_L = \sqrt{3} \times I_P$$

Note that a load connected in delta would draw three times the line current and consequently three times as much power as the same load connected in star. For this reason, induction motors are sometimes connected in **star–delta**.

This means they start off in a star connection (with a reduced starting current) and are then switched to delta. In doing so we reduce the heat that would otherwise be generated in the windings.

Star connection

Although we can have a balanced load connected in star (three-wire), we tend to use the star connection when we have an unbalanced load, i.e. one where the current in each of the phases is different. In this circumstance, one end of each of the three star connected loops is connected to a central point and it is then from this point that we take our neutral connection, which in turn is normally connected to earth. This is the three-phase four-wire system.

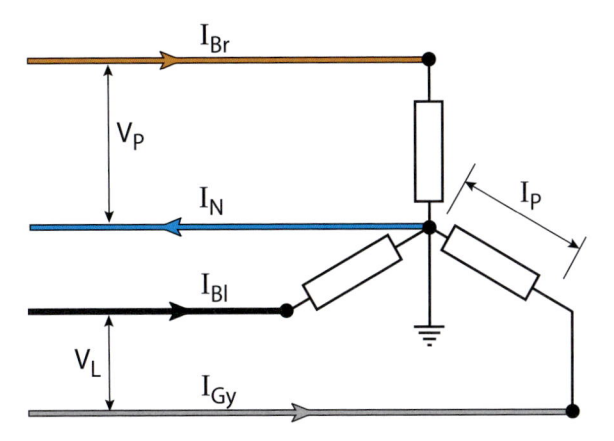

Figure 8.74 Star connection

Another advantage of the star connected system is that it allows us to have two voltages – one when we connect between any two phases (400 V) and another when we connect between any phase and neutral (230 V). You should note that we will also have 230 V between any phase and earth.

In Figure 8.74 we have shown a three-phase load that has been star connected.

As with delta, we refer to the connection made between phases as the line voltage and have shown this on the drawing as V_L. However, unlike delta, the **phase voltage** exists between any phase conductor and the neutral conductor and we have shown this as V_P. Our line currents have been represented by I_{Br}, I_{Gy} and I_{Bl} with the phase currents being represented by I_P.

In a star connected load, the line currents and phase currents are the same, but the line voltage (400 V) is greater than the phase voltage (230 V).

In a star connected load, we are therefore able to state the following formulae:

$$I_L = I_P \qquad V_L = \sqrt{3} \times V_P$$

Using the star connected load we have access to a 230 V supply, which we use in most domestic and low load situations.

Example 1

A three-phase star connected supply feeds a delta-connected load as shown in the diagram below.

If the star-connected phase voltage is 230 V and the phase current is 20 A, calculate the following:

- the line voltages and line currents in the star connection
- the line and phase voltages and currents in the delta connection.

The star connection

In a star system the line current (I_L) is equal to the phase current (I_P). Therefore if we have been given I_P as 20 A, then I_L must also be 20 A.

We find line voltage in a star connection using the formula:

$$V_L = \sqrt{3} \times V_P$$

$\sqrt{3}$, the square root of 3, is a constant having the value 1.732. Therefore if we substitute our values, we get:

$V_L = 1.732 \times 230 =$ **398 V**

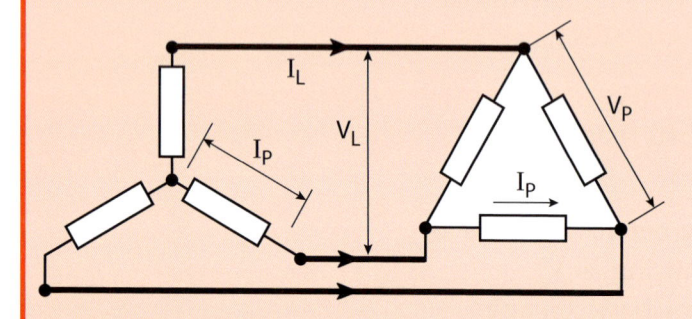

The delta connection

In a delta system, the line current (I_L) is 1.732 times greater than the phase current (I_P). We calculate this using the formula: $I_L = \sqrt{3} \times I_P$

However, we know that I_L is 20 A, so if transpose our formula and substitute our values, we get:

$$I_P = \frac{I_L}{\sqrt{3}} = \frac{20}{1.732} = 11.5 A$$

We know that for a delta connection line voltage and phase voltage have the same values.
Therefore: $V_L = V_P =$ **398 V**

Example 2

Three identical loads of 30 Ω resistance are connected to a 400 V three-phase supply. Calculate the phase and line currents if the loads were connected:

(a) in star (b) in delta.

Star connection

First we need to establish the phase voltage. If $V_L = \sqrt{3} \times V_P$ and $\sqrt{3} = 1.732$ then by transposition:

$$V_P = \frac{V_L}{\sqrt{3}} = \frac{400}{1.732} = 230.9 \text{ V}$$

Using Ohm's Law: $I_P = \frac{V_P}{Z} = \frac{230.9}{30} = \textbf{7.7 A}$

However, in a star-connected load, $I_P = I_L$, therefore I_L will also = 7.7 A.

Delta connection

In a delta connection $V_L = V_P$ and therefore we know the phase voltage will be 400 V.

Using Ohm's Law: $I_P = \frac{V_P}{Z} = \frac{400}{30} = \textbf{13.33 A}$

However, the line current:

$I_L = \sqrt{3} \times I_P = 1.732 \times 8 = \textbf{23.09 A}$

As can be seen from this example, the current drawn from a delta connected load (13.33 A) is three times that of a star-connected load (7.7 A).

Neutral currents

As we have already discussed, where we have a balanced load, we can have a three-phase system with three wires. However, in truth it is more likely that we will find an unbalanced system and will therefore need to use a three-phase four-wire system.

In such a system, we are saying that each line (Br, Gy, Bl) will have an unequal load and therefore the current in each line can be different. It therefore becomes the job of the neutral conductor to carry the out of balance current. If we used **Kirchhoff's Law** in this situation, we would find that the current in the neutral is normally found by the phasor addition of the currents in the three lines.

> **Key term**
>
> **Kirchhoff's Law** – the sum of the voltage drops around a closed loop in the network must equal zero

Example 3

For a three-phase four-wire system, the line currents are found to be $I_R = 30$ A and in phase with V_{Br}, $I_{Gy} = 20$ A and leading V_{Gy} by 20° and $I_{Bl} = 25$ A and lagging V_{Bl} by 10°. Calculate the current in the neutral by phasor addition.

The phasor diagram for this example has been provided. However, you should note that in order to establish the current in the neutral (I_N), you would need to draw two parallelograms. The first should represent the resultant currents I_{Br} and I_{Gy}. The second, I_N, should be drawn between this resultant and the current in I_{Bl}.

When this has been done to scale, we should find a current in the neutral of 18 A – look at Figure 8.75 below.

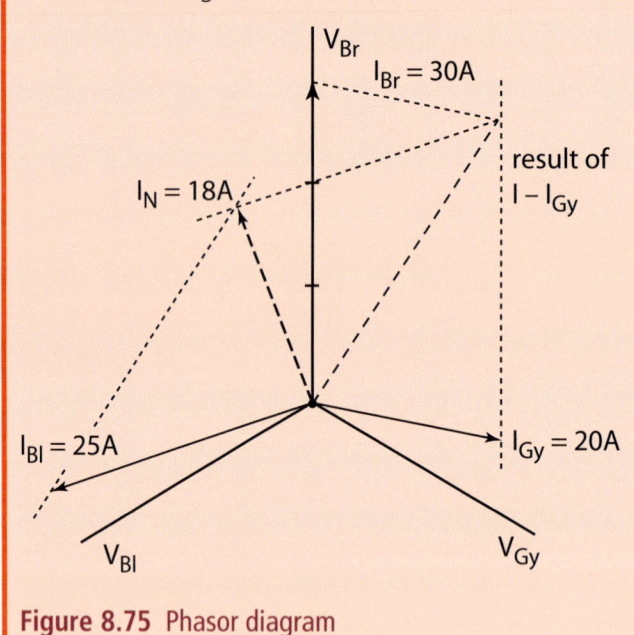

Figure 8.75 Phasor diagram

Load balancing

The regional electricity companies (now commonly known as District Network Operators – DNO) also require load balancing as a condition of their electricity supply, because it is important to try to achieve balanced currents in the mains distribution system.

In order to design a three-phase four-wire electrical installation for both efficiency and economy, it needs to be subdivided into load categories. By doing this, the maximum demand can be assessed and items of equipment can be spread over all three phases of the supply to achieve a balanced system. This section looks at load balancing on three-phase systems. The designer needs to make a careful assessment of the various installed loads, which in turn leads to the proper sizing of the main cable and associated switchgear.

Standard circuit arrangements exist for many final circuits operating at 230 V. For example, a ring final circuit is rated at 30–32 A, a lighting circuit at 5–6 A, and on a cooking appliance is rated at 30–45 A. Where more than one standard circuit arrangement is present, such as three ring final circuits and/or two cooking appliances, then a diversity allowance can be applied.

Once the designer has made these allowances for diversity, the single-phase loads can be evenly spread over all three phases of the supply so that each phase takes approximately the same amount of current. If this is done carefully, minimum current will flow along the neutral conductor; the sizes of cables and switchgear can be kept to a minimum, thus reducing costs, and the system is therefore said to be reasonably well balanced.

Example

A small guesthouse with 10 bedrooms is supplied with a 400/230 V, three-phase four-wire supply. It has the following installed loads:

20 × filament lighting points each rated at 100 W
6 × ring final circuits supplying 13 socket outlets
6 × 4 kW showers (instantaneous type)
3 × 3 kW immersion heaters
2 × 10 kW cookers.

Apply diversity as required by the IET On-site Guide Table 1B and 'spread' the loads evenly over the three-phase supply to produce the most effective load-balanced situation.

Filament lamps

$$\frac{2000 \text{ W}}{230 \text{ V}} = 8.69 \text{ A @ } 75\% = 6.5 \text{ A (after diversity)}$$

Ring final circuits

Spread evenly at two per phase on each phase:

First ring = 100% = 30 A
Second ring = 50% = 15 A (after diversity)
Total = 45 A

Showers

Spread evenly at two per phase both at 100%:

$$\frac{4000 \text{ W}}{230 \text{ V}} = 17.4 \text{ A} \times 2 = 34.8 \text{ A}$$

Example continued

Immersion heaters

Spread as required over the phases (no diversity):

$$\frac{3000 \text{ W}}{230 \text{ V}} = 13 \text{ A}$$

Cookers

If both cookers are on the same phase then:

$$100\% \text{ and } 80\% = \frac{10000 \text{ W}}{230 \text{ V}} = 43 \text{ A for } 100\%$$

$$= 35 \text{ A for } 80\% \text{ (after diversity)}$$

Or if both on different phases both are at 100% (this is what will be used).

Based on these calculations, a suggested load balancing is shown below. Note the immersion heaters have all been allocated to the Blue phase.

	Brown	Grey	Black
Ring circuits	45 A	45 A	45 A
Shower	34.8 A	34.8 A	34.8 A
Immersion			13 A
			13 A
			13 A
Cooker	43 A	43 A	
Lighting			6.5 A
Totals	122.8 A	122.8 A	125.3 A

Also, should one phase be 'lost', only a portion of the lighting will fail. If this is the case then care must be taken, as there could be three-phase voltage values present in multi-gang switches. It is also advisable to position distribution boards in large factories and commercial premises as near to the load centre as possible. This will help reduce voltage drop and make the installation more cost-effective.

The load balancing shown above provides satisfactory balancing over three phases, but it may be recommended that if discharge lighting is to be used, then it should be spread over all three phases because of the stroboscopic effect.

Power in three-phase supplies

As we have seen in the previous section, we can find the power in a single-phase a.c. circuit by using the following formula:

Power = V × I × cos Ø

Logically, you might assume that for three-phase we could multiply this formula by three. Although this is not far from the truth we must remember that this could only apply where we have a balanced three-phase load. We can therefore say, that for any three-phase balanced load, the formula to establish power is:

$$\text{Power} = \sqrt{3} \times (V_L \times I_L \times \cos \varnothing)$$

However, in the case of an unbalanced load, we need to calculate the power for each separate section and then add them together to get total power (see pages 330–3 for calculating power).

Remember

We have already looked at determining the neutral current in a three-phase supply and the values of voltage and current in star and delta connected systems on pages 335–340.

Example 1

A balanced load of 10 Ω per phase is star connected and supplied with 400 V 50 Hz at unity power factor. Calculate the following:

a. Phase voltage

b. Line current

c. Total power consumed

a. Phase voltage

$$V_L = \sqrt{3} \times V_P$$

Therefore by transposition

$$V_P = \frac{V_L}{\sqrt{3}} = \frac{400}{1.732} = \mathbf{231\ V}$$

b. Line current

$$I_L = I_P$$

and therefore

$$I_P = \frac{V_P}{R_P} = \frac{231}{10} = \mathbf{23.1\ A}$$

c. Total power consumed

In a balanced system

Power = $\sqrt{3} \times V_L \times I_L \times \cos \varnothing$

Power = 1.732 × 400 × 23.1 × 1 = **16 kW**

Example 2

Three coils of resistance 40 Ω and inductive reactance 30 Ω are connected in delta to a 400 V 50 Hz three-phase supply. Calculate the following:

a. **Current in each coil**

b. **Line current**

c. **Total power**

a. **Current in each coil**

We must first find the impedance of each coil:

$$Z = \sqrt{R^2 + X_L^2} = \sqrt{40^2 + 30^2} = \sqrt{2500} = 50\,\Omega$$

The current in each coil (I_P) can then be found by applying Ohm's Law:

This gives $I_P = \dfrac{V}{Z} = \dfrac{400}{50} =$ **8 A**

b. **Line current**

For a delta connected system

$$I_L = \sqrt{3} \times I_P$$

Therefore $I_L = 1.732 \times 8 =$ **13.86 A**

c. **Total power**

We must first find the power factor using the formula:

$$\cos\varnothing = \dfrac{R}{Z}$$

This gives us $\cos\varnothing = \dfrac{40}{50} = 0.8$

And for a delta system $V_L = V_P$

Therefore we can now use the power formula of:

$$P = \sqrt{3} \times V_L \times I_L \times \cos\varnothing$$

P = 1.732 × 400 × 13.86 × 0.8

P = **7682 W** or **7.682 kW**

Example 3

A small industrial estate is fed by a 400 V, three-phase, 4-wire TN-S system. On the estate there are three factories connected to the system as follows:

Factory A taking 50 kW at unity power factor
Factory B taking 80 kVA at 0.6 lagging power factor
Factory C taking 40 kVA at 0.7 leading power factor

Calculate the overall kW, kVA, kVar and power factor for the system.

To clarify, we are trying to find values of P (true power), S (apparent power) and Q (reactive power). First, we need to work out the situations for each factory.

Factory A

We know that power factor

$$\cos\varnothing = \frac{\text{True power}(P)}{\text{Apparent power}(S)}$$

We also know that the power factor is 1.0 and that P = 50 kW. Therefore, by transposition:

$$S = \frac{P}{\cos\varnothing} = \frac{50}{1} = 50 \text{ kVA}$$

And with unity power factor for Factory A, Q = 0

Factory B

Using the same logic, we need to find true power and reactive power.

Therefore P = cos Ø × S = 0.6 × 80 kW = 48 kW

Reactive component (Q) = S × sin Ø = 80 × 0.8 = 64 kVAr

Factory C

P = S × cos Ø = 40 kW × 0.7 = 28 kW
Q = S × sin Ø = 40 × 0.714 = 28.6 kVAr

We can now find the **total kW** by addition: 50 + 48 + 28 = **126 kW**

We can find the **total kVAr** as the difference between the reactive power components, the larger one, Factory B, lagging and the smaller one, Factory C, leading:

64 kVAr − 28.6 kVAr = **35.4 kVAr**

We can use Pythagoras' Theorem to find the total kVA

$$S = \sqrt{P^2 + Q^2} = \sqrt{126^2 + 35.4^2} = \textbf{131 kVA}$$

Consequently, the overall power factor will be:

$$\cos\varnothing = \frac{P}{S} = \frac{126}{131} = \textbf{0.96 lagging}$$

Power factor correction (PFC)

As we have seen throughout this section, in the eyes of the suppliers it is desirable to have the power factor of a system as close to perfection (1.0) as possible.

Power factor is adjusted by supplying reactive power of the opposite form. For example, if power factor is being dragged down below 1.0 due to the inductive effect of motors, we can add capacitors locally to control the effect. If the load has a capacitive value, then we add inductors to correct the power factor.

We can therefore generalise to say that inductors remove reactive power from a system and capacitors add reactive power to the system.

Power factor correction may be applied by an electrical distributor to improve the stability and efficiency of the transmission network; a high power factor is generally required in a transmission system to reduce transmission losses and improve voltage regulation at the load.

However, in some installations, an automatic power factor correction unit is used to improve power factor. These normally consist of a bank of capacitors switched in via contactors when the power factor is detected as being higher than a pre-set value.

A power factor correction unit usually consists of a number of capacitors that are switched by contactors. These contactors are controlled by a regulator that measures power factor in an electrical network. To be able to measure power factor, the regulator uses a current transformer to measure the current in one phase.

Large industrial plants also use the effect of the synchronous motor to cancel out the effect of the induction motor, as it provides a leading power factor. Taken one step further, a synchronous condenser is a synchronous motor without a load, which when introduced into the system can compensate either a leading or lagging power factor, by absorbing or supplying reactive power to the system, which in turn enhances voltage regulation.

Operating principles, applications and limitations of transformers

The **transformer** is one of the most widely used pieces of electrical equipment and can be found in situations such as

> **Did you know?**
>
> Power factor correction is a mandatory requirement on large industrial plants and sports stadiums. Failure to produce the required running levels of power factor correction can result in heavy fines from supply authorities.

electricity distribution, construction work and electronic equipment. Its purpose, as the name implies, is to transform something – the something in this case being the voltage, which can enter the transformer at one level (input) and leave at another (output).

When the output voltage is higher than the input voltage we say that we have a step-up transformer and when the output voltage is lower than the input, we say that we have a step-down transformer.

In this section we will be looking at the following areas:

- mutual inductance
- transformer types
- step-up and step-down transformers.

Mutual inductance

In their operation, transformers make use of an action known as **mutual inductance**. We have looked at inductance on pages 317–319 but we will revisit the subject here to help you try to understand the concept clearly. Let us look at the situation in Figure 8.76. Two coils, primary and secondary, are placed side by side, but not touching each other. The primary coil has been connected to an a.c. supply and the secondary coil is connected to a load, such as a resistor.

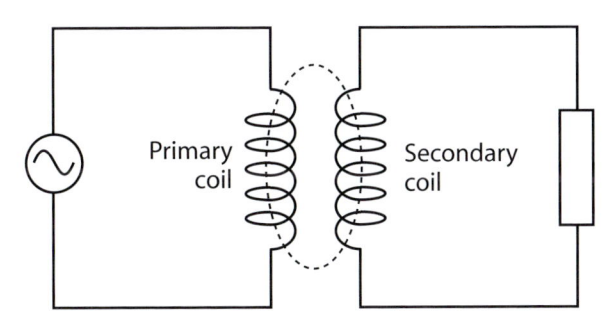

Figure 8.76 Mutual inductance

If we now allow current to flow in the primary coil, we know that current flow will create a magnetic field. Therefore, as the current in the primary coil increases up to its maximum value, it creates a changing magnetic flux and, as long as there is a changing magnetic flux, there will be an e.m.f. 'induced' into the secondary coil, which would then start flowing through the load.

This effect, where an alternating e.m.f. in one coil causes an alternating e.m.f. in another coil, is known as mutual inductance.

> **Did you know?**
>
> There is something called self-inductance. This is when a conductor that is formed into a coil, carries an alternating current. The magnetic field created around the coil grows and collapses with the changing current. This will induce into the coils an e.m.f. voltage that is always in opposition to the voltage that creates it.

In transformers we are only really interested in mutual inductance, and it is the rising and falling a.c. current that causes the change of magnetic flux. In other words, **we need an a.c. supply to allow transformers to operate correctly**.

If the two coils were now wound on an iron core, we would find that the level of magnetic flux is increased and consequently the level of mutual inductance is also increased.

Transformer types

Core-type transformers

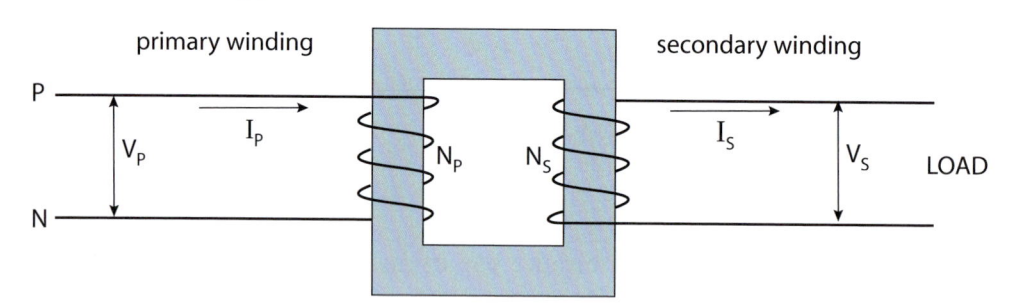

Figure 8.77 Double wound, core-type transformer

In Figure 8.77 the supply is wound on one side of the iron core (primary winding) and the output is wound on the other (secondary winding). In other words, double wound means that there is more than one winding.

The number of turns in each winding will affect the induced e.m.f., with the number of turns in the primary being referred to as (N_p), and those in the secondary referred to as (N_s). We call this the turns ratio. When voltage (V_p) is applied to the primary winding, it will cause a changing magnetic flux to circulate in the core. This changing flux will cause an e.m.f. (V_s) to be induced in the secondary winding.

Assuming that we have no losses or leakage (i.e. 100% efficient), then power input will equal power output and the ratio between the primary and secondary sides of the transformer can be expressed as follows:

$$\frac{V_P}{V_S} = \frac{N_P}{N_S} = \frac{I_S}{I_P}$$

(where I_P represents the current in the primary winding and I_S the current in the secondary winding).

As we can see from Figure 8.77, a transformer has no moving parts. Consequently, provided that the following general statements apply, it ends up being a very efficient piece of equipment:

- Transformers use laminated (layered) steel cores, not solid metal. In a solid metal core 'eddy currents' are induced which cause heating and power losses. Using laminated cores, reduces these eddy currents.
- Laminated cores, where each lamination is insulated, help to reduce this effect.
- Soft iron with high magnetic properties is used for the core.
- Windings are made from insulated, low resistance conductors. This prevents short circuits occurring either within the windings, or to the core.

> **Remember**
> No transformer can be 100% efficient. There will always be power losses.

The losses that occur in transformers can normally be classed under the following categories.

Copper losses

Although windings should be made from low resistance conductors, the resistance of the windings will cause the currents passing through them to create a heating effect and subsequent power loss. This power loss can be calculated using the formula:

$P_C = I^2 \times R$ watts

Iron losses

These losses take place in the magnetic core of the transformer. They are normally caused by eddy currents (small currents which circulate inside the laminated core of the transformer) and **hysteresis**. To demonstrate let's say that you push on some material and it bends. When you stop pushing, does it return to its original shape immediately? If it doesn't the material is demonstrating hysteresis. Let's look at this in context.

> **Key term**
> **Hysteresis** – a generic term meaning a lag in the effect of a change of force

Figure 8.78 shows the effect within ferromagnetic materials of hysteresis. Starting with an unmagnetised material at point A; and here both field strength and flux density are zero. The field strength increases in the positive direction and the flux begins to grow along the dotted path until we reach saturation at point B. This is called the initial magnetisation curve.

If the field strength is now relaxed, instead of retracing the initial magnetisation curve, the flux falls more slowly. In fact, even when the applied field has returned to zero, there will still be a degree of flux density (known as the **remanence**) at point C. To force the flux to go back to zero (point D), we have to reverse the applied field. The field strength here that is necessary to drive the field back to zero is known as the **coercivity**. We can then continue reversing the field to get to point E, and so on. This is known as the hysteresis loop.

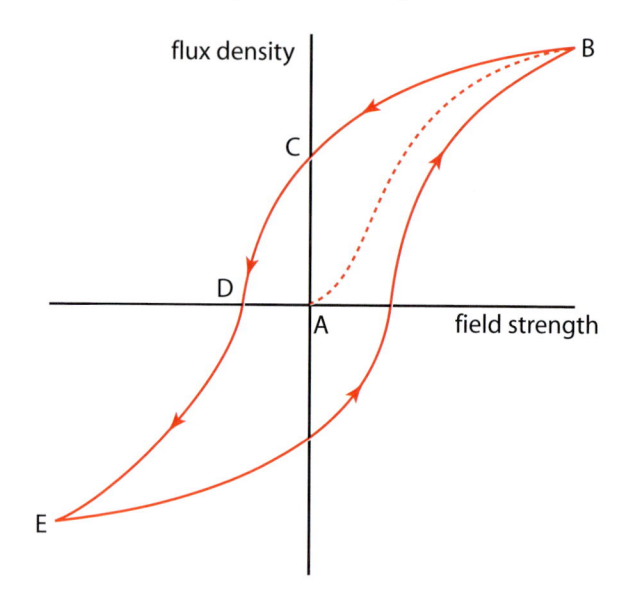

Figure 8.78 Hysteresis loop diagram

As we have already said, we can help reduce eddy currents by using a laminated core construction. We can also help to reduce hysteresis by adding silicon to the iron from which the transformer core is made.

The version of the double wound transformer that we have looked at so far makes the principle of operation easier to understand. However, this arrangement is not very efficient, as some of the magnetic flux being produced by the primary winding will not react with the secondary winding and is often referred to as 'leakage'. We can help to reduce this leakage by splitting each winding across the sides of the core (see Figure 8.79).

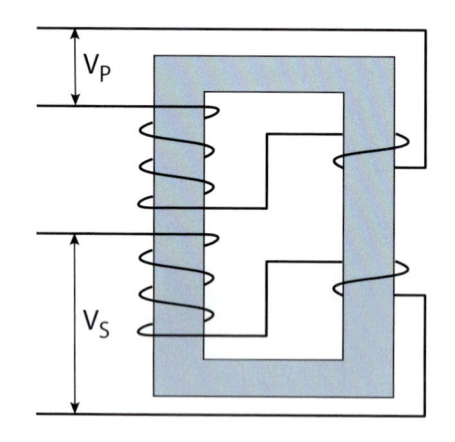

Figure 8.79 Reducing 'leakage'

Shell-type transformers

We can reduce the magnetic flux leakage a bit more, by using a shell-type transformer.

In the shell-type transformer, both windings are wound onto the central leg of the transformer and the two outer legs are then used to provide parallel paths for the magnetic flux.

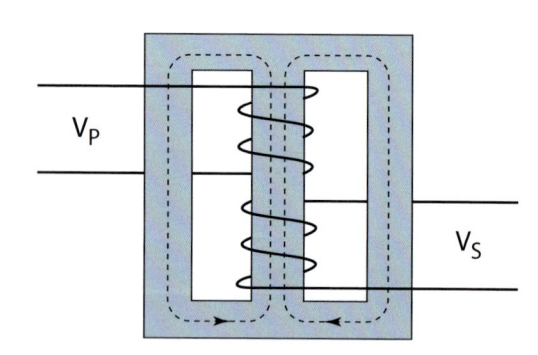

Figure 8.80 Shell-type transformer

The autotransformer

The autotransformer uses the principle of 'tapped' windings in its operation. Remember that the ratio of input voltage against output voltage will depend upon the number of primary winding turns and secondary winding turns (the turns ratio). But what if we want more than one output voltage?

Some devices are supplied with the capability of providing this, such as small transformers for calculators, musical instruments or doorbell systems. Tapped connections are the normal means by which this is achieved. A tapped winding means that we have made a connection to the winding and then brought this connection out to a terminal. Now, by connecting between the different terminals, we can control the number of turns that will appear in that winding and we can therefore provide a range of output voltages.

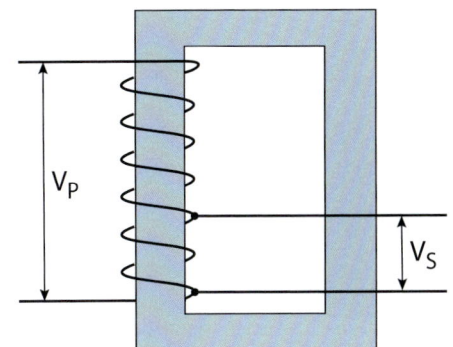

Figure 8.81 Autotransformer

An autotransformer has only one tapped winding and the position of the tapping on that winding will dictate the output voltage.

One of the advantages of the autotransformer is that, because it only has one winding, it is more economical to manufacture. However, on the down side, we have made a physical connection to the winding. Therefore, if the winding ever became broken between the two tapping points, then the transformer would not work and the input voltage would appear on the output terminals. This would then present a real hazard.

Instrument transformers

Instrument transformers are used in conjunction with measuring instruments because it would be very difficult and expensive to design normal instruments to measure the high current and voltage that we find in certain power systems. We therefore have two types of instrument transformer, both being double wound.

The current transformer

The current transformer (c.t.) normally has very few turns on its primary winding so that it does not affect the circuit to be measured, with the actual meter connected across the secondary winding.

Care must be taken when using a c.t. Never open the secondary winding while the primary is 'carrying' the main current. If this happened, a high voltage would be induced into the secondary winding. Apart from the obvious danger, this heat build-up could cause the insulation on the c.t. to break down.

Figure 8.82 Current transformer

The voltage transformer

This is very similar to our standard power transformer, in that it is used to reduce the system voltage. The primary winding is connected across the voltage that we want to measure and the meter is connected across the secondary winding.

Step-up and step-down transformers

Step-up transformers

A step-up transformer is used when it is desirable to step voltage up in value.

The primary coil has fewer turns than the secondary coil. We already know that the number of turns in a transformer is given as a ratio. When the primary has fewer turns than the secondary, voltage and impedance are stepped up. In the circuit shown in Figure 8.83, voltage is stepped up from 120 V a.c. to 240 V a.c. Since impedance is also stepped up, current is stepped down from 10 A to 5 A.

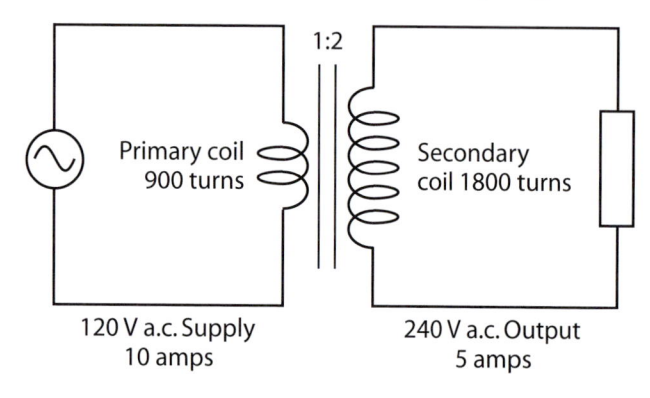

Figure 8.83 Step-up transformer

(Primary coil 900 turns, 120 V a.c. Supply, 10 amps; Secondary coil 1800 turns, 240 V a.c. Output, 5 amps; ratio 1:2)

Step-down transformers

A step-down transformer is used when it is desirable to step voltage down in value.

The primary coil has more turns than the secondary coil. The step-down ratio is 2:1. Since the voltage and impedance are stepped down, the current is stepped up in this case to 10 A.

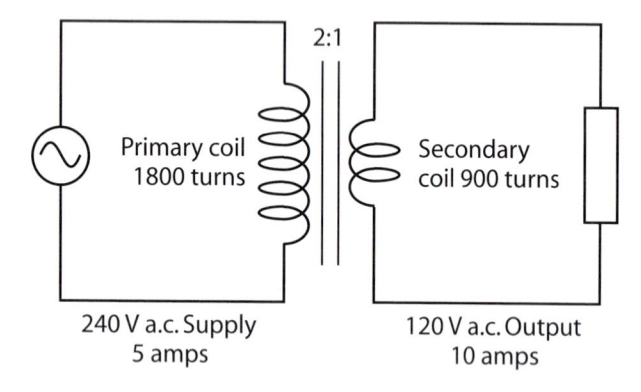

Figure 8.84 Step-down transformer

(Primary coil 1800 turns, 240 V a.c. Supply, 5 amps; Secondary coil 900 turns, 120 V a.c. Output, 10 amps; ratio 2:1)

Safety isolating transformer

Another use of a transformer is to isolate the secondary output from the supply. In a bathroom, the shower socket should be supplied from a 1:1 safety isolating transformer. The output of the transformer has no connection to earth thereby ensuring that output from the transformer is totally isolated from the supply.

In areas of installations where there is an increased risk of electric shock, the voltage is reduced to less than 50 V and is supplied from a safety isolating transformer; we know this as **Separated Extra Low Voltage** (**SELV**) supply.

Transformer ratings

Transformers are rated in kVA (kilovolt-amps). This rating is used rather than watts because loads are not purely resistive. Only resistive loads are measured in watts.

The kVA rating determines the current that a transformer can deliver to its load without overheating. Given volts and amps, kVA can be calculated. Given kVA and volts, amps can be calculated.

The kVA rating of a transformer is the same for both the primary and the secondary. At this point let us try some examples.

> **Example 1**
>
> A transformer having a turns ratio of 2:7 is connected to a 230 V supply. Calculate the output voltage.
>
> When we give transformer ratios, we give them in the order primary, then secondary. Therefore in this example we are saying that for every two windings on the primary winding, there are seven on the secondary. If we therefore use our formula:
>
> $$\frac{V_P}{V_S} = \frac{N_P}{N_S}$$
>
> We should now transpose this to get:
>
> $$V_S = \frac{V_P \times N_S}{N_P}$$
>
> However, we do not know the exact number of turns involved.
>
> But do we need to, if we know the ratio? Let us find out.
>
> The ratio is 2:7, meaning for every 2 turns on the primary, there will be 7 turns on the secondary. Therefore, if we had 6 turns on the primary, this would give us 21 turns on the secondary, but the ratio of the two has not changed. For every 2 on the primary we are still getting 7 on the secondary.
>
> This means we can just insert the ratio rather than the individual number of turns into our formula:
>
> $$V_S = \frac{V_P \times N_S}{N_P} = \frac{230 \times 7}{2} = \mathbf{805\ V}$$
>
> Now, to prove our point about ratios, let us say that we know the number of turns in the windings to be 6 in the primary and 21 in the secondary (which is still giving us a 2:7 ratio). If we now apply this to our formula we get:
>
> $$V_S = \frac{V_P \times N_S}{N_P} = \frac{230 \times 21}{6} = \mathbf{805\ V}$$
>
> The same answer.

Example 2

A single phase transformer, with 2000 primary turns and 500 secondary turns, is fed from a 230 V a.c. supply. Find:

a. the secondary voltage

b. the volts per turn.

a. Secondary voltage

$$\frac{V_P}{V_S} = \frac{N_P}{N_S}$$

Using transposition, rearrange the formula to give:

$$V_S = \frac{N_P \times N_S}{N_P}$$

$$V_S = \frac{230 \times 500}{2000} = \frac{115000}{2000} = \mathbf{57.5\ V}$$

b. Volts per turn

This is the relationship between the volts in a winding and the number of turns in that winding. To find volts per turn, we simply divide the voltage by the number of turns.

Therefore, in the primary:

$$\frac{V_P}{N_P} = \frac{230}{2000} = \mathbf{0.115\ volts\ per\ turn}$$

In the secondary

$$\frac{V_S}{N_S} = \frac{57.5}{500} = \mathbf{0.115\ volts\ per\ turn}$$

Example 3

A single-phase transformer is being used to supply a trace heating system. The transformer is fed from a 230 V 50 Hz a.c. supply and needs to provide an output voltage of 25 V. If the secondary current is 150 A and the secondary winding has 50 turns, find:

a. the output kVA of the transformer

b. the number of primary turns

c. the primary current

d. the volts per turn.

a. The output kVA

$$kVA = \frac{volts \times amperes}{1000} = \frac{V_S \times I_S}{1000} = \frac{25 \times 150}{1000} = \textbf{3.75 kVA}$$

b. The number of primary turns

If: $\frac{V_P}{V_S} = \frac{N_P}{N_S}$ then by transposition

$$N_P = \frac{V_P \times N_S}{V_S} = \frac{250 \times 50}{25} = \textbf{460 turns}$$

c. The primary current

If: $\frac{V_P}{V_S} = \frac{I_S}{I_P}$ then by transposition

$$I_P = \frac{V_S \times I_S}{V_P} = \frac{25 \times 150}{230} = \textbf{16 A}$$

d. The volts per turn

In the primary: $\frac{V_P}{N_P} = \frac{230}{460} = \textbf{0.5 volts per turn}$

In the secondary: $\frac{V_S}{N_S} = \frac{25}{50} = \textbf{0.5 volts per turn}$

Example 4

A step-down transformer, having a ratio of 2:1, has an 800 turn primary winding and is fed from a 400 V a.c. supply. The output from the secondary is 200 V and this feeds a load of 20 Ω resistance. Calculate:

a. the power in the primary winding

b. the power in the secondary winding.

Well, we know that the formula for power is: $P = V \times I$

And we also know that we can use Ohm's Law to find current: $I =$

Therefore, if we insert the values that we have, we can establish the current in the secondary winding:

$$I_S = \frac{V_S}{R_S} = \frac{200}{20} = 10 \text{ A}$$

Now that we know the current in the secondary winding, we can use the power formula to find the power generated in the secondary winding:

$P = V \times I = 200 \times 10 = 2000$ watts = **2 kW**

We now need to find the current in the primary winding. To do this we can use the formula:

$$\frac{V_P}{V_S} = \frac{I_S}{I_P}$$

However, we need to transpose the formula to find I_P. This would give us:

$$I_P = \frac{I_S \times V_S}{V_P}$$

Which, if we now insert the known values to, gives us:

$$I_P = \frac{I_S \times V_S}{V_P} = \frac{10 \times 200}{400} = \frac{2000}{400} = 5 \text{ A}$$

Now that we know the current in the primary winding, we can again use the power formula to find the power generated in the secondary winding:

$P = V \times I = 400 \times 5 = 2000$ watts = **2 kW**

Instruments and measurement

As electricians we are often responsible for the measurement of different electrical quantities. Some must be measured as part of the inspection and testing of an installation (e.g. insulation resistance). However, of the remainder, the most common quantities that we are likely to come across are shown in Table 8.10.

Property	Instrument
Current	Ammeter
Voltage	Voltmeter
Resistance	Ohmmeter
Power	Wattmeter

Table 8.10 Electrical quantities

We will look at the different instruments that we use to measure these quantities in the following section.

But before we ever use a meter, we must always ask ourselves:

- Have we chosen the correct instrument?
- Is it working correctly?
- Has it been set to the correct scale?
- How should it be connected?

Measuring current (ammeter)

An ammeter measures current by series connections. Although we know that we can use a multimeter on site, the actual name for an instrument that measures current is an ammeter. Ammeters are connected in series so that the current to be measured passes through them. The circuit diagram in Figure 8.85 illustrates this. Consequently, they need to have a very low resistance or they would give a false reading.

If we now look at the same circuit, but use a correctly set multimeter, the connections would look as in Figure 8.86.

Measuring voltage (voltmeter)

On site or for general purposes, we can use a multimeter to measure voltage, but the actual device used for measuring voltage is called a voltmeter. It measures the potential difference between two points (for instance, across the two connections of a resistor). The voltmeter must be connected in parallel across the load or circuit to be measured as shown in Figure 8.87.

If we now look at the same circuit, but use our correctly set multimeter, the connections would look as in Figure 8.88.

The internal resistance of a voltmeter must be very high if we wish to get accurate readings.

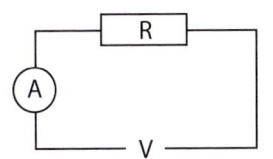

Figure 8.85 Ammeter in circuit

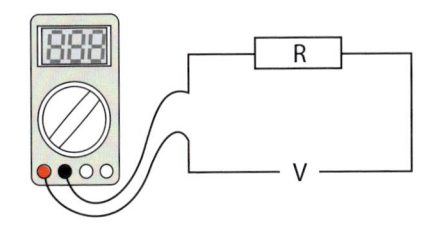

Figure 8.86 Multimeter in circuit

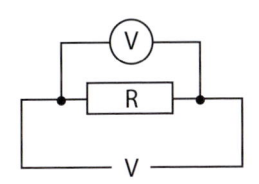

Figure 8.87 Voltmeter in circuit

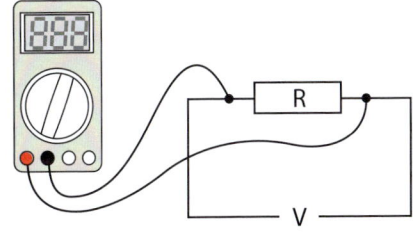

Figure 8.88 Multimeter measuring voltage

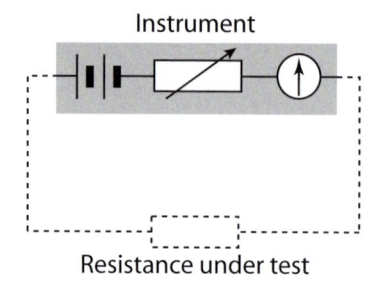

Figure 8.89 Ohmmeter

Measuring resistance (ohmmeter)

There are many ways in which we can measure resistance. However, in the majority of cases we do so by executing a calculation based upon instrument readings (ammeter/voltmeter) or from other known resistance values. Once again we can use our multimeter to perform the task, but we refer to the actual meter as being an ohmmeter.

The principle of operation involves the meter having its own internal supply (battery). The current, which then flows through the meter, must be dependent upon the value of the resistance under scrutiny.

However, before we start our measurement, we must ensure that the supply is off and then connect both leads of the meter together and adjust the meter's variable resistor until full-scale deflection (zero) is reached.

The dotted line in Figure 8.89 indicates the circuit under test. All other components are internal to the meter.

Using an ammeter and voltmeter

If we consider the application of Ohm's Law, we know that we can establish resistance by the following formula:

$$R = \frac{V(\text{voltmeter})}{I(\text{ammeter})}$$

If we therefore connect an ammeter and voltmeter in the circuit, then by using a known supply (battery) we can apply the above formula to the readings that we get.

In practice, we invariably find that we need a long 'wandering' test lead to perform such a test. Where this is the case, we must remember to deduct the value of the test lead from our circuit resistance.

Loading errors

Connecting a measuring instrument to a circuit will invariably have an effect on that circuit. In the case of the average electrical installation, this effect can largely be ignored. But if we were to measure electronic circuits, the effect may be drastic enough to actually destroy the circuit.

We normally refer to these errors introduced by the measuring instrument as loading errors, and in Figure 8.90 we will look at an example of this.

Let's assume in (a) that we have been asked to measure the voltage across resistor B in the circuit, where both items have a resistance of 100 kΩ. Note that in this arrangement, the potential difference (p.d.) across each resistor will be 115 V.

Our instrument, a voltmeter, has an internal resistance of 100 kΩ and would therefore be connected as in Figure 8.90.

In effect, by connecting our meter in this way, we are introducing an extra resistance in parallel into the circuit. We could, therefore, now draw our circuit as in Figure 8.90(b).

Now, if we find the equivalent resistance of the parallel branches:

$$\frac{1}{R_t} = \frac{1}{R_1} + \frac{1}{R_2} = \frac{1+1}{100} = \frac{2}{100} k\Omega$$

Therefore:

$$R_t = \frac{100}{2} = 50 k\Omega$$

And then if we treat the parallel combination as being in series with 100 kΩ resistor A then total resistance:

$R = R_A + R_t = 100 k\Omega + 50 k\Omega = 150 k\Omega$

If you applied Ohm's Law again, we would find we now have a current of 0.00153 A (1.53 mA) flowing in the circuit. This in turn would give a p.d. across each resistor of 76 V and not the 115 V we would expect to see.

The problem is caused by the amount of current that is flowing through the meter, and this is normally solved by using meters with a very high resistance (as resistance restricts current). So you now see why we said earlier that voltmeters must have a high internal resistance in order to be accurate.

Meter displays

There are two types of display, **analogue** and **digital**. Analogue meters have a needle moving around a calibrated scale, whereas digital tend to show results, as numeric values, via a liquid crystal or LED display.

For many years electricians used the analogue meter and indeed it is still common for continuity/insulation resistance testers to be of the analogue style. However, their use is slowly being replaced by the digital meter. Most modern digital meters are based on semiconductors and consequently have very high impedance, making them ideal for accurate readings and good for use with electronic circuits.

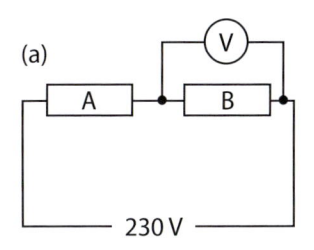

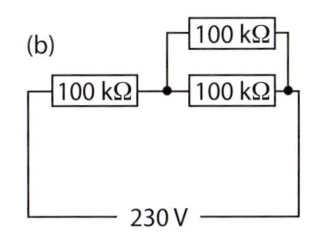

Figure 8.90 Loading errors

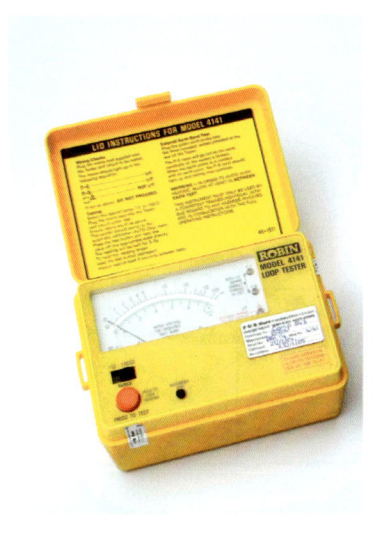

Figure 8.91 Analogue meter

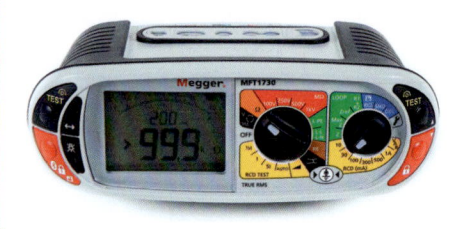

Figure 8.92 Digital meter

What sort of meter will I need?

Although instruments exist to measure individual electrical quantities, most electricians will find that a cost-effective solution is to use a digital multimeter similar to the one pictured. Meters like this are generally capable of measuring current, resistance and voltage, both in a.c. and d.c. circuits and across a wide range of values.

Is the meter working correctly?

It is important to check that your meter is functioning correctly and physically fit for the purpose. There are commercially available proving units that will help. However, on site it is more normal to use a known supply as the means of proving voltage measurement and the shorting/separating of leads to prove operation.

Measuring power

In a d.c. circuit it is possible to measure the power supplied by using a voltmeter and an ammeter. Then by using the formula $P = V \times I$ we can arrive at the power in watts.

However, in an a.c. circuit, this method only produces the apparent power, a figure that will not be accurate unless we have unity power factor. This is because components such as capacitors and inductors will cause the current to lead or lag the voltage.

For a single-phase a.c. circuit we therefore use the formula:

$P = V \times I \times \cos Ø$

This indicates how we would connect measuring instruments to establish power. For a single-phase resistive load, we can use the ammeter and voltmeter, but for a circuit containing capacitance or inductance we must use a wattmeter.

Figure 8.93 indicates how a wattmeter is connected into a single-phase circuit. Just like Ohm's Law, where to calculate power we need to know the values of voltage and current, a wattmeter also requires values of voltage and current.

The instrument shown in Figure 8.93 has a current coil, indicated between W_1 and W_2, that is wired in series with the load, and a voltage coil, shown between P_1 and P_2, wired in parallel across the supply.

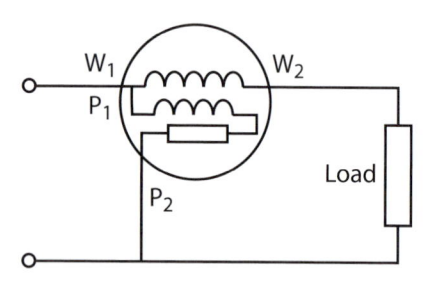

Figure 8.93 Wattmeter connected to load

This can be used to measure the power in a single-phase circuit or in a balanced three-phase load, where the total power will be equal to three times the value measured on the meter.

Measuring power in a three-phase, four-wire balanced load – one-wattmeter method

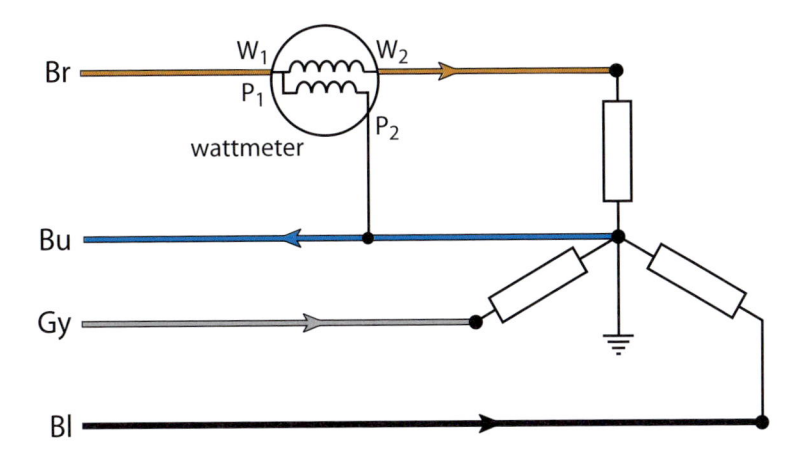

Figure 8.94 One-meter method

Measuring power in a three-phase balanced load – two-wattmeter method

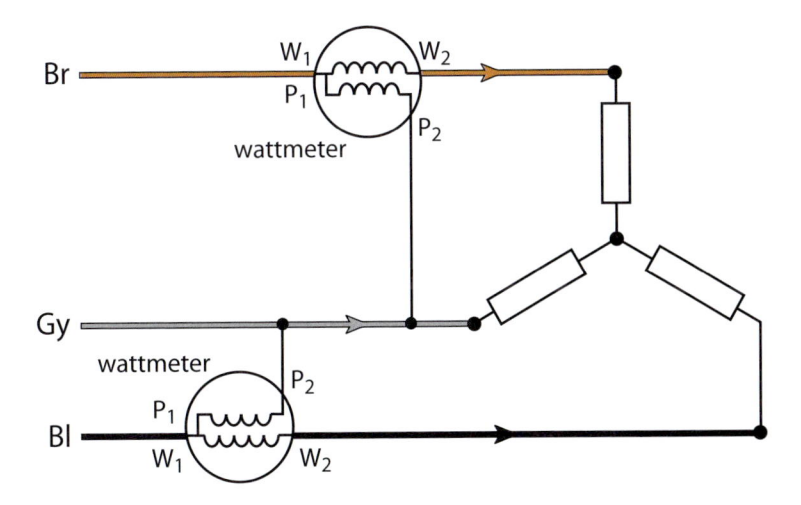

Figure 8.95 Wattmeter connected to load

In the two-wattmeter method, the total power is found by adding the two values together. At unity power factor the instruments will read the same and be half of the total load. For other power factors the instrument readings will be different, the difference in the reading could then be used to calculate the power factor.

Measuring power in an unbalanced three-phase circuit – three-wattmeter method

Here the total power will be the sum total of the readings on the three-wattmeters.

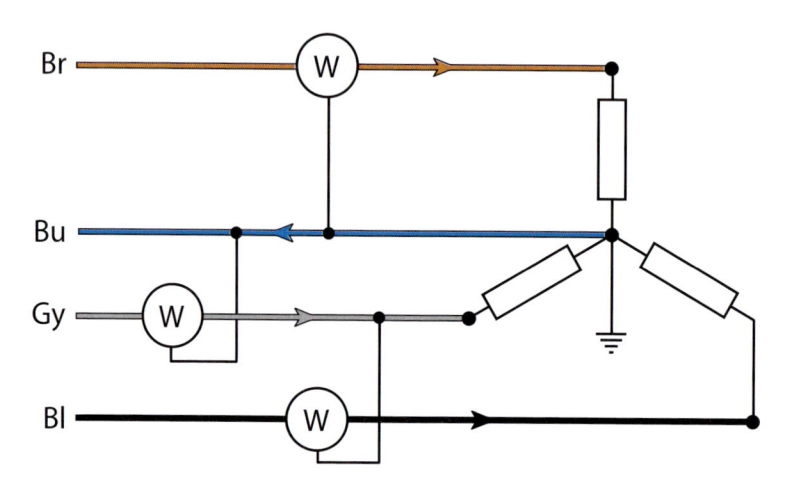

Figure 8.96 Three-phase circuit

This allows the power to be measured in a situation where the load is unbalanced, such as the three-phase supply to a large building where it is impossible to balance the load completely.

Measuring power factor

To measure power factor, there are a number of purpose-made instruments available. All of these meters include both voltage and current measurement in circuits. Most designs are based on the clamp meter idea where the meter clamps around the current-carrying conductor. However, additional leads are then required to connect the meter across the supply. Many meters are also combined with the ability to measure all aspects of power, i.e. kW, kVA and kVAr.

Figure 8.97 illustrates the connections of a power factor measuring instrument.

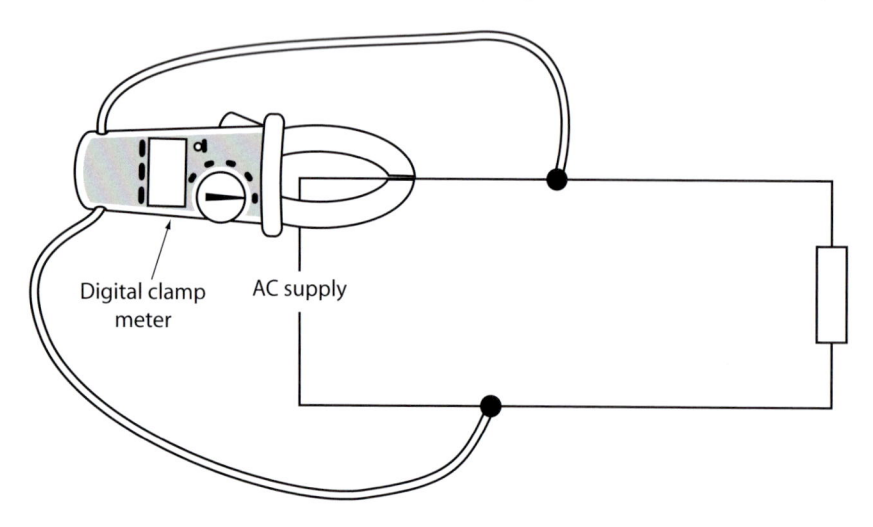

Figure 8.97 Circuit diagram with digital clamp meter

There is also an alternative method. The calculation to establish power factor is to divide true power by the apparent power (VA). As we know that a wattmeter will give us the true power, by adding an ammeter and a voltmeter to our circuit we can establish the apparent power (VA) and therefore establish the power factor. Figure 8.98 shows this arrangement.

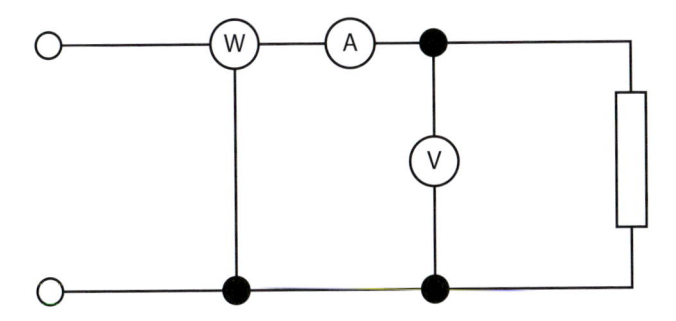

Figure 8.98 Circuit with wattmeter, ammeter and voltmeter connected

Measuring frequency

Frequency is an electrical quantity with no equivalent in d.c. circuits and can become important when an a.c. power system has been designed to run at a specific frequency.

Frequency can be measured in a variety of ways, such as by measuring the shaft speed of the generator that produced the alternating current or by using a Cathode Ray Oscilloscope.

However, one very popular method is to use a vibrating reed frequency meter as shown in Figure 8.99. These work very much like a tuning fork that is used to tune a piano, in that any object that possesses elasticity has a frequency that it likes to vibrate at. If you can remember those tuning forks, the longer ones vibrated at lower notes when struck than the shorter ones, which had a higher note when struck.

If you placed a row of such tuning forks side by side fixed to a common base material, the base could be vibrated at the frequency of the supply by means of an electromagnet. The fork that likes to vibrate at the frequency needed will therefore start to 'rattle' about, and it is the end of those forks that we can see in such a meter.

Figure 8.99 Vibrating reed frequency meter

Measuring impedance

Impedance (Z) is generally defined as the total opposition a device or circuit offers to the flow of an alternating current (a.c.) at a given frequency, and is represented as a complex quantity which is graphically shown on a vector plane. An impedance vector consists of resistance (R) and an 'imaginary' part, reactance (X), which as we discussed earlier, takes two forms: inductive (X_L) and capacitive (X_C).

To find the impedance, we therefore need to measure at least two values because impedance is a complex quantity and many modern impedance measuring instruments measure the real and the imaginary parts of an impedance vector before converting them into the desired parameter.

Measurement ranges and accuracy for a variety of impedance parameters are then determined from those specified for impedance measurement. However, automated instruments allow you to make a measurement by simply connecting the component, circuit, or material to the instrument.

Quite often these meters are referred to as LRC meters because of the circuit components involved.

> **Remember**
>
> Information about how to measure energy was covered on page 286, where we looked at the kWh meter.

> **Progress check**
>
> 1. Describe briefly the main problems of having a low power factor.
> 2. Explain the difference between peak values and average values when discussing a.c. sine waves.
> 3. Identify and briefly explain the following:
>   - R
>   - X_L
>   - X_C
>   - Z
> 4. A resistor of 10 Ω is connected in series with an inductor of 0.02 H and a capacitor of 120 μF across a 240 V 50 Hz supply. Calculate the following:
>   - the impedance
>   - the supply current
>   - the power factor
> 5. Describe the main reason for utilising star–delta connected supplies.
> 6. How is power factor correction achieved in industrial installations?

K8. Understand the operating principles and application of d.c. machines and a.c. motors

In essence there are two categories of motor: those that run on direct current (d.c.) and those that run on alternating current (a.c.).

The basic concept of the motor was actually explained in our electromagnetism section. As lines of flux cannot cross, Figure 8.100 shows what the result would be if we actually placed a current-carrying conductor in a magnetic field.

> **Remember**
>
> To understand motors and generators we need to remember the basic principles of magnetic fields, current flow and induced motion and how when used together they can make a motor rotate.

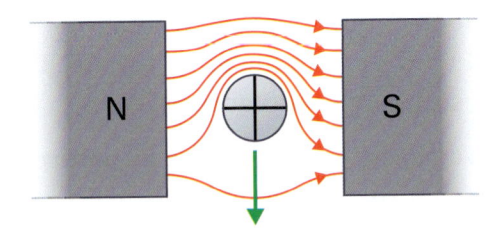

Figure 8.100 Force on a current-carrying conductor in a magnetic field

In Figure 8.100 the main field now becomes distorted. As the two fields above the conductor are in the same direction, the amount of flux is high and therefore the force will move the conductor downwards. It is this basic principle of a force being exerted on a current-carrying conductor in a magnetic field that is used in the construction of motors and moving coil instruments.

Basic types, applications and operating principles of d.c. machines

The operating principle of a d.c. motor is fairly straightforward. Looking at Figure 8.101, if we take a conductor, form it into a loop that is pivoted at its centre, place it within a magnetic field and then apply a d.c. supply to it, a current will pass around the loop. This will cause a magnetic field to be produced around the conductor, and this magnetic field will interact with the magnetic field between the two poles (**pole pair**) of the magnet.

The magnetic field between the two poles (pole pair) becomes distorted and, as mentioned earlier, this

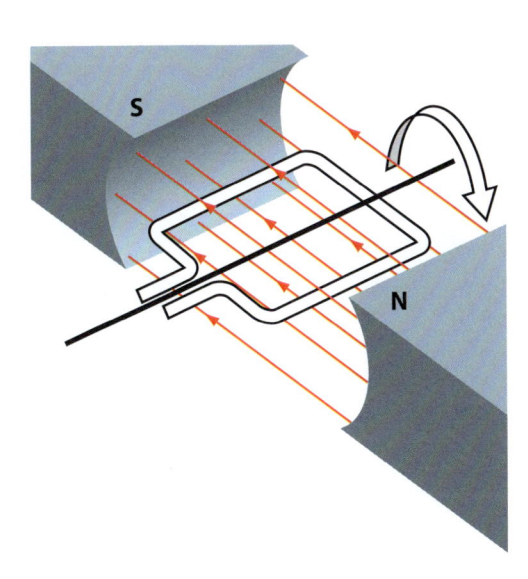

Figure 8.101 Single current-carrying loop in a magnetic field

interaction between the two magnetic fields causes a bending or stretching of the lines of force. The lines of force behave a bit like an elastic band and consequently are always trying to find the shortest distance between the pole pair. These stretched lines, in their attempt to return to their shortest length, therefore exert a force on the conductor that tries to push it out of the magnetic field, and thus they cause the loop to rotate.

In reality we don't have just one loop, we have many, with them being wound on to a central rotating part of the d.c. motor known as the **armature**. When the armature is positioned so that the loop sides are at right angles to the magnetic field, a turning force is exerted. But we have a problem when the coil has rotated 180°, as the magnetic field in the loop is now opposite to that of the field, and this will tend to push the armature back the way it came, thus stopping the rotating motion.

The solution is to reverse the current in the armature every half rotation so that the magnetic fields will work together to maintain a continuous rotating motion. The device that we use to achieve this switching of polarity is known as the **commutator**.

The commutator

Figure 8.102 now gives a different perspective of the same arrangement, showing the device that enables the supply to be connected to the loop, enabling the loop to rotate continuously. This is the commutator. This simplified version has only two segments, connecting to either side of the loop, but in actual machines there may be any number. Large machines would have in excess of 50, and the numbers of loops can be in the hundreds. The commutator is made of copper, with the segments separated from each other by insulation.

Figure 8.103 shows the direction of current around the loop when connected to a d.c. supply. The brushes remain in a fixed position, against the copper segments of the commutator. Note the direction of current in those sections indicated by X and Y in the diagram.

Figure 8.104 shows the wire loop having rotated 180°. X and Y have changed positions but, as can be seen from the arrows, current flow remains as it was for the previous diagram. As the poles of the armature electromagnet pass

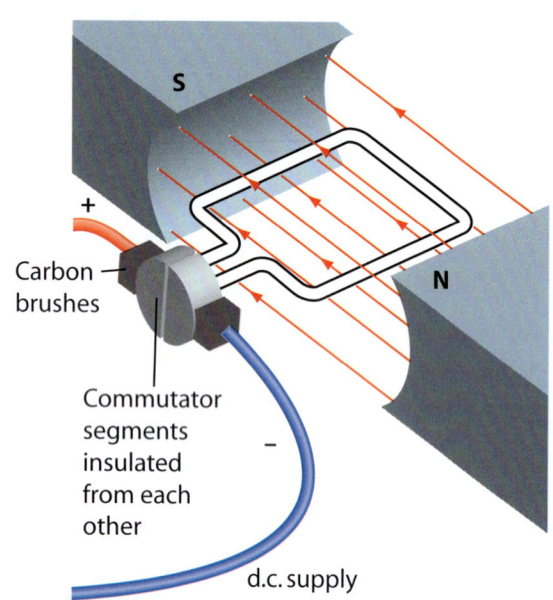

Figure 8.102 Single loop with commutator

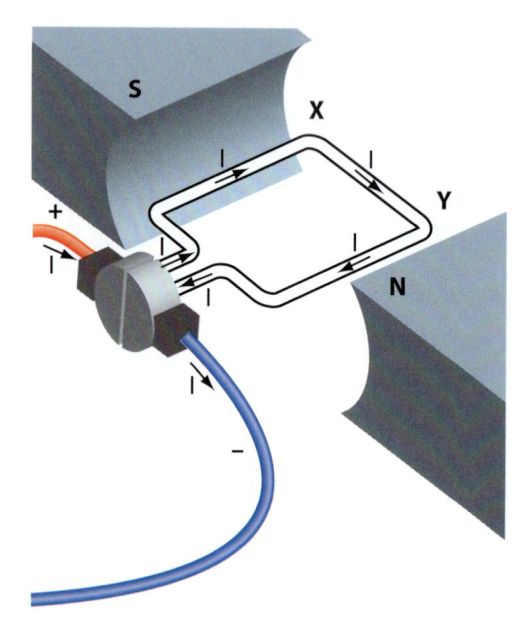

Figure 8.103 Single loop with d.c. flowing

the poles of the permanent magnets, the commutator reverses the polarity of the armature electromagnet. In that instant of switching polarity, inertia keeps the motor going in the proper direction and thus the motor continues to rotate in one direction.

In the previous drawings, we have shown the armature rotating between a pair of magnetic poles. Practical d.c. motors do not use permanent magnets, but use electromagnets instead. The electromagnet has two advantages over the permanent magnet:

1. By adjusting the amount of current flowing through the wire the strength of the electromagnet can be controlled.
2. By changing the direction of current flow the poles of the electromagnet can be reversed.

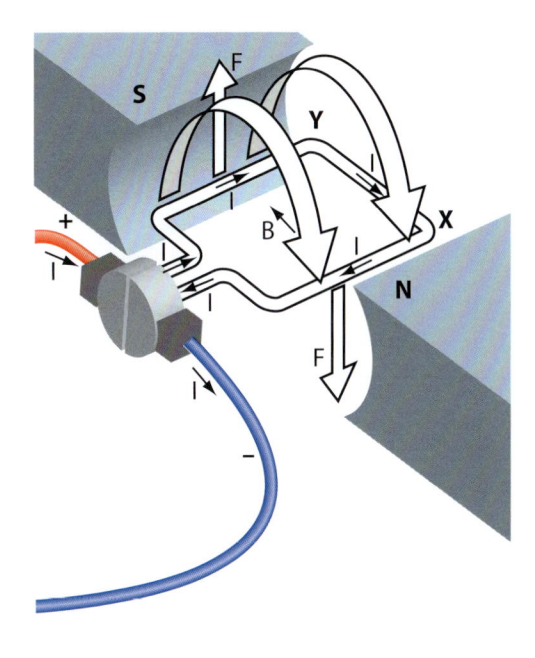

Figure 8.104 Single loop rotated 180°

Reversing a d.c. motor

The direction of rotation of a d.c. motor may be reversed by either:

- reversing the direction of the current through the field hence changing the field polarity
- reversing the direction of the current through the armature.

Common practice is to reverse the current through the armature, and this is normally achieved by reversing the armature connections only.

Types of d.c. motor

There are three basic forms of d.c. motor:

- series
- shunt
- compound.

They are very similar to look at, the difference being the way in which the field coil and armature coil circuits are wired.

The series motor

As the name suggests, this motor has the field winding wired in series with the armature. It is also called a 'universal motor' as it can run in both d.c. and a.c. situations.

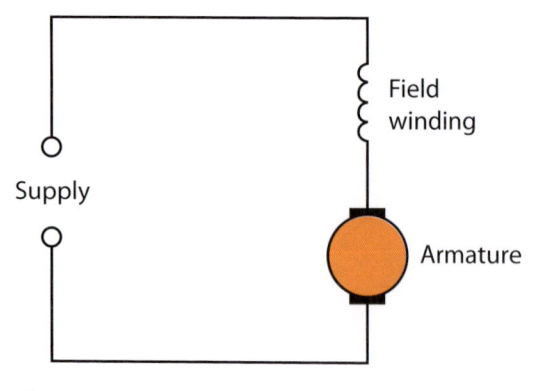

Figure 8.105 Series motor

The large conductors used in the field windings and armature allow the series motor to develop a large magnetic field and consequently a high starting **torque** (rotational force). This means that it is able to move a large load connected to the shaft when first switched on.

As the armature begins to rotate and gather speed the current and torque reduces. However, should the motor ever lose its load (e.g. the shaft breaks or the drive belt breaks) then the load current falls, which reduces the amount of back e.m.f. produced by the armature. As the armature is no longer producing sufficient back e.m.f. and the load is no longer exerting a force on the shaft, the armature will continue to 'runaway'. In other words, it will increase in speed until it self-destructs. We therefore tend to use sensors to disconnect the machine should the **revolutions per minute** (rpm) exceed a set level.

As these motors have a very high starting current, they are started using an external resistance placed in series with the armature. As we saw earlier, the speed of the motor is inversely proportional to the strength of the magnetic field and therefore we can control speed by connecting a variable resistor in the field circuit.

The shunt motor

Electrically, a shunt is something connected in parallel. In the shunt motor it is the field winding that is connected in parallel with the armature.

This motor has a low starting torque and therefore the load on the shaft has to be quite small. The armature's torque increases as the motor gains speed due to the fact that the torque is directly proportional to the armature current. Consequently, when the motor is starting and speed is very low, the motor has very little torque. Once the motor reaches full rpm, its torque is at its highest level.

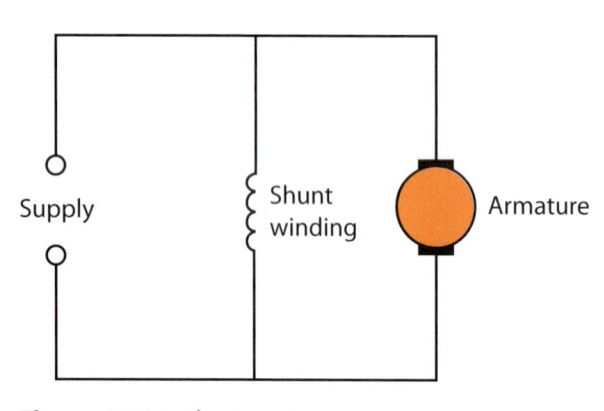

Figure 8.106 Shunt motor

The shunt motor's speed can be controlled by varying the amount of current supplied to the shunt field winding, which will allow the rpm to be changed by up to 20%. We can reverse the direction of rotation by changing the polarity of either the armature coil or the field coil.

The compound motor

A compound is when we produce something by combining two or more parts. The compound motor therefore gets its name by combining the characteristics of the series and shunt motors. There are actually two main types of compound motor – the most commonly used cumulative compound and the rarely used differential compound.

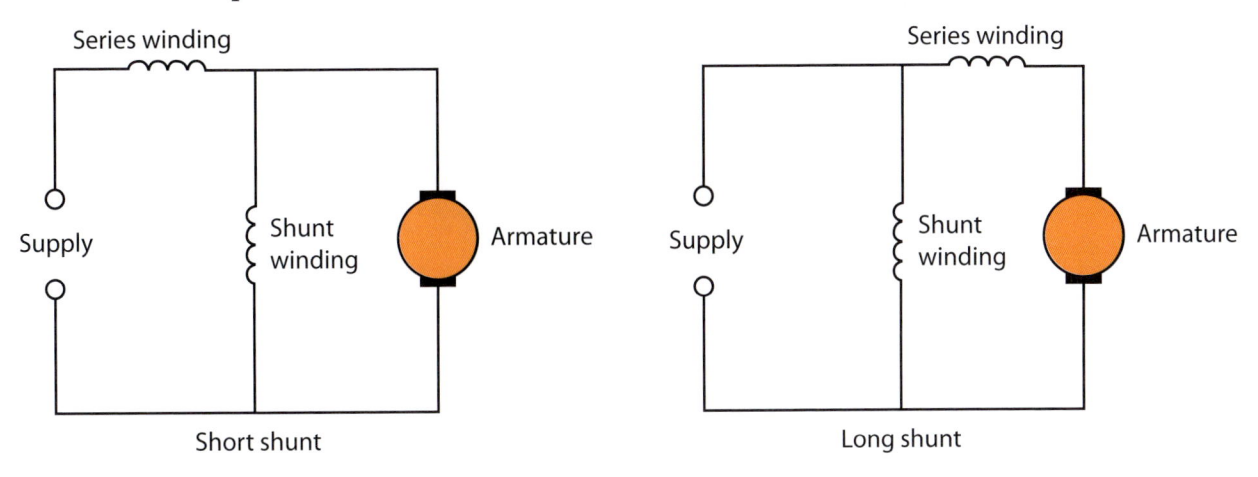

Figure 8.107 Compound motor

Figure 8.107 shows the short and long shunt version of the cumulative compound motor. This is called cumulative because the polarity of the shunt field is the same as that of the armature and thus the shunt field aids the magnetic fields of the series field and armature. The shunt winding can be wired as a long shunt or as a short shunt.

In the case of the differential compound motor, the polarity of the shunt field is reversed with the negative terminal of the shunt field being connected to the positive terminal of the armature and as a result it opposes the flux of the armature and series field.

The cumulative compound motor is probably the most common d.c. motor because it provides high starting torque and good speed regulation. As a result it tends to be used in situations where a constant speed is needed with varying loads.

Speed control is achieved by voltage regulation of the shunt field and reversal of shaft rotation can be achieved by changing the polarity of the armature winding.

Operating principles, basic types, applications and limitations of a.c. motors

Figure 8.108 A simple a.c. motor

As we mentioned earlier, there are three general types of d.c. motor. However, there are many types of a.c. motor, each one having a specific set of operating characteristics such as torque, speed, single-phase or three-phase, and this determines their selection for use. We can essentially group them in to two categories: single-phase and three-phase.

In a d.c. motor, electrical power is conducted directly to the armature through brushes and a commutator. Due to the nature of an alternating current, an a.c. motor doesn't need a commutator to reverse the polarity of the current. Whereas a d.c. motor works by changing the polarity of the current running through the armature (the rotating part of the motor), the a.c. motor works by changing the polarity of the current running through the stator (the stationary part of the motor).

The series-wound (universal) motor

We will begin the discussion of a.c. motors by looking at the series-wound (universal) motor because it is different in its construction and operation from the other a.c. motors considered here, and also because it is constructed as we have previously discussed in the Types of d.c. motor section on page 367, having field windings, brushes, commutator and an armature.

As can be seen in Figure 8.109, because of its series connection, current passing through the field windings also passes through the armature. The turning motion (torque) is produced as a result of the interaction between the magnetic field in the field windings and the magnetic field produced in the armature.

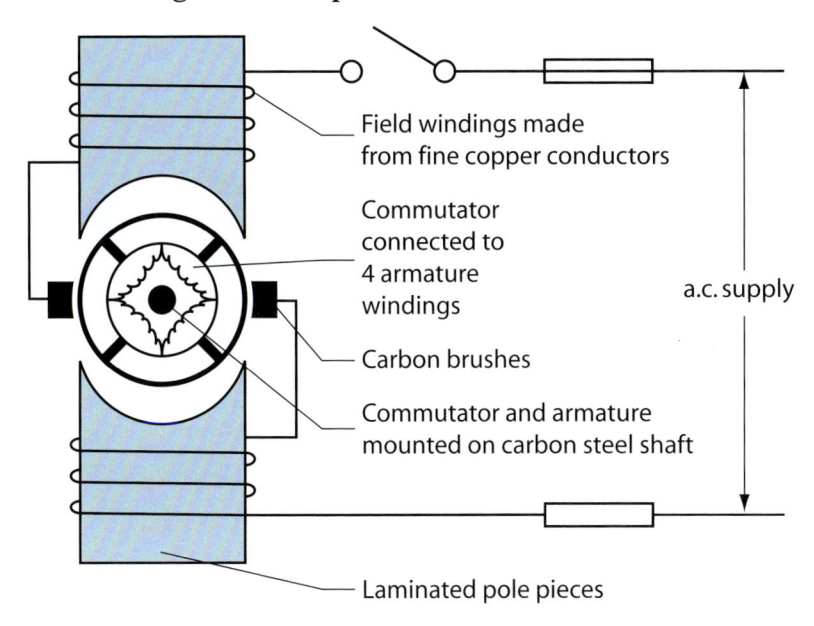

Figure 8.109 Series universal motor

For this motor to be able to run on an a.c. supply, modifications are made both to the field windings and armature **formers**. These are heavily laminated to reduce eddy currents and I^2R losses, which reduces the heat generated by the normal working of the motor, thus making the motor more efficient.

This type of motor is generally small (less than a kilowatt) and is used to drive small hand tools such as drills, vacuum cleaners and washing machines.

A disadvantage of this motor is that it relies on contact with the armature via a system of carbon brushes and a commutator. It is this point that is the machine's weakness, as much heat is generated through the arcing that appears across the gap between the brushes and the commutator. The brushes are spring-loaded to keep this gap to a minimum, but even so the heat and friction eventually cause the brushes to wear down and the gap to increase. These then need to be replaced, otherwise the heat generated as the gap gets larger will eventually cause the motor to fail.

The advantages of this machine are:

- more power for a given size than any other normal a.c. motor
- high starting torque
- relative cheapness to produce.

Three-phase a.c. induction motors

Figure 8.110 Three-phase induction motor showing component parts

Induction motors operate because a moving magnetic field induces a current to flow in the rotor. This current in the rotor then creates a second magnetic field, which combines with the field from the stator windings to exert a force on the rotor conductors, thus turning the rotor.

Production of the rotating field

Figure 8.111 shows the stator of a three-phase motor to which a three-phase supply is connected. The windings in the diagram are in star formation and two windings of each phase are wound in the same direction.

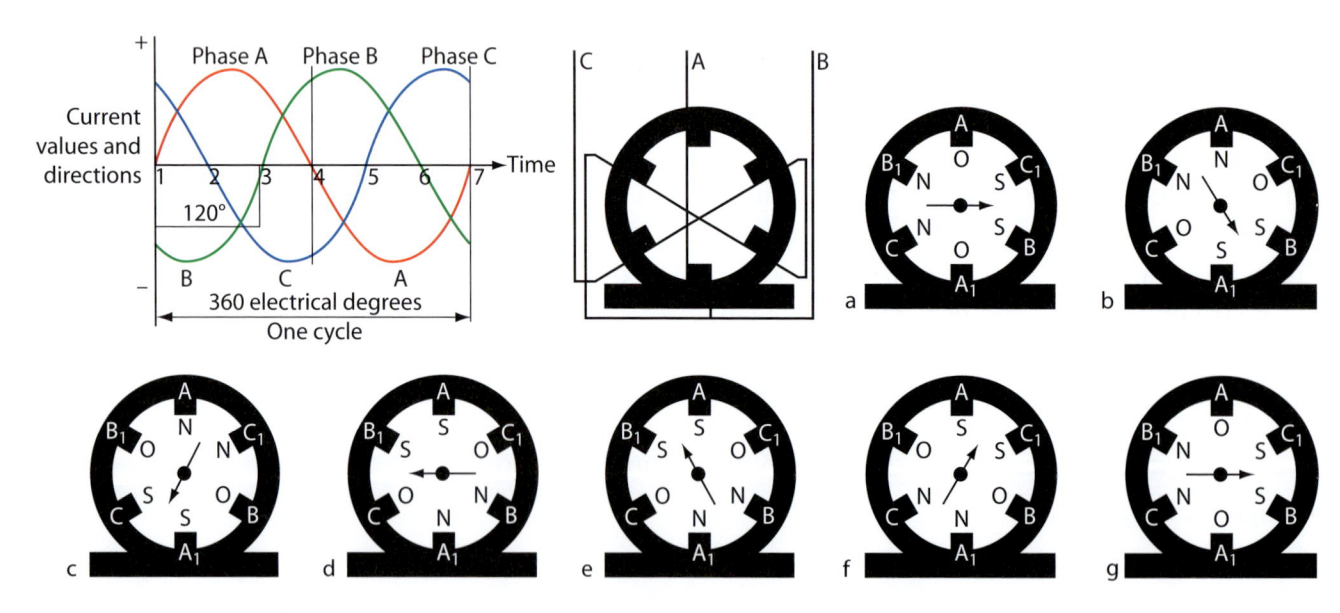

Figure 8.111 Stator of three-phase motor with three-phase supply connection

Each pair of windings will produce a magnetic field, the strength of which will depend upon the current in that particular phase at any instant of time. When the current is zero, the magnetic field will be zero. Maximum current will produce the maximum magnetic field.

As the currents in the three phases are 120° out of phase (see graph in Figure 8.111) the magnetic fields produced will also be 120° out of phase. The magnetic field set up by the three-phase currents will therefore give the appearance of rotating clockwise around the stator.

The resultant magnetic field produced by the three phases is at any instant of time in the direction shown by the arrow in the diagram, where diagrams (1) to (7) in Figure 8.111 show how the direction of the magnetic field changes at intervals of 60° through one complete cycle. The speed of rotation of the magnetic field depends upon the supply frequency and the number of 'pole pairs', and is referred to as the **synchronous** speed. We will discuss synchronous speed later in this unit.

The direction in which the magnetic field rotates is dependent on the sequence in which the phases are connected to the windings. Reversing the connection of any two incoming phases can therefore reverse rotation of the magnetic field.

Stator construction

As shown in Figure 8.112, the stator (stationary component) comprises the field windings, which are many turns of very fine copper wire wound on to formers, which are then fixed to the inside of the stator steel frame (sometimes called the yoke).

The formers have two roles:

1. to contain the conductors of the winding
2. to concentrate the magnetic lines of flux to improve the flux linkage.

The formers are made of laminated silicon steel sections to reduce eddy currents, thereby reducing the I^2R losses and reducing heat. The number of poles fitted will determine the speed of the motor.

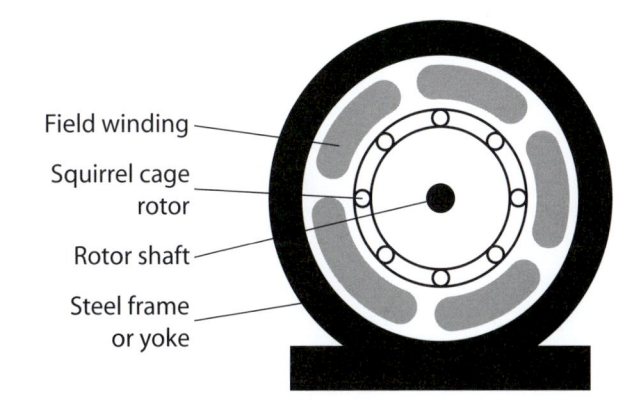

Figure 8.112 Stator construction

Figure 8.113 Stator field winding

Squirrel-cage rotor

In the squirrel cage (see Figure 8.114), the bars of the rotor are shorted out at each end by 'end rings' to form the shape of a cage. This shape creates numerous circuits within it for the induced e.m.f. and resultant current to flow and thus produce the required magnetic field.

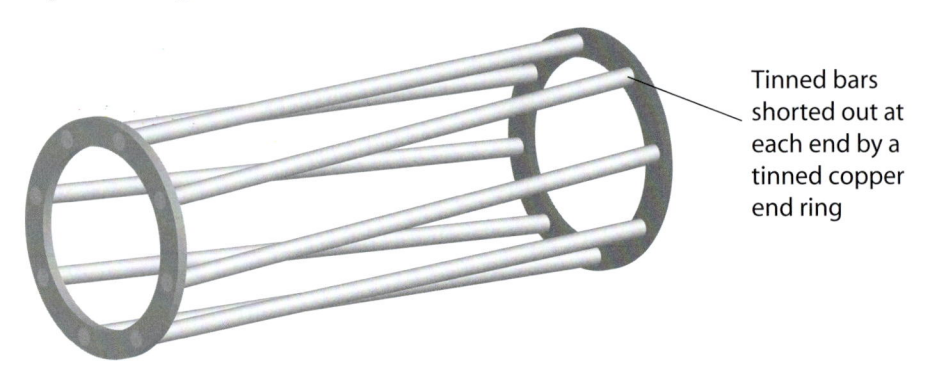

Tinned bars shorted out at each end by a tinned copper end ring

Figure 8.114 Squirrel-cage rotor

Figure 8.115 shows the cage fitted to the shaft of the motor. The rotor bars are encased within many hundreds of very thin laminated (insulated) segments of silicon steel and are skewed to increase the rotor resistance.

Carbon steel shaft

Tinned copper rotor bars encased in thin 0.5 mm laminated steel segments, to reduce losses and skewed to assist starting

Low-friction bearings

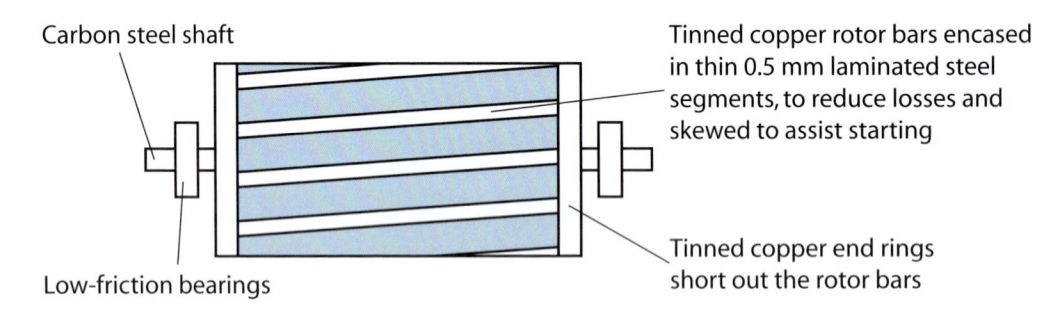

Tinned copper end rings short out the rotor bars

Figure 8.115 Cage fitted to shaft and motor

> **Did you know?**
>
> Squirrel cage induction motors are sometimes referred to as the 'workhorse' of the industry, as they are inexpensive and reliable and suited to most applications.

On the shaft will be two low-friction bearings which enable the rotor to spin freely. The bearings and the rotor will be held in place within the yoke of the stator by two end caps which are normally secured in place by long nuts and bolts that pass completely through the stator.

When a three-phase supply is connected to the field windings, the lines of magnetic flux produced in the stator, rotating at 50 revolutions per second, cut through the bars of the rotor, inducing an e.m.f. into the bars.

Faraday's Law states that 'when a conductor cuts, or is cut, by a magnetic field, then an e.m.f. is induced in that conductor the magnitude of which is proportional to the rate at which the conductor cuts or is cut by the magnetic flux'.

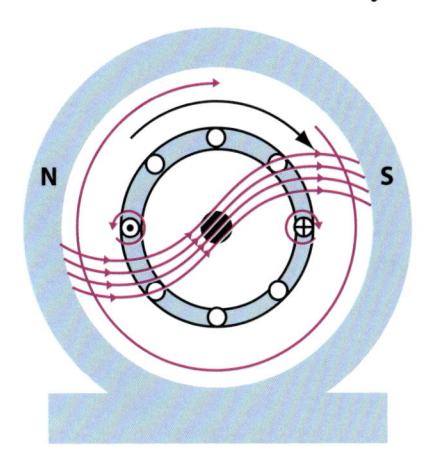

Figure 8.116 Rotating field

The e.m.f. produces circulatory currents within the rotor bars, which in turn result in the production of a magnetic field around the bars. This leads to the distortion of the magnetic field as shown in Figure 8.116. The interaction of these two magnetic fields results in a force being applied to the rotor bars, and the rotor begins to turn. This turning force is known as a torque, the direction of which is always as Fleming's left-hand rule indicates.

Wound rotor

In the wound rotor type of motor, the rotor conductors form a three-phase winding, which is starred internally. The other three ends of the windings are brought out to slip rings mounted on the shaft. Thus it is possible through brush connections to introduce resistance into the rotor circuit, albeit this is normally done on starting only to increase the starting torque. This type of motor is commonly referred to as a slip-ring motor.

Figure 8.117 shows a completed wound-rotor motor assembly. Although it looks like a squirrel-cage motor, the difference is that the rotor bars are exchanged for heavy conductors that run through the laminated steel rotor, the ends then being brought out through the shaft to the slip rings on the end.

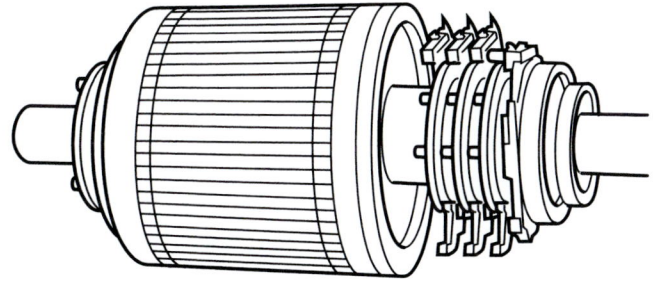

Figure 8.117 Wound-rotor motor assembly

The wound-rotor motor is particularly effective in applications where using a squirrel-cage motor may result in a starting current that is too high for the capacity of the power system. The wound-rotor motor is also appropriate for high-inertia loads having a long acceleration time. This is because you can control the speed, torque and resulting heating of the wound-rotor motor. This control can be automatic or manual. It's also effective with high-slip loads as well as adjustable-speed installations that do not require precise speed control or regulation. Typical applications include conveyor belts, hoists and elevators.

Single-phase induction motors

If we were to construct an induction motor as shown in Figure 8.118, we would find that, on connecting a supply to it, it would not run. However, if we were then to spin the shaft with our fingers, we would find that the motor *would* continue to run. Why is this?

When an a.c. supply is connected to the motor, the resulting current flow, and therefore the magnetic fields produced in the field windings, changes polarity, backwards and forwards, 100 times per second. Therefore no lines of flux cut through the rotor bars. If no lines of flux cut through the rotor bars then there is no e.m.f. being produced in them and there is therefore no magnetic field for the stator winding to interact with.

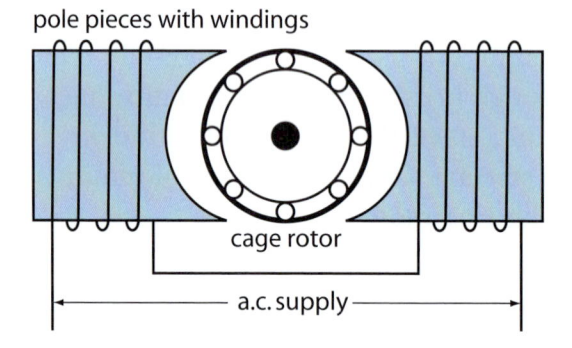

Figure 8.118 Single-phase induction motor

However, if we spin the rotor, we create the effect of the rotor bars cutting through the lines of force, hence the process starts and the motor runs up to speed. If we stop the motor and then connect the supply again, the motor will still not run. If this time we spin the rotor in the opposite direction, the motor will run up to speed in the new direction. So how can we get the rotor to turn on its own?

If we think back to the three-phase motor we discussed previously, we did not have this problem, because the connection of a three-phase supply to the stator automatically produced a rotating magnetic field. This is what is missing from the motor in Figure 8.118: we have no rotating field.

The split-phase motor (induction start/induction run)

We can overcome this problem if we add another set of poles, positioned 90° around the stator from our original wiring, as shown in Figure 8.119.

Now when the supply is connected, both sets of windings are energised, both windings having resistive and reactive components to them – resistance as every conductor has and also inductive reactance because the conductors form a coil. These are known as the 'start' and 'run' windings.

The start winding is wound with fewer turns of smaller wire than the main winding, so it has a higher resistance. This extra resistance creates a small phase shift that results in the current in the start winding lagging that in the run winding by approximately 30°, as shown in Figure 8.120.

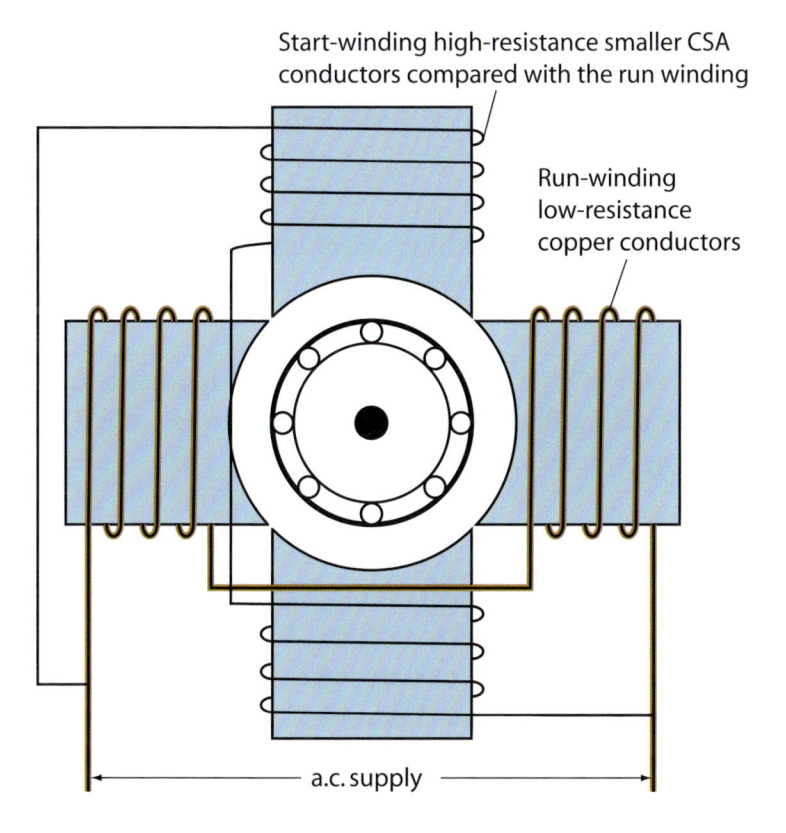

Figure 8.119 Split-phase induction motor

Strong magnetic field in the run winding, weaker field in the start winding

Strong magnetic field in the start winding 30° later, weaker field in the run winding

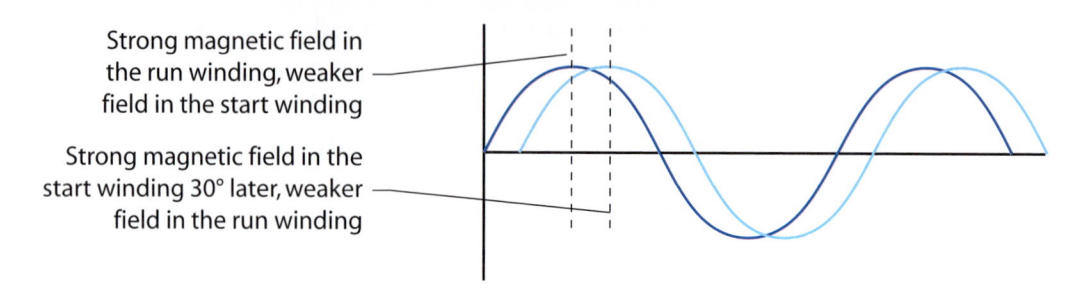

Figure 8.120 Run and start winding phases

Therefore the magnetic flux in each of the windings is growing and collapsing at different periods in time, so that for example as the run winding is having a strong north/south on the face of its pole pieces, the start winding will only have a weak magnetic field.

In the next instance, the run winding's magnetic field has started to fade, but the start winding's magnetic field is now strong, presenting to the rotor an apparent shift in the lines of magnetic flux.

In the next half of the supply cycle, the polarity is reversed and the process repeated, so there now appears to be a rotating magnetic field. The lines of force cutting through the rotor bars induce an e.m.f. into them, and the resulting current flow now produces a magnetic field around the rotor bars: the interaction between the magnetic fields of the rotor and stator takes place and the motor begins to turn.

It is because the start and run windings carry currents that are out of phase with each other that this type of motor is called the 'split-phase'.

Once the motor is rotating at about 75% of its full load speed, the start winding is disconnected by the use of a device called a centrifugal switch, which is attached to the shaft; see Figure 8.121.

This switch works by centrifugal action, in that sets of contacts are held closed by a spring, and this completes the circuit to the start winding. When the motor starts to turn, a little weight gets progressively thrown away from the shaft, forcing the contacts to open and thus disconnecting the start winding. It's a bit like the fairground ride known as 'the Rotor', in which you are eventually held against the sides of the ride by the increasing speed of the spinning wheel. Once the machine has disconnected the start winding, the machine continues to operate from the run winding.

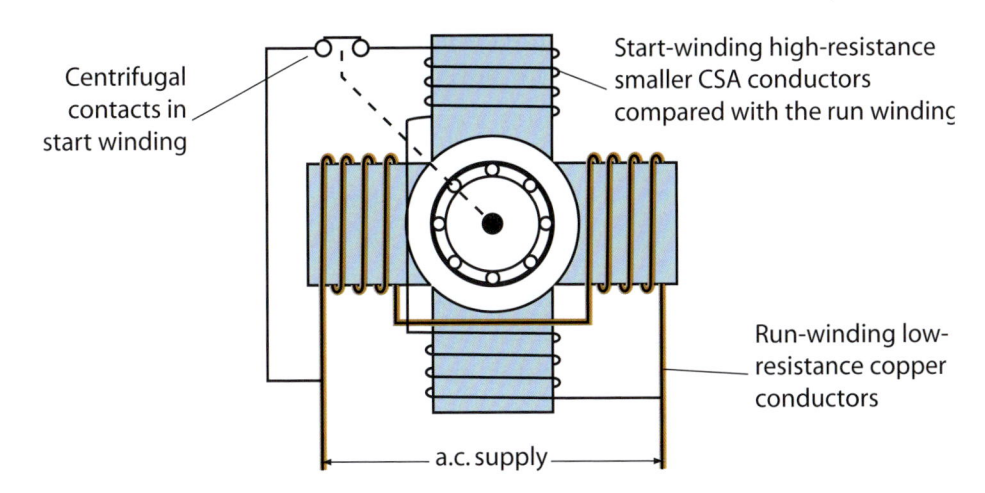

Figure 8.121 Split-phase induction motor with centrifugal switch

The split-phase motor's simple design makes it typically less expensive than other single-phase motors. However, it also limits performance. Starting torque is low at about 150% to 175% of the rated load. Also, the motor develops high starting currents of about six to nine times the full load current. A lengthy starting time can cause the start winding to overheat and fail, and therefore this type of motor shouldn't be used when a high starting torque is needed. Consequently it is used on light-load applications such as small hand tools, small grinders and fans, where there are frequent stop/starts and the full load is applied after the motor has reached its operating speed.

Reversal of direction

If you think back to the start of this section, we talked about starting the motor by spinning the shaft. We also said that we were able to spin it in either direction and the motor would run in that direction. It therefore seems logical that in order to change the direction of the motor, all we have to concern ourselves with is the start winding. We therefore need only to reverse the connections to the start winding to change its polarity, although you may choose to reverse the polarity of the run winding instead. The important thing to remember is that if you change the polarity through both the run and start windings, the motor will continue to revolve in the same direction.

The capacitor-start motor (capacitor start/induction run)

Normally perceived as being a wide-ranging industrial motor, the capacitor-start motor is very similar to the split-phase motor discussed previously. Indeed, it probably helps to think of this motor as being a split-phase motor but with an enhanced start winding that includes a capacitor in the circuit to help out with the start process. If we look at Figure 8.122 we can see the capacitor mounted on top of the motor case.

In this motor the start winding has a capacitor connected in series with it, and since this gives a phase difference of nearly 90° between the two currents in the windings, the starting performance is improved. We can see this represented in the sine waves shown in Figure 8.123.

Figure 8.122 Capacitor-start motor

Strong magnetic field in the run winding, weaker field in the start winding

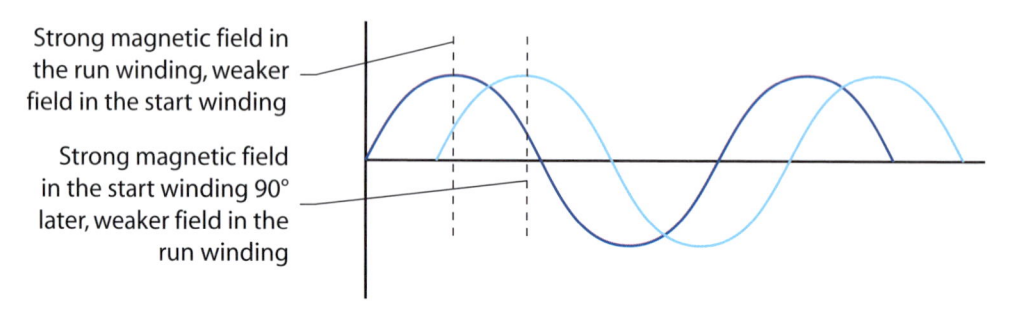

Strong magnetic field in the start winding 90° later, weaker field in the run winding

Figure 8.123 Magnetic field in capacitor-start motor

In this motor the current through the run winding lags the supply voltage due to the high inductive reactance of this winding, and the current through the start winding leads the supply voltage due to the capacitive reactance of the capacitor. The phase displacement in the currents of the two windings is now approximately 90°. Figure 8.124 shows the winding connections for a capacitor-start motor.

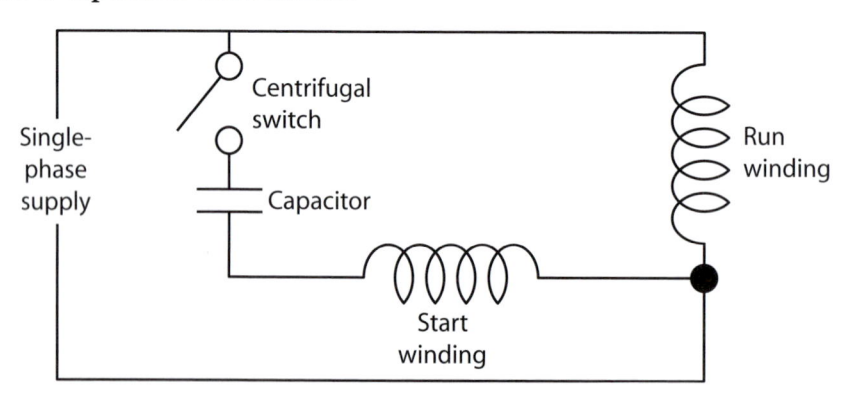

Figure 8.124 Winding connections for capacitor-start split-phase motor

The magnetic flux set up by the two windings is much greater at starting than in the standard split-phase motor, and this produces a greater starting torque. The typical starting torque for this type of motor is about 300% of full-load torque, and a typical starting current is about five to nine times the full-load current.

The capacitor-start motor is more expensive than a comparable split-phase design because of the additional cost of the capacitor. But the application range is much wider because of higher starting torque and lower starting current. Therefore, because of its improved starting ability, this type of motor is recommended for loads that are hard to start, so we see this type of motor used to drive equipment such as lathes, compressors and small conveyor systems.

As with the standard split-phase motor, the start windings and the capacitor are disconnected from the circuit by an automatic centrifugal switch when the motor reaches about 75% of its rated full-load speed.

Reversal of direction

Reversing the connections to the start winding will only change its polarity, although we may choose to reverse the polarity of the run winding instead.

Permanent split capacitor (PSC) motors

Permanent split capacitor (PSC) motors look exactly the same as capacitor-start motors. However, a PSC motor doesn't have either a starting switch or a capacitor that is strictly used for starting. Instead, it has a run-type capacitor permanently connected in series with the start winding, and the second winding is permanently connected to the power source. This makes the start winding an auxiliary winding once the motor reaches running speed. However, because the run capacitor must be designed for continuous use, it cannot provide the starting boost of a starting capacitor.

Typical starting torques for thus type of motor are low, from 30% to 150% of rated load, so these motors are not used in difficult starting applications. However, unlike the split-phase motor, PSC motors have low starting currents, usually less than 200% of rated load current, making them excellent for applications with high cycle rates.

PSC motors have several advantages. They need no starting mechanism and so can be reversed easily, and designs can easily be altered for use with speed controllers. They can also be designed for optimum efficiency and high power factor at rated load.

Permanent split capacitor motors have a wide variety of applications depending on the design. These include fans, blowers with low starting torque and intermittent cycling uses such as adjusting mechanisms, gate operators and garage-door openers, many of which also need instant reversing.

Capacitor start-capacitor run motors

In appearance we can distinguish this motor because of the two capacitors that are mounted on the motor case.

This type of motor is widely held to be the most efficient single-phase induction motor, as it combines the best of the capacitor-start and the permanent split capacitor designs and is able to handle applications too demanding for any other kind of single-phase motor.

As shown in Figure 8.125, it has a start capacitor in series with the auxiliary winding like the capacitor-start motor, and this allows for high starting torque. However, like the PSC motor, it also has a run capacitor that remains in series with the auxiliary winding after the start capacitor is switched out of the circuit.

Figure 8.125 Capacitor start-capacitor run motor

Another advantage of this type of motor is that it can be designed for lower full-load currents and higher efficiency, which means that it operates at a lower temperature than other single-phase motor types of comparable horsepower. Typical uses include woodworking machinery, air compressors, high pressure water pumps, vacuum pumps and other high-torque applications.

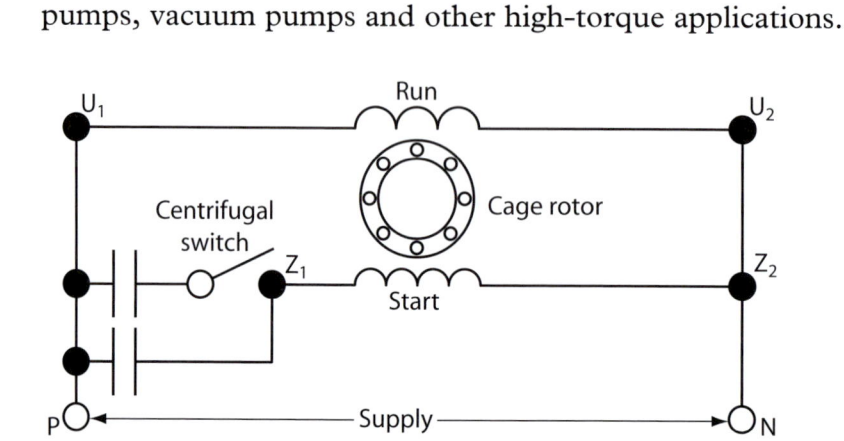

Figure 8.126 Capacitor start-capacitor run 'split-phase' motor

Shaded-pole motors

One final type of single-phase induction motor that is worthy of mention is the shaded-pole type. We cover this last as, unlike all of the previous single-phase motors we have discussed, shaded-pole motors have only one main winding and no start winding.

Starting is by means of a continuous copper loop wound around a small section of each motor pole. This 'shades' that portion of the pole, causing the magnetic field in the ringed area to lag the field in the non-ringed section. The reaction of the two fields then starts the shaft rotating.

Because the shaded pole motor lacks a start winding, starting switch or capacitor, it is electrically simple and inexpensive. In addition, speed can be controlled merely by varying voltage or through a multi-tap winding.

The shaded pole motor has many positive features, but it also has several disadvantages. As the phase displacement is very small, it has a low starting torque, typically in the region of 25% to 75% of full-load torque. Also, it is very inefficient, usually below 20%.

Low initial costs make shaded pole motors suitable for light-duty applications such as multi-speed fans for household use and record turntables.

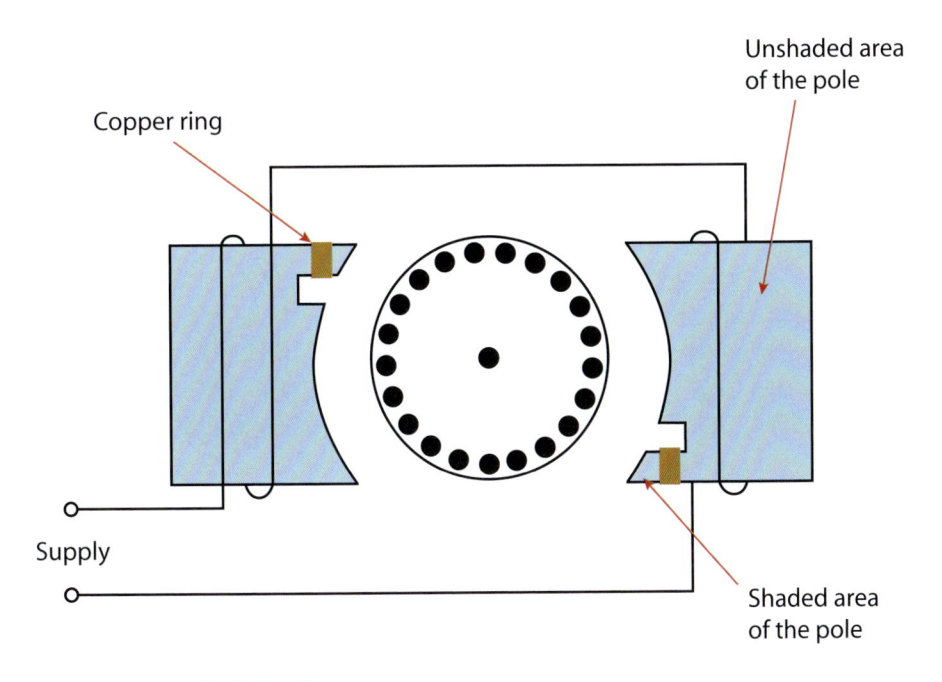

Figure 8.127 Shaded pole motor

Construction of a three-phase squirrel-cage induction motor

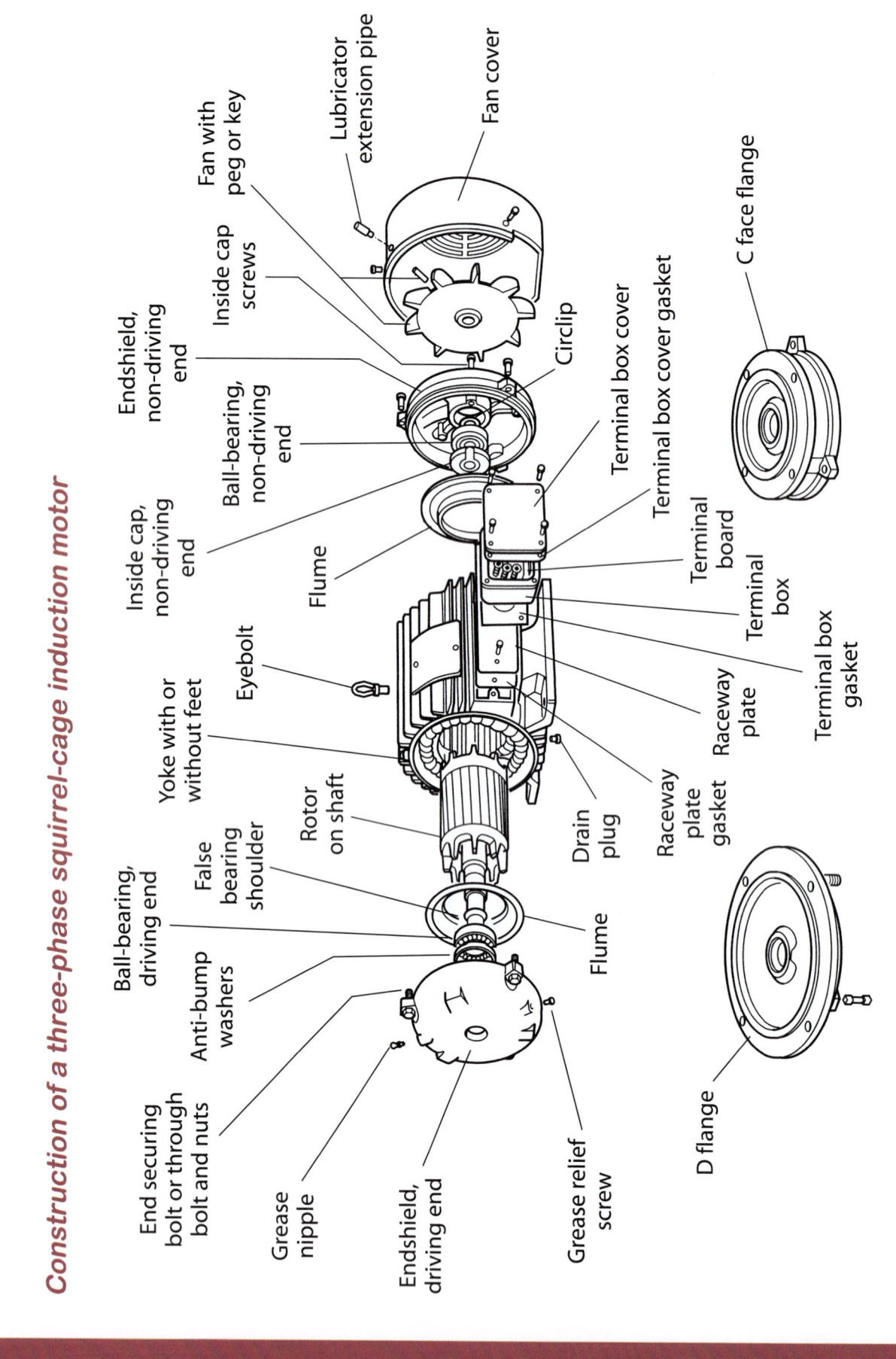

- Lubricator extension pipe
- Fan cover
- Fan with peg or key
- Inside cap screws
- Endshield, non-driving end
- Ball-bearing, non-driving end
- Circlip
- Terminal box cover
- Terminal box cover gasket
- C face flange
- Inside cap, non-driving end
- Flume
- Terminal board
- Terminal box
- Eyebolt
- Raceway plate
- Terminal box gasket
- Yoke with or without feet
- Rotor on shaft
- Drain plug
- Raceway plate gasket
- Ball-bearing, driving end
- False bearing shoulder
- Flume
- Anti-bump washers
- End securing bolt or through bolt and nuts
- Grease nipple
- Endshield, driving end
- Grease relief screw
- D flange

Figure 8.128 Construction of a three-phase squirrel-cage induction motor

Construction of a single-phase, capacitor start, induction motor

This motor consists of a laminated stator wound with single-phase windings arranged for split-phase starting, and a cage rotor. The cage rotor is a laminated framework with lightly insulated longitudinal slots into which copper or aluminium bars are fitted.

The bars are then connected together at their ends by metal end-rings. No electrical connections are made to the rotor.

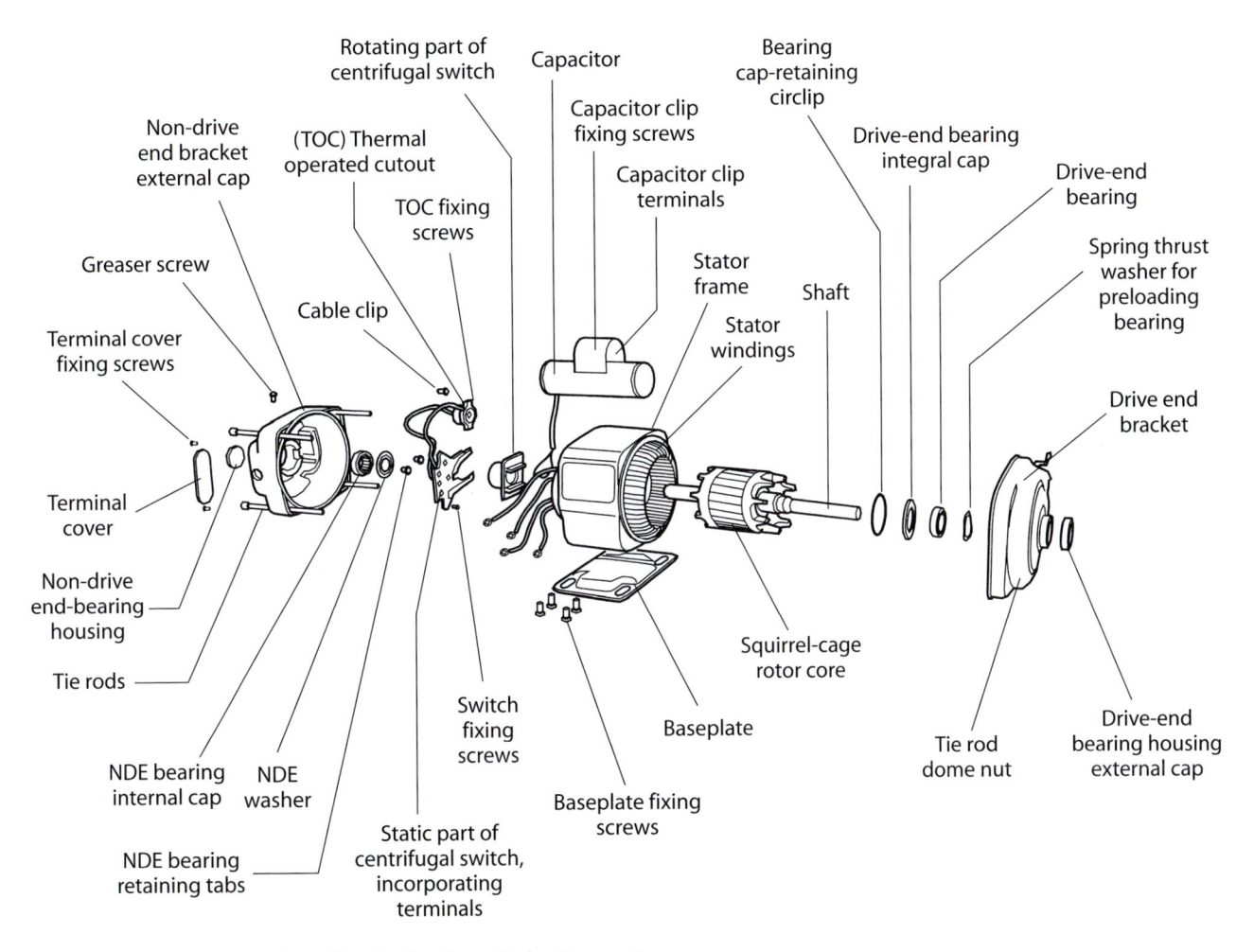

Figure 8.129 Construction of a single-phase induction motor

Motor speed and slip calculation

There are essentially two ways to express the speed of a motor.

- **Synchronous speed**. For an a.c. motor this is the speed of rotation of the stator's magnetic field. Consequently, this is really only a theoretical speed, as the rotor will always turn at a slightly slower rate.

- **Actual speed**. This is the speed at which the shaft rotates. The nameplate on most a.c. motors will give the actual motor speed rather than the synchronous speed.

The difference between the speed of the rotor and the synchronous speed of the rotating magnetic field is known as the **slip**, which can be expressed either as a unit or in percentage terms. Because of this, we often refer to the induction motor as being an asynchronous motor.

Remember, the speed of the rotating magnetic field is known as the synchronous speed, and this will be determined by the frequency of the supply and the number of pairs of poles within the machine. The speed at which the rotor turns will be between 2% and 5% slower, with an average of 4% being common.

The reduced speed of the rotor is due to it having to overcome friction, as during the turning movement there is friction from the bearings in addition to any friction deriving from the load that the motor is connected to. Another factor that comes into play in determining the speed of a motor is **windage**. This means that within the enclosure there is a certain amount of air, and as the rotor rotates it has to move the air, which in turn contributes to the slowing down of the rotor. Of course there are no moving parts involved with the rotating magnetic field, so the rotor will never catch up. However, were some miracle to happen and the rotor reached the synchronous speed, we would have a different problem.

We already know that when a conductor passes at right angles through a magnetic field current is induced in the conductor. The direction of the induced current is dependent on the direction of movement of the conductor, and the strength of the current is determined by the speed at which the conductor moves. If the rotating magnetic field and the rotor are now revolving at the same speed, there will be no lines of magnetic flux cutting through the rotor bars, no induced e.m.f. and consequently no resultant magnetic field around the rotor bars to interact with the rotating magnetic field of the stator. The motor will immediately slow down and, having slowed down, will then start to speed up as the lines of magnetic flux start to cut through the rotor bars again – and so the process would continue.

Standard a.c. induction motors therefore depend on the rotor trying, but never quite managing, to catch up with the stator's magnetic field. The rotor speed is just slow enough to cause the

> **Key term**
>
> **Windage** – the air resistance of a moving object or the force of the wind on a stationary object

> **Remember**
>
> When a conductor passes at right angles through a magnetic field, current is induced in the conductor. The direction of the induced current will depend on the direction of movement of the conductor, and the strength of the current will be determined by the speed at which the conductor moves.

proper amount of rotor current to flow, so that the resulting torque is sufficient to overcome windage and friction losses and drive the load.

Determining synchronous speed and slip

All a.c. motors are designed with various numbers of magnetic poles. Standard motors have two, four, six or eight poles, and these poles play an important role in determining the synchronous speed of an a.c. motor.

As we said before, the synchronous speed can be determined by the frequency of the supply and the number of pairs of poles within the machine. We can express this relationship with the following formula:

Synchronous speed (n_s) in revolutions per second

$$= \frac{\text{Frequency (f) in Hz}}{\text{The number of pole pairs (p)}}$$

Example 1

Calculate the synchronous speed of a four-pole machine connected to a 50 Hz supply.

$$n_s = \frac{f}{p}$$

As we know the motor has four poles, this means it has two pole pairs. We can therefore complete the calculation as:

$$n_s = \frac{50}{2} \quad \text{therefore, } n_s = 25 \text{ revolutions per second (rps)}$$

To convert revolutions per second into the more commonly used revolutions per minute (rpm), simply multiply n_s by 60. This new value is referred to as N_s, which in this example will become 25 × 60, giving 1500 rpm.

We also said that we refer to the difference between the speed of the rotor and the synchronous speed of the rotating magnetic field as the slip, which can be expressed either as a unit (S) or in percentage terms (S per cent). We express this relationship with the following formula:

$$\text{per cent slip} = \frac{\text{Sychronous speed}(n_s) - \text{Rotor speed}(n_r)}{\text{Synchronous speed}(n_s)} \times 100$$

Example 2

In this example numbers have been rounded up for ease.
A six-pole cage-induction motor runs at 4% slip.

Calculate the motor speed if the supply frequency is 50 Hz.

$$S(\text{per cent}) = \frac{(n_s - n_r)}{n_s} \times 100$$

We therefore need first to establish the synchronous speed; as the motor has six poles it will have three pole pairs. Consequently:

Synchronous speed $n_s = \frac{f}{p}$ giving us $\frac{50}{3}$ and therefore $n_s = 16.7$ revs/sec

We can now put this value into our formula and then, by transposition, rearrange the formula to make n_r the subject. Consequently:

$$S(\text{per cent}) = \frac{(n_s - n_r)}{n_s} \times 100 \quad \text{giving us} \quad 4 = \frac{(16.7 - n_r)}{16.7} \times 100$$

Therefore by transposition:

$$4 = \frac{(16.7 - n_r)}{16.7} \times 100 \quad \text{give us} \quad (16.7 - n_s) = \frac{4 \times 16.7}{100}$$

When calculated:

$$(16.7 - n_r) = \frac{4 \times 16.7}{100} \quad \text{becomes} \quad (16.7 - n_r) = 0.668$$

Therefore by further transposition:

$(16.7 - n_r) = 0.668$ becomes $16.7 - 0.668 = n_r$

Therefore n_r = **16.032 rps or N_r = 962 rpm**

Working life

You are called out to check a split-phase motor which has been reported as not turning and making a humming sound.

1. What steps would you take to diagnose the problem?
2. What are the likely causes of this problem?

Synchronised motors

A synchronous motor, as the name suggests, runs at synchronous speed. Because of the problems discussed earlier, this type of motor is not self-starting and instead must be brought up to almost synchronous speed by some other means.

Three-phase a.c. synchronous motors

To understand how the synchronous motor works, assume that we have supplied three-phase a.c. power to the stator, which in turn causes a rotating magnetic field to be set up around the rotor. The rotor is then supplied via a field winding with d.c. and consequently acts a bit like a bar magnet, having north and south poles. The rotating magnetic field now attracts the rotor field that was activated by the d.c. This results in a strong turning force on the rotor shaft, and the rotor is therefore able to turn a load as it rotates in step with the rotating magnetic field.

It works this way once it's started. However, one of the disadvantages of a synchronous motor is that it cannot be started from a standstill by just applying a three-phase a.c. supply to the stator. When a.c. is applied to the stator, a high-speed rotating magnetic field appears immediately. This rotating field rushes past the rotor poles so quickly that the rotor does not have a chance to get started. In effect, the rotor is repelled first in one direction and then the other.

An induction winding (squirrel-cage type) is therefore added to the rotor of a synchronous motor to cause it to start, effectively meaning that the motor is started as an induction motor. Once the motor reaches synchronous speed, no current is induced in the squirrel-cage winding, so it has little effect on the synchronous operation of the motor.

Synchronous motors are commonly driven by transistorised variable-frequency drives.

Single-phase a.c. synchronous motors

Small single-phase a.c. motors can be designed with magnetised rotors. The rotors in these motors do not require any induced current so they do not slip backwards against the mains frequency. Instead, they rotate synchronously with the mains frequency. Because of their highly accurate speed, such motors are usually used to power mechanical clocks, audio turntables and tape drives; formerly they were also widely used in accurate timing instruments such as strip-chart recorders or telescope drive mechanisms. The shaded-pole synchronous motor is one version.

As with the three-phase version, inertia makes it difficult to accelerate the rotor instantly from stopped to synchronous speed, and the motors normally require some sort of special feature to get started. Various designs use a small induction motor (which may share the same field coils and rotor as the synchronous motor) or a very light rotor with a one-way mechanism (to ensure that the rotor starts in the 'forward' direction).

Motor windings

A motor can be manufactured with the windings internally connected. If this is the case and there are three terminal connections in the terminal block labelled U, V and W, you would expect the motor windings to be connected in a delta configuration. This is shown in Figure 8.130.

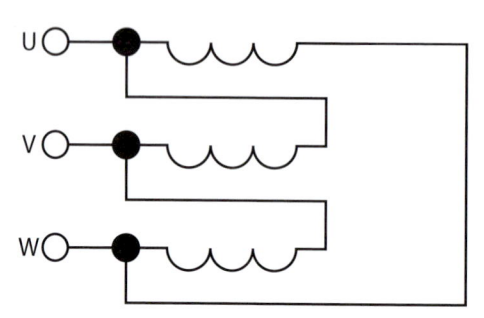

Figure 8.130 Motor windings with delta connection

However, there may be four connections in the terminal box labelled U, V, W and N. If this is the case, the windings would be arranged to give a star configuration, as shown in Figure 8.131.

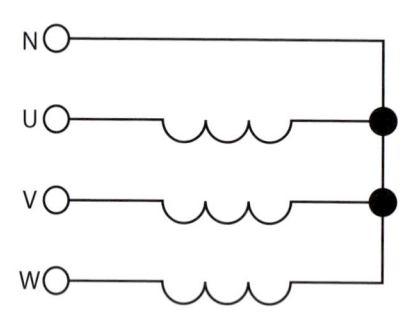

Figure 8.131 Motor windings with four connections

> **Remember**
>
> Some older motors may have different markings in the terminal block, and you should therefore refer to the manufacturer's data.

Alternatively the terminal block may contain six connections: U1, U2, V1, V2, W1 and W2. This is used where both star and delta configurations are required. The terminal connections can then be reconfigured for either star or delta, starting within the terminal block. Figure 8.132 illustrates the connections that would come out to the terminal block.

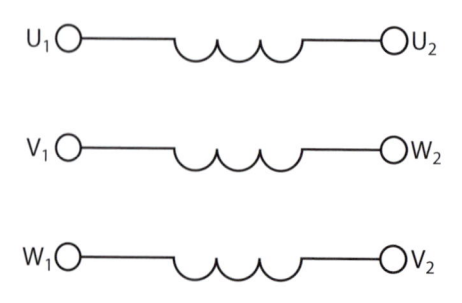

Figure 8.132 Motor windings with six connections

Operating principles, limitations and applications of motor control

A practical motor starter has to do more than just switch on the supply. It also has to have provision for automatically disconnecting the motor in the event of overloads or other faults. The overload protective device monitors the current consumption of the motor and is set to a predetermined value that the motor can safely handle. When a condition occurs that exceeds the set value, the overload device opens the motor starter control circuit and the motor is turned off. The overload protection can come in a variety of types, including solid-state electronic devices.

The starter should also prevent automatic restarting should the motor stop because of supply failure. This is called **no-volts protection** and will be discussed later in this section.

The starter must also provide for the efficient stopping of the motor by the user. Provision for this is made by the connection of remote stop buttons and safety interlock switches where required.

The Direct-On-Line (DOL) starter

This is the simplest and cheapest method of starting squirrel-cage (induction) motors.

The expression 'Direct-On-Line' starting means that the full supply voltage is directly connected to the stator of the motor by means of a contactor-starter, as shown in Figure 8.133.

Since the motor is at a standstill when the supply is first switched on, the initial starting current is heavy. This 'inrush' of current can be as high as 6 to 10 times the full load

Figure 8.133 DOL starter

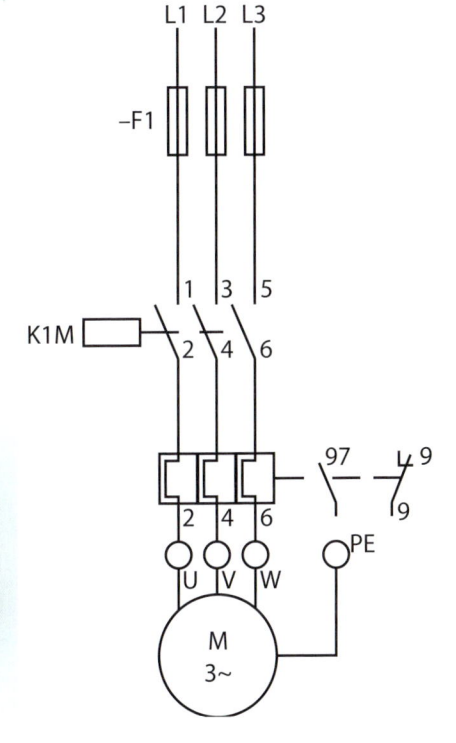

Figure 8.134 DOL starter

current, i.e. a motor rated at 10 A could have a starting current inrush as high as 60 A, and the initial starting torque can be about 150% of the full-load torque. Thus you may observe motors 'jumping' on starting if their mountings are not secure. As a result, Direct-On-Line starting is usually restricted to comparatively small motors with outputs of up to about 5 kW.

DOL starters should also incorporate a means of overload protection, which can be operated by either magnetic or thermal overload trips. These activate when there is a sustained increase in current flow.

To reverse the motor you need to interchange any two of the incoming three-phase supply leads. If a further two leads are interchanged then the motor will rotate in the original direction.

Operating principle of a DOL starter

A three-pole switch controls the three-phase supply to the starter. This switch normally includes fuses, which provide a means of electrical isolation and also short-circuit protection. We shall look at the operation of the DOL starter in stages.

Let's start by looking at the one-way switch again. In this circuit (Figure 8.135), the switch is operated by your finger and the contacts are then held in place mechanically.

We could decide that we don't want to operate the switch this way and instead use a relay. In this system (Figure 8.136), when the coil is energised it creates a magnetic field. Everything in the magnetic field will be pulled in the direction of the arrow and the metal strip will be pulled onto the contacts. As long as the coil remains energised and is a magnet, the light will stay on.

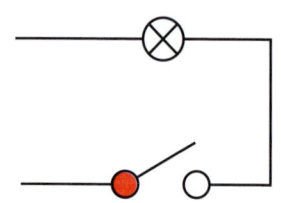

Figure 8.135 One-way switch

Looking at Figures 8.135 and 8.136, we can see that the first option works well enough for its intended purpose. However, it couldn't be used in a three-phase system as we would need one for each phase and would have to trust to luck each time we tried to hit all three switches at the same time. However, the second option does provide us with an effective method of controlling more than one thing from one switch, as long as they are all in the same magnetic field.

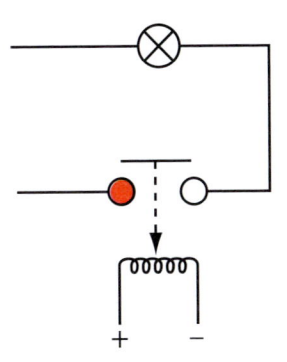

Figure 8.136 Switch relay

Let's apply this knowledge to a DOL starter. We know that we can't have three one-way switches in the starter. But it helps to try to think of the contacts: where the switches aren't operated by your fingers, but are pulled in the direction of the arrow by the magnetic effect of the coil (see Figure 8.137), items affected by the same magnetic effect are normally shown linked by a dotted line. For ease of explanation we'll use a 230 V coil.

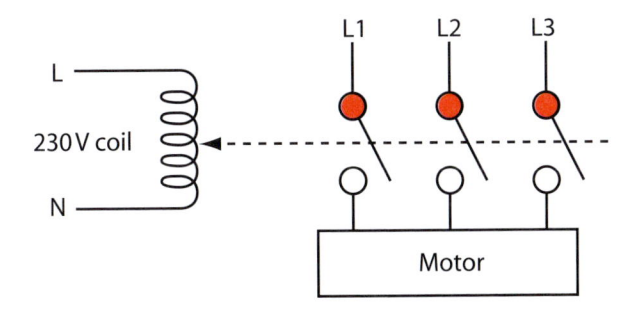

Figure 8.137 Three-switch relay

From this we can see that as soon as we put a supply onto the coil it energises, becomes a magnet and pulls the contacts in. Obviously, this would be no good for our starter, as every time the power is put on, the starter will become active and start operating whatever is connected to it. In the case of machinery this could be very dangerous. The starter design therefore goes one step further to include no-volts protection. The simple addition of a 'normally open' Start button gives this facility, as shown in Figure 8.138.

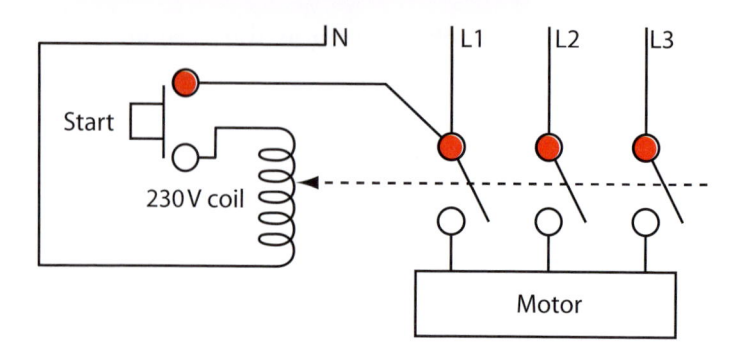

Figure 8.138 Three-switch relay with 'normally open' Start button

So far so good. We now have to make a conscious decision to start the motor.

Our next problem though, is that every time we take our finger off the Start button, the button springs back out, the supply to the coil is lost and the motor stops. This is the 'no volt' protection element in operation.

What we need for normal operation is a device called the 'hold in' or 'retaining' contact, as shown in Figure 8.139. This is a 'normally closed' contact (position 1) which is also placed in the magnetic field of the coil. Consequently, when the coil is energised, it is also pulled in the direction shown, and in this case, across and onto the terminals of the Start button (position 2).

We can now take our finger off the Start button, as the supply will continue to feed the coil by running through the 'hold in' contact that is linking out the Start button terminals (position 2).

This now means that we can only break the coil supply and stop the motor by fitting a Stop button. In the case of starters fitted with a 230 V coil, this will be a normally closed button placed in the neutral wire.

Now, for the fraction of a second that we push the Stop button, the supply to the coil is broken, the coil ceases to be a magnet and the 'hold in' contact returns to its original position (position 1). Since the Start button had already returned to its original open position when we released it, when we take our finger off the Stop button everything will be as we first started. Therefore any loss of supply will immediately break the supply to the coil and stop the motor: if a supply failure was restored, the equipment could not restart itself – someone would have to take the conscious decision to do so.

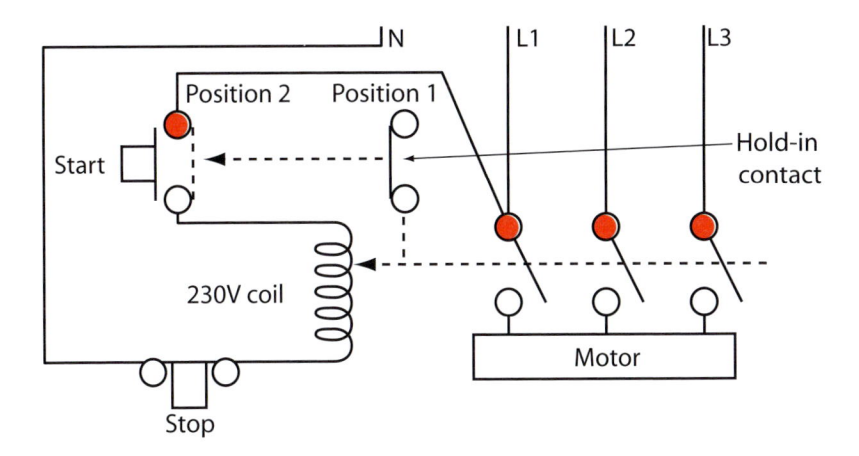

Figure 8.139 Restarting the motor

This system is basically the same as that of a contactor. In fact, many people refer to this item as the contactor starter. Such starters are also available with a 400 V coil, which is therefore connected across two of the phases.

Remote stop/start control

In the DOL starter as described, we have the means of stopping and starting the motor from the buttons provided on the starter enclosure. However, there are situations where the control of the motor needs to take place from some remote location. This could be, for example, in a college workshop where, in the case of an emergency, emergency stop buttons located throughout the workshop can be activated to break the supply to a motor. Equally, because of the immediate environment around the motor, it may be necessary to operate it from a different location.

Commonly known as a remote stop/start station, the enclosure usually houses a start and a stop button connected in series. However, depending on the circumstances it is also possible to have an additional button included to give 'inch' control of a motor.

If we use the example of our DOL starter as described in the previous diagrams, but now include the remote stop/start station, the circuit would look as in Figure 8.140, where for ease the additional circuitry has been shown in red.

As can be seen, the remote start button is effectively in parallel with the start button on the main enclosure with the supply to both of these buttons being routed via the stop button of the remote station.

If the intention is to provide only emergency stops, omit the remote station shown so that these are all connected in series with the stop button on the main enclosure.

> **Did you know?**
> Inch control gives the ability to partially start a motor without it commencing its normal operation.

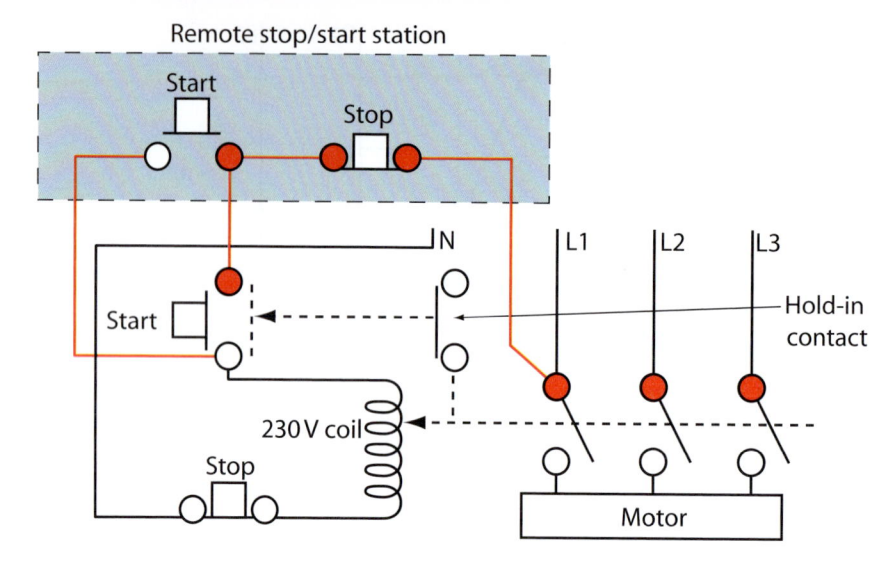

Level 3 NVQ/SVQ Diploma Installing Electrotechnical Systems and Equipment Book B

Figure 8.140 Remote stop/start control

Hand-operated star–delta starter

Remember

Although hand-operated star–delta starters are now rare, you may still come across them in your career as an electrician.

This is a two-position method of starting a three-phase squirrel-cage motor, in which the windings are connected firstly in star for acceleration of the rotor from standstill, and then secondly in delta for normal running.

The connections of the motor must be arranged for star–delta, starting with both ends of each phase winding – six in all – brought out to the motor terminal block. The starter itself, in its simplest form, is in effect a changeover switch. Figure 8.141 gives the elementary connections for both star and delta.

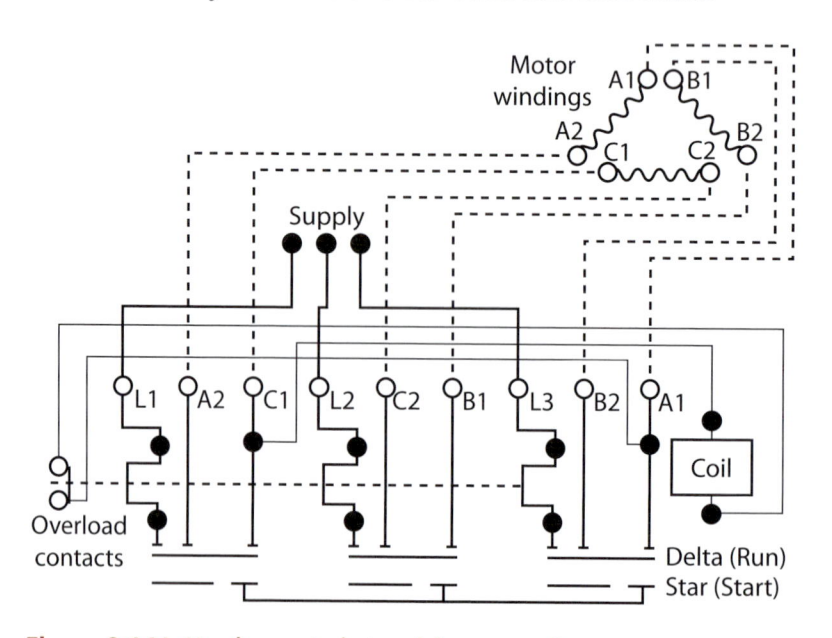

Figure 8.141 Hand-operated star–delta connections

When the motor windings are connected in star, the voltage applied to each phase winding is reduced to 58% of the line voltage, thus the current in the winding is correspondingly reduced to 58% of the normal starting value.

Applying these reduced values to the typical three-phase squirrel-cage induction motor, we would have: initial starting current from two to three-and-a-half times full-load current and initial starting torque of about 50% of the full-load value.

The changeover from star to delta should be made when the motor has reached a steady speed on star connection, at which point the windings will now receive full line voltage and draw full-rated current.

If the operator does not move the handle quickly from the start to run position, the motor may be disconnected from the supply long enough for the motor speed to fall considerably. When the handle is eventually put into the run position, the motor will therefore take a large current before accelerating up to speed again. This surge current could be large enough to cause a noticeable voltage dip. To prevent this, a mechanical interlock is fitted to the operating handle. However, in reality the handle must be moved quickly from start to run position, otherwise the interlock jams the handle in the start position.

The advantage of this type of starter is that it is relatively cheap. It is best suited for motors against no load or light loads, and it also incorporates no-volts protection and overload protection.

Automatic star-delta starter

Bearing in mind the user actions required of the previous hand-operated starter, the fully automatic star-delta contactor starter (as shown in Figure 8.143) is the most satisfactory method of starting a three-phase cage-induction motor. The starter consists of three triple-pole contactors, one employing thermal overload protection, the second having a built-in time-delay device, and the third providing the star points.

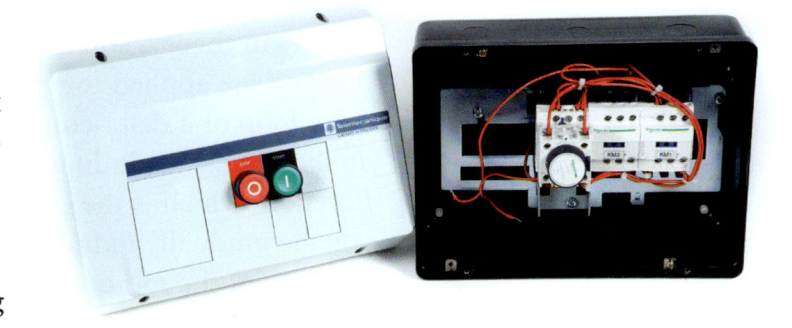

Figure 8.142 Star–delta starter

The changeover from star to delta is carried out automatically by the

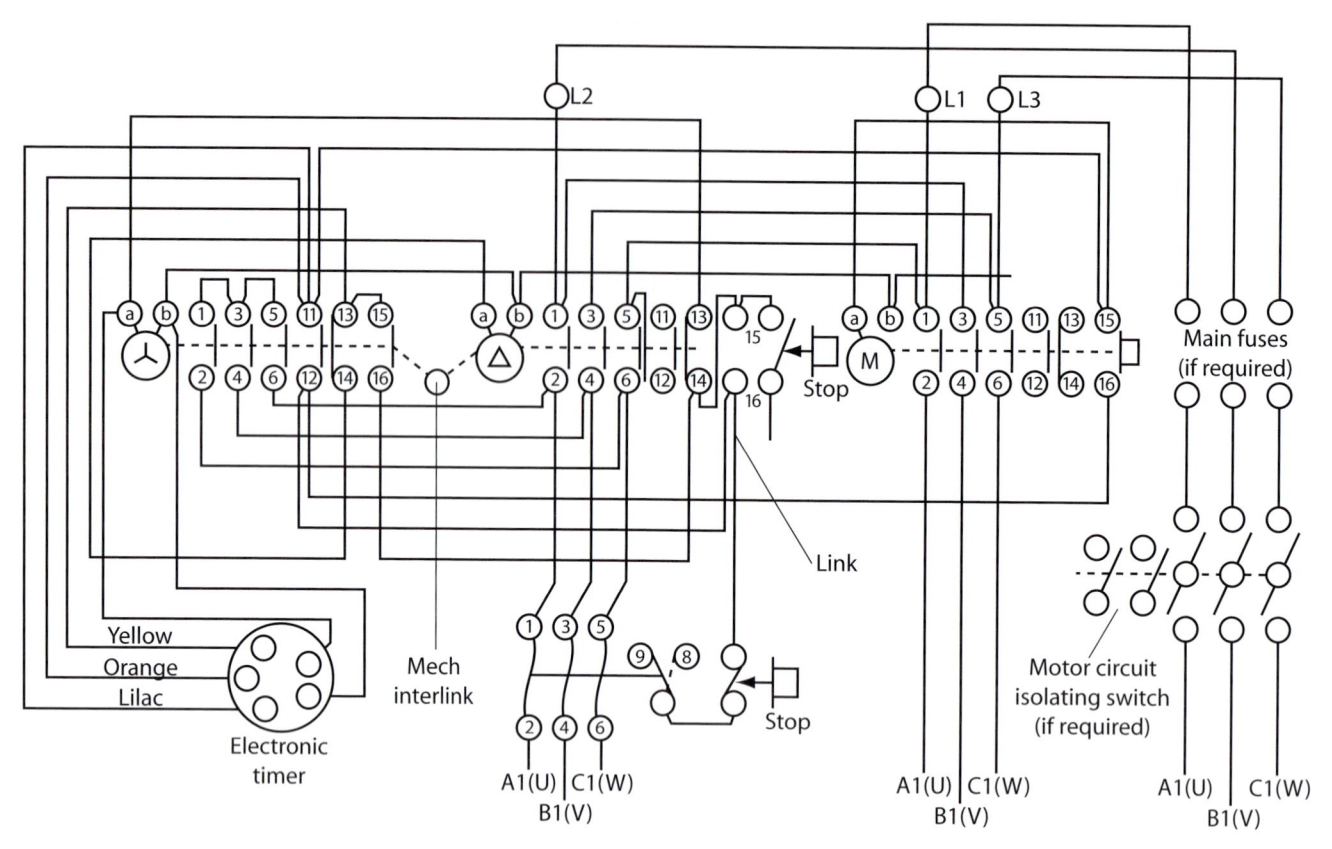

Figure 8.143 Automatic star–delta contactor starter

timing device, which can be adjusted to achieve the best results for a particular situation.

Soft starters

A soft starter is a type of reduced-voltage starter that reduces the starting torque for a.c. induction motors. The soft starter is in series with the supply to the motor, and uses solid-state devices to control the current flow and therefore the voltage applied to the motor. In theory, soft starters can be connected in series with the line voltage applied to the motor, or can be connected inside the delta loop of a delta-connected motor, controlling the voltage applied to each winding. Soft starters can also have a soft-stop function, which is the exact opposite to soft start, and sees the voltage gradually reduced and thus a reduction in the torque capacity of the motor.

The auto-transformer starter

This method of starting is used when star–delta starting is unsuitable, either because the starting torque would not be sufficiently high using that type of starter, or because only three terminals have been provided at the motor terminal box – a practice commonly found within the UK water industry.

This again is a two-stage method of starting three-phase squirrel-cage induction motors, in which a reduced voltage is applied to the stator windings to give a reduced current at start. The reduced voltage is obtained by means of a three-phase auto transformer, the tapped windings of which are designed to give 40%, 60% and 75% of the line voltage respectively. Although there are a number of tappings, only one tapping is used for the initial starting, as the reduced voltage will also result in reduced torque. Once this has been selected for the particular situation in which the motor is operating, it is left at that position and the motor is started in stages – much like the star-delta starter in that once the motor has reached sufficient speed the changeover switch moves onto the run connections, thus connecting the motor directly to the three-phase supply. Figure 8.144 illustrates the connections for an auto-transformer starter.

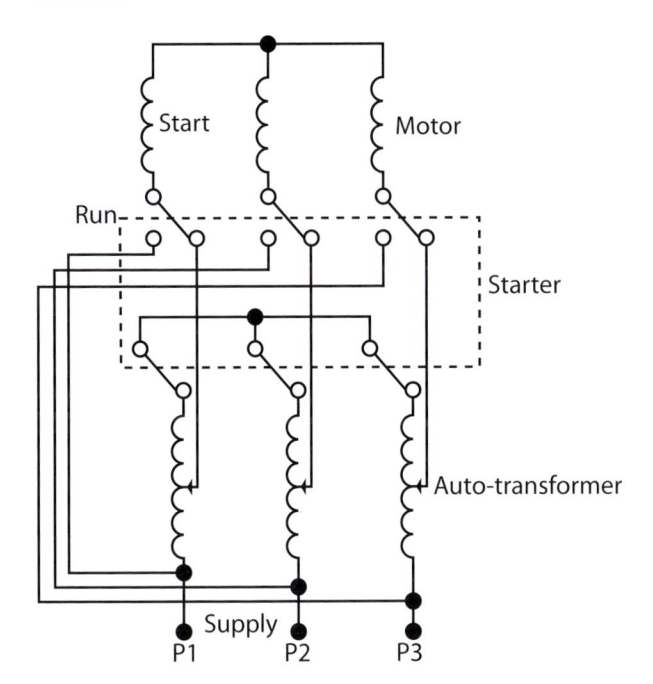

Figure 8.144 Connections for an auto-transformer starter

The rotor-resistance starter

This type of starter is used with a slip-ring wound-rotor motor. These motors and starters are primarily used where the motor will start against full load, as an external resistance is connected to the rotor windings through slip rings and brushes, which serves to increase the starting torque.

When the motor is first switched on, the external rotor resistance is at a maximum. As the motor speed increases, the resistance is reduced until at full speed, when the external resistance is completely eliminated and the machine runs as a squirrel-cage induction motor.

The starter is provided with no-volts and overload protection and an interlock to prevent the machine being switched on with no rotor resistance connected. (For clarity these are not shown in Figure 8.145, since the purpose of the diagram is to show the principle of operation.)

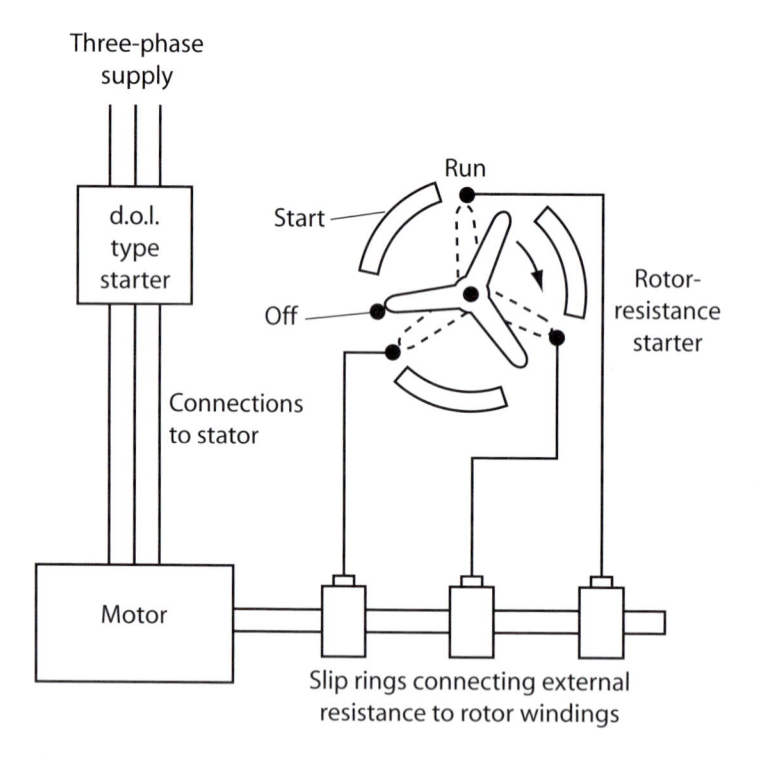

Figure 8.145 Rotor-resistance starter

Motor speed control

Speed control of d.c. machines

We said at the beginning of this chapter that there are three types of d.c. motor: series, shunt and compound, and that one of the advantages of the d.c. machine is the ease with which the speed may be controlled.

Some of the more common methods used to achieve speed control on a d.c. machine are described below.

Armature resistance control

With this system of control we are effectively reducing the voltage across the armature terminals by inserting a variable resistor into the armature circuit of the motor. In effect we are creating the illusion of applying a lower-than-rated voltage across the armature terminals.

The disadvantages of this method of control are that we see much of the input energy dissipated in the variable resistor, a loss of efficiency in the motor and poor speed regulation in the shunt and compound motors.

Although not discussed in this book, the principle of applying a lower-than-rated voltage across the armature terminals forms the basis of the Ward-Leonard system of speed control, which in essence provides a variable voltage to the armature terminals by controlling the field winding of a separate generator. Although expensive, this method gives excellent speed control and therefore finds use in situations such as passenger lifts.

Field control

This method works on the principle of controlling the magnetic flux in the field winding. This can be controlled by the field current and as a result controls the motor speed. As the field current is small, the power dissipated by the variable resistor is reasonably small. We can control the field current in the various types of motor as follows:

- **Series motor** – Place a variable resistor in parallel with the series field winding.
- **Shunt motor** – Place a variable resistor in series with the field shunt winding.

This method of speed control is not felt to be suitable for compound machines, as any reduction in the flux of the shunt is offset by an increase in flux from the series field because of an increase in armature current.

Pulse width modulation (PWM)

We know all about the problem with the variable resistor: although it works well, it generates heat and hence wastes power. PWM d.c. motor control uses electronics to eliminate this problem. It controls the motor speed by driving the motor with short pulses. These vary in duration to change the speed of the motor. The longer the pulses, the faster the motor turns, and vice versa. The main disadvantages of PWM circuits are their added complexity and the possibility of generating radio frequency interference (RFI), although this can be minimised by the use of short leads or additional filtering on the power supply leads.

Speed control of a.c. induction motors

We have already established that synchronous speed is directly proportional to the frequency of the supply and inversely proportional to the number of pole pairs. Therefore the speed of an induction motor can be changed by varying the frequency and/or the number of poles. We can also control the speed by changing the applied voltage and the armature resistance.

For adjustable speed applications, variable speed drives (VSD) use these principles by controlling the voltage and frequency delivered to the motor. This gives control over motor torque and reduces the current level during starting. Such drives can control the speed of the motor at any time during operation.

The phrase 'a.c. drive' has different meanings to different people. To some it means a collection of mechanical and electro-mechanical components (the variable frequency inverter and motor combination), which, when connected together, will move a load. More commonly – and also for our purposes – an a.c. drive should be considered as being a variable-frequency inverter unit (the drive) with the motor as a separate component. Manufacturers of variable-frequency inverters will normally refer to these units as being a variable-frequency drive (VFD).

A variable-frequency drive is a piece of equipment that has an input section – the converter – which contains diodes (see page 463–471 for more information on diodes) arranged in a bridge configuration. This converts the a.c. supply to d.c. The next section of the VFD, known as the constant-voltage d.c. bus, takes the d.c. voltage and filters and smoothes out the wave form. The smoother the d.c. wave form is, the cleaner the output wave form from the drive.

The d.c. bus then feeds the final section of the drive, the inverter. This section of the drive will invert the d.c. back to a.c. using Insulated Gate Bipolar Transistors (IGBT), which create a variable a.c. voltage and frequency output.

Working life

You are asked to investigate a problem in the plant room of a hospital, as the estates maintenance manager has reported a noisy motor on the air-conditioning system. When you look at the motor, you find that as well as it being noisy there is a vibration.

1. Do you think this is a real problem? If so, what actions would you recommend?

Progress check

1. Describe how an artificial means of starting a single-phase motor is achieved.
2. Explain briefly the difference between synchronous speed and actual speed for an a.c. motor.
3. What is the major advantage of using star-delta starting rather than direct-on-line?

K9. Understand the operating principles of different electrical components

There are several electrical components you will need to be familiar with when working with electrical installations. Many of these have been covered elsewhere in this unit or in this qualification. For more information on the following turn to these pages:

- contactors (Unit ELTK 07, pages 194–95)
- relays (pages 292–94)
- solenoids (pages 297–98)
- over-current protection devices (Unit ELTK 07, pages 195–96)
- RCDs and RCBOs (Unit ELTK 04a, pages 27–30 and 36–37).

K10. Understand the principles and applications of electrical lighting systems

Did you know?

The first edition of the Regulations, introduced in 1882, was entitled 'Rules and Regulations for the Prevention of Fire Risks Arising from Electric Lighting'.

Illumination by means of electricity has been available for over 100 years. In that time it has changed in many ways, though many of the same ideas are still in use. The first type of electric lamp was the 'arc lamp', which used electrodes to draw an electrode through the air, this is now known as discharge lighting. This was quite an unsophisticated use of electricity, and many accidents and fires resulted from it. Regulations had to be developed to control discharge lighting installations, and the use of electrical lighting systems has changed dramatically over the centuries.

This section will look at the basic principles of illumination and the applications of the different methods used to calculate lighting requirements, after introducing the key operating principles, types and applications of common luminaires you will use.

Operating principles, types, limitations and applications of luminaires

Remember

A luminaire is more commonly known as a lighting fitting and refers to the fully assembled enclosure, lamp, control gear and reflector etc.

Before we look at the types of lamp available, we should look at lamp caps. The cap is that part of the lamp that allows an electrical connection to be made with the supply. There are many different types, some of which are shown below.

Light emitting diodes are also types of luminaires. More information can be found about these later in this unit on pages 467–68.

The bayonet cap

The bayonet cap (BC) is probably the lamp you have come across most often. It is 22 mm diameter and has two locating lugs, the electrical contact made over two pins on the base of the cap. Two popular variations of this are the SBC cap which is 15 mm in diameter and the SCC cap. The SCC cap only has one contact on the base, the other contact being the cap itself.

The Edison Screw cap

Most Edison Screw (ES) lamps are represented as the letter 'E' followed by a number. This number denotes the diameter of the cap. The most popular types for domestic use in the UK are E14

Figure 8.146 Bayonet cap

(14 mm and also known as SES) and E27. There is also a version used in street lighting and industrial situations with mercury fluorescent lamps, the E40 or GES (Giant Edison Screw).

Halogen lamp caps

There are three common types of halogen lamp camp:

- **Halogen capsule lamps** are generally designated by the distance in millimetres between the pins of the lamp. The most common of these, G4 (2 pins – 4 mm apart) is used in low voltage applications such as desk lamps.

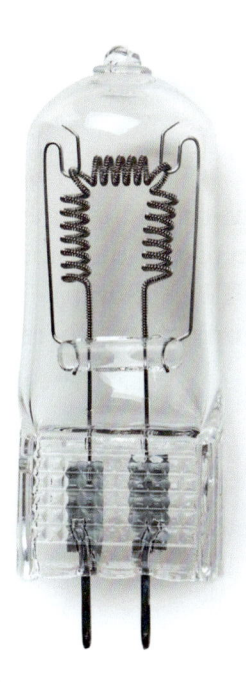

Figure 8.147 Edison Screw cap

Figure 8.148 Halogen capsule lamp

- **Linear tungsten halogen lamps** are mains voltage and normally seen in security lights, floodlights and some up-lighters. They have what is known as an R7 cap at each end of a thin gas-filled quartz tube.

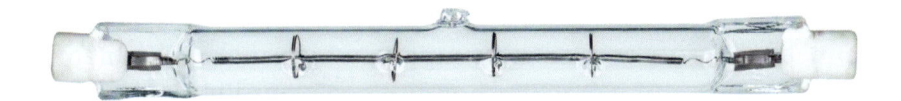

Figure 8.149 Linear tungsten halogen capsule lamp

- **Halogen spotlights** have become ever more popular and are seen in many domestic applications such as bathrooms, dining rooms and kitchens or commercially in display applications. The most common is the GU10 version shown in Figure 8.150.

Figure 8.150 Halogen spotlights

Low pressure mercury (fluorescent) caps)

Using their common name, fluorescent tubes have a bi-pin cap at both ends of the gas-filled tube. Diameters of the tube range from T5 (16 mm) through T8 (25 mm) to T12 (38 mm). T8 and T12 tubes normally have pins that are 13 mm apart, whereas the T5 tube has pins that are 5 mm apart.

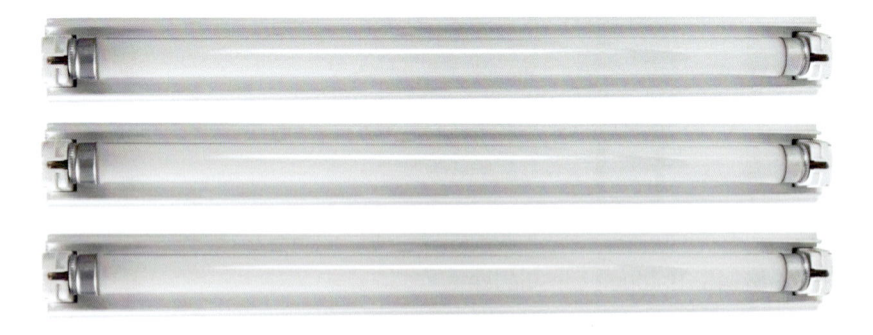

Figure 8.151 Low pressure mercury (fluorescent) caps

Incandescent lamps – GLS and tungsten halogen

In this method of creating light a fine filament of wire is connected across an electrical supply, which makes the filament wire heat up until it is white-hot and gives out light. The filament wire reaches a temperature of 2500–2900°C. These lamps are very inefficient and only a small proportion of the available electricity is converted into light; most of the electricity is converted into heat as infrared energy. The light output of this type of lamp is mainly found at the red end of the visual spectrum, which gives an overall warm appearance.

> **Remember**
>
> The filament wire in an incandescent lamp reaches a temperature of about 2500–2900°C.

Operation of GLS lamps

The General Lighting Service (GLS) lamp is one type of incandescent lamp and is commonly referred to as the 'light bulb'. It has at its 'core' a very thin tungsten wire that is formed into a small coil and then coiled again.

A current is passed through the tungsten filament, which causes it to reach a temperature of 2500°C or more so that it glows brightly. At these temperatures, the oxygen in the atmosphere would combine with the filament to cause failure, so all the air is removed from the glass bulb and replaced by gases such as nitrogen and argon. Nitrogen is used to minimise the risk of arcing and argon is used to reduce the evaporation process. On low-power lamps such as 15 and 25 watt, the area inside the bulb remains a vacuum. The efficiency of a lamp is known as the efficacy. It is expressed in lumen per watt, lm/w. For this type of lamp the efficacy is between 10 and 18 lumens per watt. This is low compared with other types of lamp, and its use is limited. However, although now being phased out, it is the most familiar type of light source used and has many advantages including:

- comparatively low initial costs
- immediate light when switched on
- no control gear
- it can easily be dimmed.

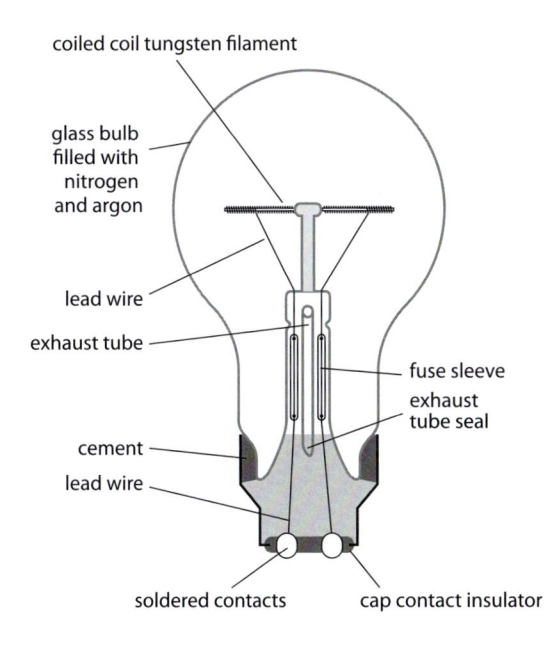

Figure 8.152 GLS lamp

> **Did you know?**
>
> The first lamp that was developed for indoor use was the carbon-filament lamp. Although this was a dim lamp by modern standards it was cleaner and far less dangerous than the exposed 'arc lamp'.

> **Did you know?**
>
> The average life of this type of lamp is 1 000 hours, after which the filament will rupture.

When a bulb filament finally fails it can cause a very high current to flow for a fraction of a second – often sufficient enough to operate a 5 or 6 amp miniature circuit breaker which protects the lighting circuit. High-wattage lamps, however, are provided with a tiny integral fuse within the body of the lamp to prevent damage occurring when the filament fails.

If the lamp is run at a lower voltage than that of its rating this results in the light output of the lamp being reduced at a greater rate than the electricity used by the lamp, and the lamp's efficacy is poor. This reduction in voltage, however, increases the lifespan and can be useful where lamps are difficult to replace or light output is not the main consideration.

It has been calculated that an increase in 5% of the supply voltage can reduce the lamp life by half. However, if the input voltage is increased by just 1% this will produce an increase of 3.5% in lamp output (lumens). When you consider that the Electricity Supply Authority is allowed to vary its voltage up to and including 6% it is easy to see that if this was carried on for any length of time the lamps would not last very long.

Tungsten halogen lamps

These types of lamps were introduced in the 1950s. For their operation the tungsten filament is enclosed in a gas-filled quartz tube together with a carefully controlled amount of halogen such as iodine. Figure 8.153 illustrates the linear tungsten halogen lamp.

> **Did you know?**
>
> Running the lamp at a higher voltage than it was designed to do results in a shorter lamp life.

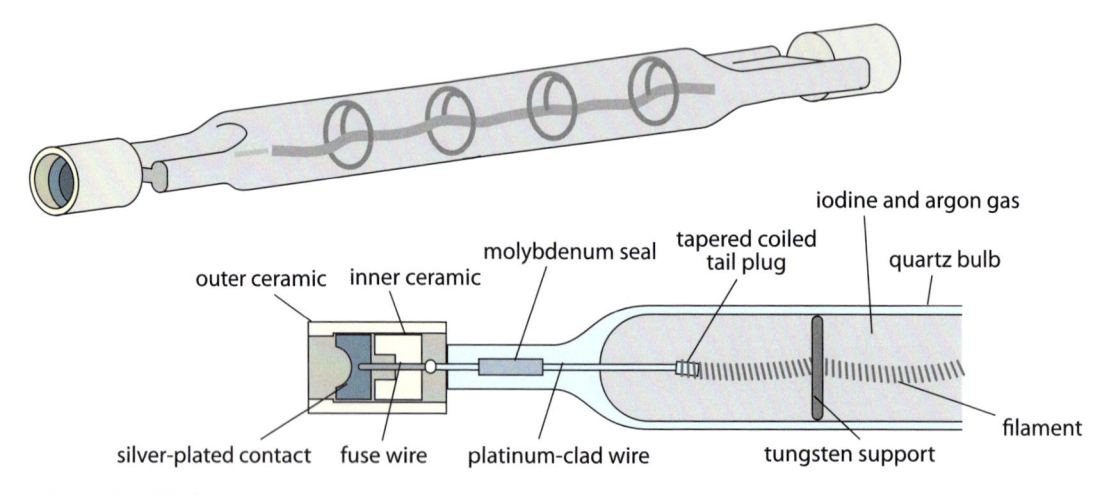

Figure 8.153 Linear tungsten lamp

Operation of tungsten halogen lamps

The inclusion of argon and iodine in the quartz tube allows the filament to burn at a much higher temperature than the incandescent lamp. The inclusion of the halogen gas produces a regeneration effect which prolongs the life of the lamp.

As small particles of tungsten fall away from the filament, they combine with the iodine passing over the face of the quartz tube, forming a new compound. Convection currents in the tube cause this new compound to rise, passing over the filament. The intense heat of the filament causes the compound to separate into its component parts, and the tungsten is deposited back on the filament.

The lamp should not be touched with bare fingers as this would deposit grease on the quartz glass tube; this would lead to small cracks and fissures in the tube when the lamp heats rapidly, causing the lamp to fail. If accidentally touched on installation, the lamp should be cleaned with methylated spirit before being used.

The linear type of lamp must be installed within 4° of the horizontal to prevent the halogen vapour migrating to one end of the tube, causing early failure. These types of lamps have many advantages, which include:

- increased lamp life (up to 2000 hours)
- increase in efficacy (up to 23 lumens per watt)
- reduction in lamp size.

There are two basic designs that have been produced: the double-ended linear lamp and the single-ended lamp, with both contacts embedded in the seal at one end (see Figure 8.154). This type of lamp has been produced to work on extra-low voltages: they are

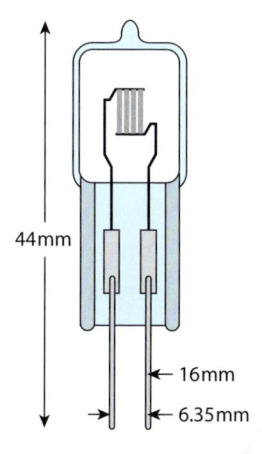

Figure 8.154 Single-ended filament lamp

used extensively in the automobile industry for vehicle headlamps. They may also be used for display spotlights where extra-low voltage is desirable. They may be supplied from an in-built 230 V/12 V transformer.

Discharge lighting

This is a term that refers to illumination derived from the ionisation of gas. This section looks at:

- low pressure mercury vapour lamps
- the glow-type starter circuit
- semi-resonant starting
- high frequency
- stroboscopic effect
- other methods of starting the fluorescent tube
- other discharge lamps.

Low pressure mercury vapour lamps

The fluorescent lamp, or, more correctly, the low pressure mercury vapour lamp, consists of a glass tube filled with a gas such as krypton or argon and a measured amount of mercury vapour. The inside of the glass tube has a phosphor coating, and at each end there is a sealed set of oxide-coated electrodes, known as cathodes.

When a voltage is applied across the ends of a fluorescent tube the cathodes heat up, and this forms a cloud of electrons, which ionise the gas around them. The voltage to carry out this ionisation must be much higher than the voltage required to maintain the actual discharge across the lamp. Manufacturers use several methods to achieve this high voltage, usually based on a transformer or **choke**. This ionisation is then extended to the whole length of the tube so that the arc strikes and is then maintained in the mercury, which evaporates and takes over the discharge. The mercury arc, being at low pressure, emits little visible light but a great deal of ultraviolet, which is absorbed by the phosphor coating and transformed into visible light.

The cathodes are sealed into each end of the tube and consist of tungsten filaments coated with an electron-emitting material. Larger tubes incorporate cathode shields – iron strips bent into an oval shape to surround the cathode. The shield traps material given off by the cathodes during the tube life and thereby prevents the lamp-ends blackening.

Did you know?

The type of phosphorus used on the lamp's inner surface will determine the colour of light given out by the lamp.

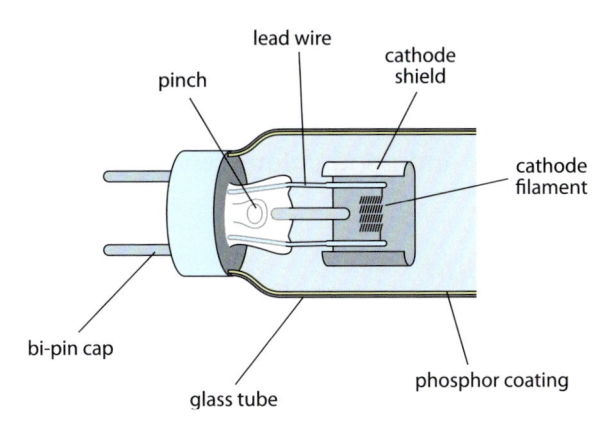

Figure 8.155 Detail of one end of a fluorescent tube

The gas in standard tubes is a mercury and argon mix, although some lamps (the smaller ones and the new slim energy-saving lamps) have krypton gas in them. The phosphor coating is a very important factor affecting the quantity and quality of light output. When choosing different lamps there are three main areas to be considered:

- lamp efficacy
- colour rendering
- colour appearance.

Lamp efficacy

This refers to the lumen output for a given wattage. For fluorescent lamps this varies between 40 and 90 lumens per watt.

Colour rendering

This describes a lamp's ability to show colours as they truly are. This can be important depending on the building usage. For example, it would be important in a paint shop but less so in a corridor of a building. The rendering of colour can affect people's attitude to work etc. – quite apart from the fact that in some jobs true colour may be essential. By restoring or providing a full colour range the light may also appear to be better or brighter than it really is.

Colour appearance

This is the actual look of the lamp, and the two ends of the scale are warm and cold. These extremes are related to temperatures: the higher the temperature, the cooler the lamp. This is important for the overall effect, and generally warm lamps are used to give a relaxed atmosphere while cold lamps are used where efficiency and businesslike attitudes are the priorities. The subject of lamp

> **Did you know?**
>
> The most economic tube life is limited to around 5000 or 6000 hours. In industry, tubes are changed at set time intervals and all the tubes, whether still working or not, are replaced. This saves money on maintenance, stoppage of machinery and scaffolding erection etc.

> **Remember**
>
> The capacitor connected across the supply terminals is to correct the poor power factor that has been created by this type of light.

> **Remember**
>
> Once the tube strikes, current no longer passes through the starter; therefore the starter takes no further part in the circuit.

choice has become very complicated, and a programme of lamp rationalisation has begun. The intention is that the whole range currently available will be reduced. Also, new work has resulted in lamps with high lumen outputs and good colour rendering possibilities.

The glow-type starter circuit

In the starter, a set of normally open contacts is mounted on bi-metal strips and enclosed in an atmosphere of helium gas. When switched on, a glow discharge takes place around the open contacts in the starter, which heats up the bi-metal strips, causing them to bend and touch each other. This puts the electrodes at either end of the fluorescent tube in circuit and they warm up, giving off a cloud of electrons; simultaneously an intense magnetic field is building up in the choke, which is also in circuit.

The glow in the starter ceases once the contacts are touching so that the bi-metal strips now cool down and they spring apart again. This momentarily breaks the circuit, causing the magnetic field in the choke to rapidly collapse. The high back-e.m.f. produced provides the high voltage required for ionisation of the gas and enables the main discharge across the lamp to take place. The voltage across the tube under running conditions is not sufficient to operate the starter and so the contacts remain open.

The resistance of the ionised gas gets lower and lower as it warms up and conducts more current. This could lead to disintegration of the tube. However, the choke has a secondary function as a current-limiting device: the impedance of the choke limits the current through the lamp, keeping it in balance. This is one reason why it is often referred to as ballast.

This type of starting may not succeed first time and can result in the characteristic flashing on/off, when initially switching it on.

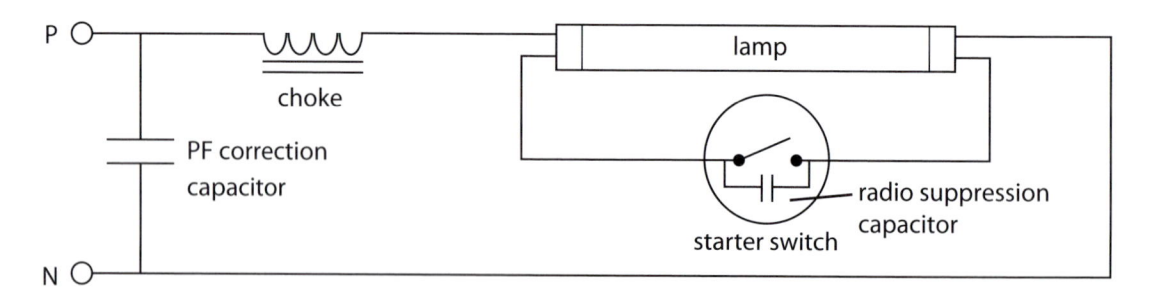

Figure 8.156 Glow-type starter circuit

Semi-resonant starting

In this circuit the place of the choke is taken by a specially wound transformer. Current flows through the primary coils to one cathode of the lamp and then through the secondary coil, which is wound in opposition to the primary coil. A fairly large **capacitor** is connected between the secondary coil and the second cathode of the lamp (the other end of which is connected to the neutral).

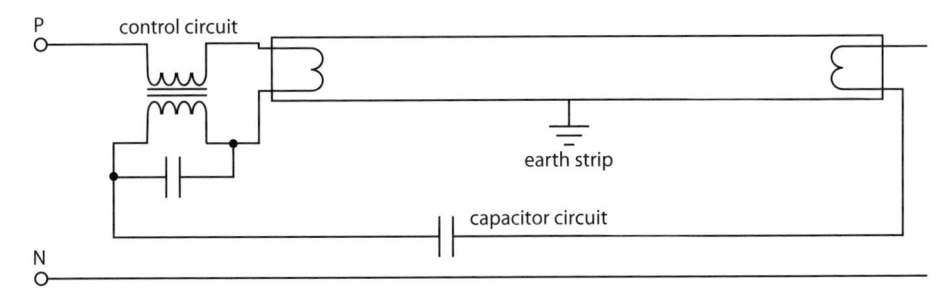

Figure 8.157 Semi-resonant circuit

The current that flows through this circuit heats the cathodes and, as the circuit is predominantly capacitive, the pre-start current leads the main voltage. Owing to the primary and secondary windings being in opposition, the voltages developed across them are 180° out of phase, so that the voltage across the tube is increased, causing the arc to strike. The primary windings then behave as a choke, thus stabilising the current in the arc. The circuit has the advantage of high power factor and easy starting at low temperatures.

High frequency

Standard fluorescent circuits operate on a mains supply frequency of 50 Hz; however, high-frequency circuits operate on about 30 000 Hz. There are a number of advantages to high frequency:

- higher lamp efficacy
- first-time starting
- noise free
- the ballast shuts down automatically on lamp failure.

The higher efficacy for this type of circuit can lead to savings of at least 10%, and in some large installations they may be as high as 30%. Many high-frequency electronic ballasts will operate on a wide range of standard fluorescent lamps. The high-frequency circuit will switch on the lamp within 0.4 of a second and there should be no flicker (unlike glow-type starter circuits).

Stroboscopic effect is not a problem as this light does not flicker on/off due to its high frequency.

However, a disadvantage of this type of circuit is that supply cables installed within the luminaire must not run adjacent to the leads connected to the ballast output terminals as interference may occur. Also, the initial cost of these luminaires is greater than traditional glow-type switch starts.

Stroboscopic effect

A simple example of this effect is that, when watching the wheels rotate on a horse-drawn cart on television, it appears that the wheels are stationary or even going backwards. This phenomenon is brought about by the fact that the spokes on the wheels are being rotated at about the same number of revolutions per second as the frames per second of the film being shot. This effect is known as the 'stroboscopic effect' and can also be produced by fluorescent lighting.

The discharge across the electrodes is extinguished 100 times per second, producing a flicker effect. This flicker is not normally observable but it can cause the stroboscopic effect, which can be dangerous. For example, if rotating machinery is illuminated from a single source it will appear to have slowed down, changed direction of rotation or even stopped. This is a potentially dangerous situation to any operator of rotating machines in an engineering workshop.

However, this stroboscopic effect can be harnessed – for example, to check the speed of a CD player or the speed of a motor vehicle for calibration purposes.

By using one of the following methods, the stroboscopic effect can be overcome or reduced. The first three maintain light falling on the rotating machine. The fourth makes the effective flicker at a different frequency from the operating frequency of the machines.

(i) Tungsten filament lamps can be fitted locally to lathes, pillar-drilling machines etc. This will lessen the effect but will not eliminate it completely.

(ii) Adjacent fluorescent fittings can be connected to different phases of the supply. Because in a three-phase supply the phases are 120° out of phase with each other, the light falling on the machine will arrive from two different sources.

> **Did you know?**
>
> Certain frequencies of stroboscopic flash can induce degrees of drowsiness, headaches, eye fatigue and, in extreme cases, disorientation. Some television shows have warnings about stroboscopic lights before their transmission.

Each of these will be flickering at a different time and will interfere with each other, reducing the stroboscopic effect.

(iii) Twin lamps can be wired on lead-lag circuits, thus counteracting each other. The lead-lag circuit, as the name implies, is a circuit that contains one lamp in which the **power factor** leads the other – hence the other lags. Using the leading current effect of a capacitor and the lagging current effect of an inductor produces the lead-lag effect. The lagging effect is produced naturally when an inductor is used in the circuit as shown in Figure 8.158. The leading effect uses a series capacitor, which has a greater effect than the inductor in the circuit. When these two circuits are combined as shown there is no need for further power factor correction as one circuit will correct the other.

(iv) The use of high-frequency fluorescent lighting reduces the effect by about 60%.

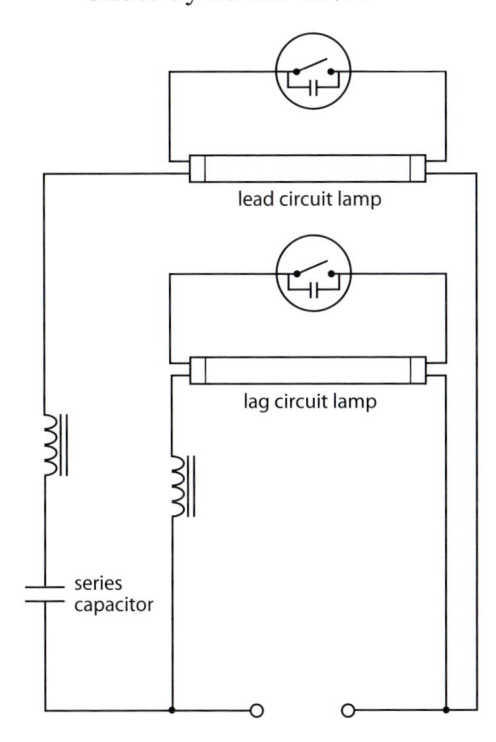

Figure 8.158 Lead lag circuit

Other methods of starting the fluorescent tube
Quick start

The electrodes of this type of circuit are rapidly pre-heated by the end windings of an autotransformer so that a quick start is possible. The method of ionisation of the gas is the same as in the semi-resonant circuit. Difficulties may occur in starting if the voltage is low.

Thermal starter circuit

This type of circuit has waned in popularity over the years. However, there are still thousands of these fittings in service, so it is worth describing them. In this starter, the normally closed contacts are mounted on a bi-metal strip. A small heater coil heats one of these when the supply is switched on. This causes the strip to bend and the contacts to open, creating the momentary high voltage and starting the circuit discharge. The starter is easily recognised as it has four pins instead of the usual two, the extra pins being for the heater connection.

Other discharge lamps

Low pressure sodium vapour

Low pressure sodium lamps have a gas discharge tube containing solid sodium and a small amount of neon and argon gas mixture to start the gas discharge. When the lamp is initially turned on it emits a dull red light as the sodium metal is warmed before becoming a bright orangey-yellow once the sodium has vaporised.

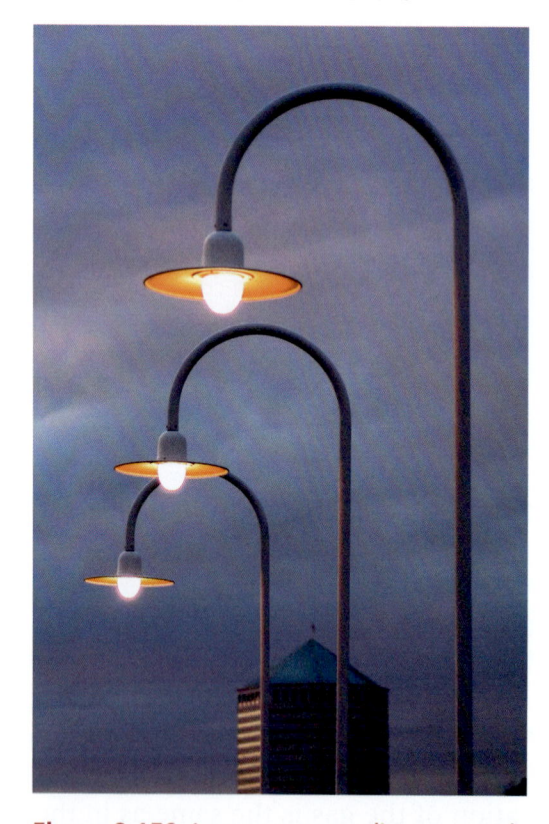

Figure 8.159 Low pressure sodium vapour lamp

These lamps have an outer glass vacuum envelope around the inner discharge tube for thermal insulation, which improves their efficiency along with coating the glass envelope with an infrared reflecting layer of metal oxide, thus resulting in their common name of SOX (**so**dium and o**x**ide) lamps.

As a result of their colour when lit and the fact that they glow rather than glare, you will most commonly find them used on street or motorway lighting where accurate colour rendition of objects is not important.

High pressure sodium vapour

High pressure sodium lamps are smaller than SOX lamps and use low pressure xenon as the starter gas required to being the sodium and mercury vaporisation.

When the lamp is initially turned on it emits a dull red light as the sodium metal is warmed before becoming a blue-white light once the sodium and mercury have vaporised. Because of this they are used where good colour rendering is considered important. As they are reasonably efficient, they are also used in street and security lighting.

As they were felt to be replicating sunlight, they are commonly referred to as SON lamps.

Figure 8.160 High pressure sodium vapour lamp

High pressure mercury vapour

When a mercury vapour lamp is first turned on, the voltage initiates an argon glow discharge between the main and the auxiliary electrode pair(s), causing a sufficient number of mercury atoms to be ionised. This in turn initiates a low pressure discharge between the two main electrodes and it will produce a dark blue glow because only a small amount of the mercury is ionised and the gas pressure in the arc tube is very low. As the main arc strikes and the gas heats up and increases in pressure, the light appears nearly white to the human eye.

Metal halide

By adding rare earth metal salts to the mercury vapour lamp, metal halide lamps produce high light-output for their size, making them a powerful and efficient light source.

Since the lamp is small compared to a fluorescent or incandescent lamp of the same light level, relatively small reflective luminaires can be used to direct the light for different applications (flood lighting outdoors, or lighting for warehouses or industrial buildings).

Metal halide lamps are used both for general lighting purposes, and for very specific applications that require specific UV or blue-frequency light. Because of their wide spectrum, they are used for indoor growing applications.

Like all discharge lamps, metal halide lamps produce light by passing an electric arc through a mixture of gases. In a metal halide lamp, the compact arc tube contains a high pressure mixture of argon, mercury, and a variety of metal halides. The argon gas in the lamp is easily ionised, and facilitates striking the arc across the two electrodes when voltage is first applied to the lamp. The heat generated by the arc then vaporises the mercury and metal halides, which produce light as the temperature and pressure increases.

Compact (energy saving) fluorescent lamps

The incandescent lamp is simply a very hot piece of wire inside a glass container. It is the heat that created the light, thus making it an inefficient device. In fact it wastes about 90% of the electricity it uses. The problem is, just about every domestic property in the UK uses them in pendant or table light fittings. The typical incandescent lamp used in the home is either 60 watts or 100 watts, which means that if 10×100 watt lamps are on then you are using 1 kW of power.

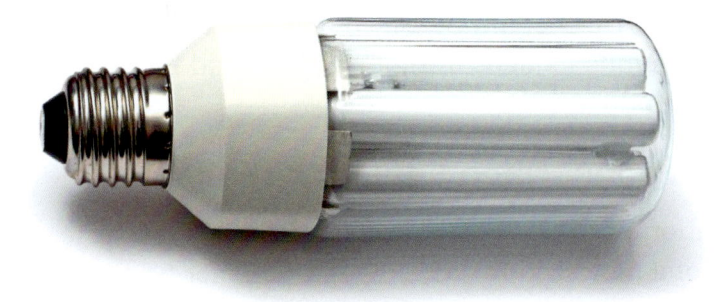

Figure 8.161 Energy-saving light bulbs

Energy-saving lights are simply a compact version of the fluorescent tube mentioned earlier; in other words they are a low pressure mercury vapour lamp. The compact design is necessary to allow domestic users to replace their old incandescent lamps without having to replace shades or fittings. Additionally, these low energy equivalents typically use only 9 or 11 watts each even though they give the same amount of light. So using our same example, instead of using 1 kW of power our demand would be 10 × 11 watt or 110 watts in total – quite an energy saving.

There are two types of compact fluorescent lamp:

- Integrated lamps combine a tube, an electronic ballast and either an Edison screw or a bayonet fitting in a single unit. Most domestic users will be familiar with these.
- Non-integrated lamps have the ballast permanently installed in the luminaire, and therefore only the lamp bulb has to be changed when it fails. Since the ballast is in the fitting itself, they tend to be larger but last longer when compared to the integrated lamp.

Standard compact fluorescents are not suitable for dimming applications and special lamps are required instead.

LED lighting

Light emitting diodes (LEDs) have been around for many years and are used for a range of items, from lighting digital clocks to Christmas tree lights.

> **Working life**
>
> Your customer has installed glow-type fluorescents in the kitchen area of a large house. They have reported that one of them continues to flicker as though trying to start but doesn't.
>
> 1. What should you do to determine the facts?
> 2. What are the most likely causes of this fault?
> 3. How would you proceed?

As with the compact fluorescent, the LED lamp saves energy against traditional incandescent lamps and they are likely to last about 30 000 hours.

Although we will cover the LED in more depth later in this unit (pages 467–8), an LED could be called 'solid-state lighting' technology as it emits its light from a piece of solid matter, namely a semiconductor.

Regulations concerning lighting circuits

The 17th Edition of the IET Regulations (BS 7671) saw the introduction of a new section (559) within Chapter 55 that applies to:

- the selection and erection of luminaires and lighting installations that are intended to be part of the fixed installation (559.1)
- highway power supplies and street furniture (559.1).

Particular requirements are specified for the following:

- fixed outdoor lighting installations.
- extra low voltage installations supplied at up to 50 V a.c. or 120 V d.c.
- lighting for display stands.

The definitions within Regulation 559.3 mean that lighting installations for roads, parks, car parks, gardens, places open to the public, sporting areas, illuminating monuments, floodlighting, telephone kiosks, bus shelters, advertising panels, road signs and road traffic signalling systems are now all covered by BS 7671.

However, BS 7671 does not cover the following:

- high voltage signs supplied at low voltage (e.g. neon tubes)
- signs and luminous discharge tube installations in excess of 1 kV.

Although Section 559 sees the introduction of some 44 new regulations, we're only going to look at the most appropriate ones in the following text.

Regulation 559.4 – requires that all luminaires and track systems comply with the relevant standards for manufacture.

Regulation 559.5 – requires that the selection and erection process considers the thermal effect of radiant and convected energy on to surroundings. This means considering the power of lamps and the fire resistance of nearby materials.

Regulation 559.6 requires that at each fixed lighting point, one of the following shall be used:

- ceiling rose
- luminaire
- luminaire support coupler
- batten lampholder or pendant set
- suitable socket outlet/connection unit
- suitable lighting distribution unit
- device for connecting a luminaire (DCL).

Regulation 559.6 also states that:

- a batten lampholder/ceiling rose for a filament lamp shall not be used for a circuit in excess of 250 V
- a ceiling rose can't be used for attaching more than one pendant unless it has been designed for that purpose
- luminaire support couplers provide for connection to, and support of, the luminaire. They must not be used to connect any other item of equipment
- adequate means of fixing the luminaire must be provided
- lighting circuits that include B15, B22, E14, E27 or E40 lampholders shall have a protective device for the circuit of no greater than 16 A
- except for E14 and E27 lampholders to BS EN 60238, in circuits on a TN or TT system, the outer contact of every ES lampholder or single-centre bayonet cap lampholder shall be connected to the neutral conductor
- the installation of through wiring in a luminaire is only permitted if the luminaire has been designed accordingly
- if through wiring is permitted then any cabling must be selected in accordance with the temperature information on that luminaire. In the absence of such information then heat resistant cables and/or conductors must be used
- groups of luminaires divided between the three line conductors of a three-phase system but having a common neutral, must have a linked circuit breaker that simultaneously disconnects all the line conductors.

Did you know?

E14 (SES) and E27 (ES) holders are in common use. The reason for this exemption regarding polarity is that the structure of the lampholder to BS EN 60238 ensures that with the bulb in place it is impossible to touch any live metal, and on removing the lamp, before the metal threaded part begins to show, the centre contact has already been broken.

It is still good practice to connect the outer threaded contact to Neutral, however for this type of lampholder it isn't mandatory. This is now reflected in inspection and testing as well.

Regulation 559.9 recognises the stroboscopic effect mentioned on page 414.

Regulations 559.10.1 and 559.10.2 generally state that for outdoor lighting installations, highway power supplies and street furniture, the protective measures 'placing out of reach', 'obstacles' and 'earth-free-equipotential bonding' shall not be used.

Regulation 559.10.3 states that:

- where the protective measure ADOS is used, then all live parts of equipment shall be protected by insulation or barriers/enclosures providing basic protection. However, a door in street furniture used for access to electrical equipment cannot be classed as a barrier/enclosure
- for every accessible enclosure, all live parts must only be accessible by using a key/tool unless the enclosure is located where it can only be accessed by a skilled person
- for a luminaire mounted less than 2.8 m above ground level, access shall only be possible after removing a barrier/enclosure that requires the use of a tool
- for an outdoor lighting installation, any nearby metallic structure such as a fence that is not part of the installation, need not be connected to the main earthing terminal
- it is recommended that equipment within the lighting arrangements of places such as telephone kiosks, bus shelters etc., be provided with additional protection in the form of an RCD
- a maximum disconnection time of 5 seconds shall apply to all circuits feeding fixed equipment in highway power supplies.

Regulation 559.10.6.1 states that as long as certain measures are taken, a suitable rated fuse carrier may be used as the means of isolation. However, **559.10.6.2** states that if the distributor's cut out is used as the means of isolation, then distributor permission must be obtained.

Regulation 559.11.5 states that the protective measure FELV cannot be used. The introduction of this new section in the Regulations also sees the introduction of new symbols and these are contained in Table 55.2 as detailed in Figure 8.162.

Description	Symbol	Description	Symbol
Luminaire with limited surface temperature (BS EN 60598 series)	D	Luminaire for use with high pressure sodium lamp having an internal starting device	I
Luminaire suitable for direct mounting on normally flammable surfaces	F	Replace any cracked protective shield	
Luminaire suitable for direct mounting on non-combustible surfaces only	(F crossed out)	Luminaire designed for use with self shielded tungsten halogen lamps only	
Luminaire suitable for mounting in/on normally flammable surfaces where a thermal insulating material may cover the luminaire. Note: The marking of the symbols corresponding to IP numbers is optional.	F	Transformer – short-circuit proof (both inherently and non-inherently)	
Use of heat-resistant supply cables, interconnecting cables or external wiring. (The number of cores shown is optional)	t......°C	Electronic convertor for an extra-low voltage lighting installation	110
Luminaire designed for use with bowl mirror lamps		A "class P" thermally protected ballast(s)/transformer(s)	P
Rated maximum ambient temperature	t_a .....°C	A temperature declared thermally protected ballast(s)/transformers(s) with a marked value equal to or below 130°C	...
Warning against the use of cool beam lamps	COOL BEAM	The generally recognised symbol is of an independent ballast of EN60417	
Minimum distance from lighted objects (m)	---m	Note: These symbols are referenced within BS EN 60598-1:2004. However, some of these symbols at the time of going to press, are the subject of change; the reader is advised to consult the latest edition of BS EN 60598 for current luminaire marking requirements	
Rough service luminaires	T		
Luminaire for use with high pressure sodium lamps that require an external ignitor (to the lamp)	E		

Figure 8.162 Explanations of symbols used in luminaires in control gear for luminaires and in installation of luminaires (Table 55.2)

Basic principles of illumination

Before we can perform any calculations, it is probably a good idea to know what we are trying to calculate and what we can use to achieve it.

Simply, two early things to consider are:

- How powerful is something at source? (How bright is the light?)
- How much light has landed on an object a certain distance away from the light?

Measuring light

We refer to the 'brightness' of a source (the power of the light) as **luminous intensity** and this is given the symbol I and is measured in candelas (cd).

Luminous flux is the measure of the flow (or amount) of light being emitted from that source and one of the factors used when designing lighting systems is **illuminance**.

Formerly called illumination, illuminance is our measure of the amount of light falling on a surface. This is defined as 'the density of the luminous flux striking a surface'.

Using the symbol E, illuminance has the unit of measurement (Lux), with one Lux being the illuminance at a point on a surface that is one metre from, and perpendicular to, a uniform point source of one candela.

Let's explore that a bit more. Take a ruler 1 metre long and place it flat on the floor with one end touching a 1 m^2 wall. Fix a candle to the other end of it and then light the candle. If we assume that the candle has a luminous intensity of one candela, then the amount of light hitting the wall is one Lux. In other words, one lumen uniformly distributed over one square metre of wall surface provides an illuminance of 1 Lux (1 Lux = 1 lumen/square metre).

If we were now to move the ruler and candle further away from the wall, then the wall will appear less brightly lit. However, the amount coming from the candle has remained the same. This is inversely proportional.

As a concept what we are saying is the closer you are to a luminaire, the brighter that luminaire is. Or, put another way, if we can't change how much light comes out of the luminaire, to make more light land on an object we either have to move the luminaire closer, or add more luminaires.

Other factors that affect illuminance

Whatever type of luminaires we eventually decide to install will be affected by age, collection dust etc. All of these factors will affect our level of illuminance and are grouped under an overall title of **Maintenance Factor**.

Maintenance Factor (MF) is the ratio of the illuminance provided by an installation after a period of use against its initial

> **Did you know?**
>
> A luminaire is more commonly known as a lighting fitting and refers to the fully assembled enclosure, lamp, control gear and reflector etc.

> **Did you know?**
>
> As with the compact flourescent the LED lamp saves energy against traditional incandescent lamps and they are likely to last about 30 000 hours!

illuminance when it first started use. This is expressed as a number or percentage and has no unit.

As a simple example, let's say that a shift manager's office in a garage workshop has a ceiling luminaire with one 65 W fluoresent lamp inside it installed from new. When first installed the lamp had a lumen output of 1000 lm, but when measured again after six months in operation the output had fallen to 850 lm.

The output has decreased by a ratio of: $\dfrac{850}{1000} = 0.85$

We therefore have a Maintenance Factor of 0.85.

The Maintenance Factor is based on how often the lights are cleaned and replaced. It takes into account such factors as decreased efficiency with age, accumulation of dust within the fitting itself and the depreciation of reflectance as walls and ceiling age. It is fully represented by the following formula:

MF = LLMF × LSF × LMF × RSMF

Where:

- LLMF (lamp lumen maintenance factor) – the reduction in lumen output after specific burning hours
- LSF (lamp survival factor) – the percentage of lamp failures after specific burning hours
- LMF (luminaire maintenance factor) – the reduction in light output due to dirt deposited on or in the luminaire
- RSMF (room surface maintenance factor) – the reduction in reflectance due to dirt deposition in the room surfaces.

As a rough guide, for convenience, MF is usually taken as being around the following values:

- Good = 0.70
- Medium = 0.65
- Poor = 0.55

One other consideration is the **utilisation factor** (UF), once referred to as the coefficient of utilisation (CU). Using tables available from manufacturers, it is possible to determine the utilisation factor for different lighting fittings if the reflectance of both the walls and ceiling is known, the room index has been determined and the type of luminaire is known.

In other words, UF is a number used to represent the amount of luminous flux emitted by a lamp that reaches a working surface.

Factors that make up the UF include:

- the light output of luminaire
- the flux distribution of the luminaire
- Room Index (room dimensions and spacing and mounting height of luminaires)
- room reflectances.

When checking existing light levels in a building, you should be aware that, as the light output of lamps varies depending on their operating temperature, it is essential that the luminaires have been operating under normal thermal conditions before checking. This may require, for example, both lighting and heating or air conditioning systems to be switched on for long enough to achieve steady conditions.

Where lamps are known to be new, they should be run for about 100 hours under normal operating conditions. Also, where possible the line voltage supply to the lighting circuit should also be monitored, as fluctuations in lumen output are caused by variations in supply voltage.

Let's now look at the calculations.

The lumen method

This method of calculation is only applicable in square or rectangular rooms with a uniform array of luminaires as shown in Figure 8.163.

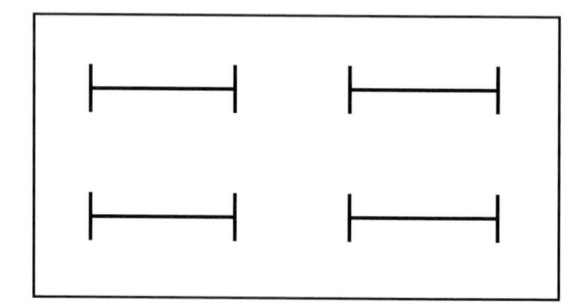

Figure 8.163 A uniform array of luminaires in a room

Using the factors mentioned previously, it is the simplest method of calculating the overall illumination level for such areas. It is accurate enough for the majority of purposes, and is the calculation most used by lighting engineers when determining the number of luminaires for a given lighting level. The formula is as follows:

$$E = \frac{F \times n \times N \times MF \times UF}{A}$$

Where:

E = average illuminance
F = initial lamp lumens
n = number of lamps in each luminaire
N = number of luminaires
MF = maintenance factor
UF = utilisation factor
A = area

Example

You have been given the following information and asked to calculate how many luminaires we need to give 300 lux at desk level within a room in a primary school.

UF = 0.44

MF = 0.85 (as the building is clean and without any air conditioning system)

n = 4 lamps per luminaire

F = 2350 lumens for the fluorescent tube

E = 300 lux at the level of the table (good for such a school)

A = 9 m × 4 m = 36 m²

By transposition of the formula:

$$N = \frac{A \times E}{F \times n \times MF \times UF}$$

Therefore:

$$N = \frac{300 \times 36}{2350 \times 4 \times 0.85 \times 0.44} = \frac{10800}{3515} = 3.07$$

Therefore 3 × 4 tube luminaires are required.

The inverse square law

In physics, an inverse square law is any physical law stating that some physical quantity or strength is inversely proportional to the square of the distance from the source of that physical quantity.

Simply, what this means is that an object that is twice the distance from a point source of light will receive only a quarter of the illumination. Or put another way, if you moved an object from 3 m to 6 m (**twice** the distance) away from a light source, you would need four times (2^2) the amount of light to maintain the same level of illumination.

We can see this in real life very easily. Consider a campfire at night – a pool of light surrounded by darkness. Or a torch being shone into the night sky – a bright beam of light that rapidly fades to nothing. You might think that when you double the distance from a light source you are now getting half as much light, but it doesn't work like that – you actually get just a quarter as much light. Figure 8.164 shows how the inverse square law works.

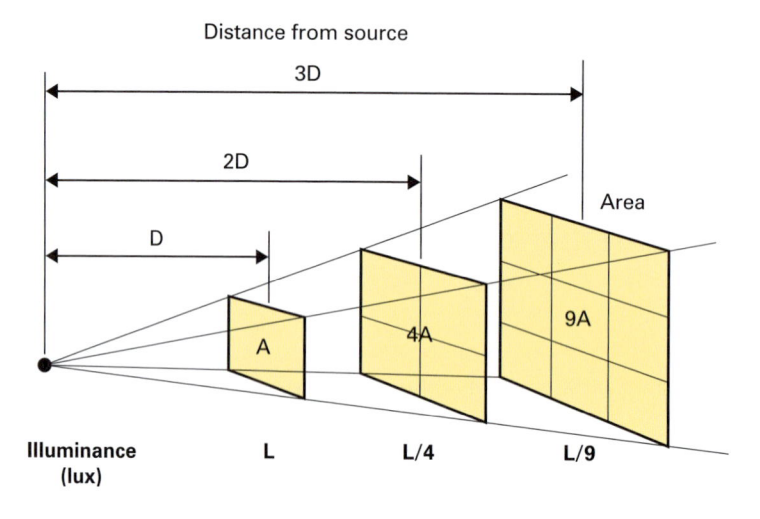

Figure 8.164 The inverse square law

Notice in Figure 8.164 how as the distance from the source increases to three times the original distance from the light source (3D), that the intensity of illumination at the new distance is nine times less ($3D^2$). This is because the amount of illumination is inversely proportional to the distance from the source.

We express this phenomenon with the following formula and apply it when the light source is directly above a surface:

Illuminance in lux $E = \dfrac{I \text{ (luminous intensity in candela)}}{d^2 \text{ (distance between source and point of measurement in m}^2\text{)}}$

> **Example**
>
> A luminaire producing a luminous intensity of 1000 cd is installed 3 m above a surface, what is the illuminance on that surface directly beneath the luminaire?
>
> $$E = \frac{I}{d^2} = \frac{1000}{3^2} = \frac{1000}{9} = \textbf{111.1 lux}$$
>
> If the luminaire was now installed 1 m higher, what would be the new level of illuminance on the surface?
>
> $$E = \frac{I}{d^2} = \frac{1000}{4^2} = \frac{1000}{16} = \textbf{62.5 lux}$$

Lambert's cosine law

Inverse square law applies when the light source is directly above the work surface and the measurement of illumination applied to a straight line beneath that luminaire. Lambert's cosine law (commonly referred to as the cosine rule) allows us to measure the illumination on the work surface but at an angle to the light source.

The law states that the illuminance on any surface varies as the cosine of the angle of incidence (the angle of incidence is the angle between the normal to the surface and the direction of the incident light).

When light strikes a surface normally (perpendicular to the surface), it gives a certain illumination level. As the angle changes from 90°, the same amount of light is spread out over a larger area, so the illumination level goes down.

If we call this angle from the perpendicular x, then the illumination level is proportional to cos x. This is demonstrated in Figure 8.165.

In the environmental context, this is why it's cold in winter and warm during the summer. During the winter the Sun's rays hit the Earth at a steep angle. The light does not spread out as much, thus increasing the amount of energy hitting any given spot. But during the winter, the Sun's rays hit the Earth at a shallow angle. The rays are therefore more spread out, which minimises the amount of energy that hits any given spot.

> **Remember**
>
> Do not confuse the Lambert's cosine law with the cosine calculations we looked at earlier in this unit (see page 238).

> **Remember**
>
> A normal is a line at right-angles to the plane.

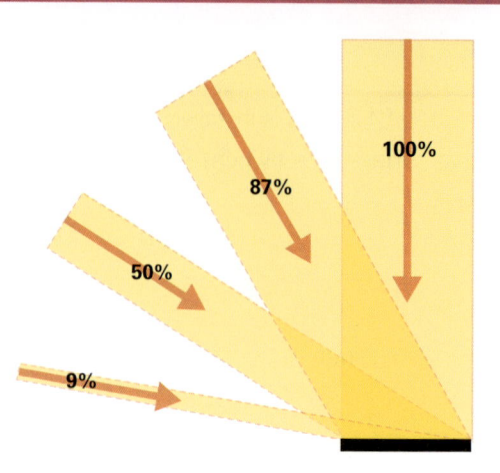

Level 3 NVQ/SVQ Diploma Installing Electrotechnical Systems and Equipment Book B

Figure 8.165 Lambert's cosine law

From our lighting perspective, if we use the pendant light in Figure 8.166 below, then we can see that the level of illuminance at point A must be higher than it is at point B, and that this reduced level at point B depends on the cosine of the angle.

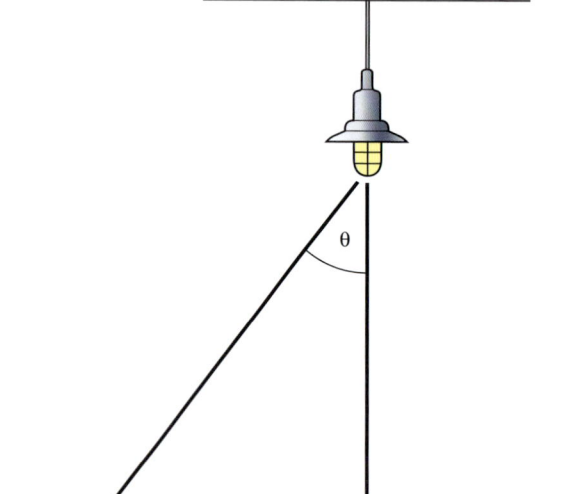

Figure 8.166 Illuminance at a series of points

We express this with the following formula, which is a composite of this and the inverse square law, namely:

$$E = \frac{I \times \cos\theta}{d^2}$$

Progress check

1. Describe how an incandescent lamp works and why it is regarded as inefficient.
2. Describe the operation of a low pressure mercury vapour discharge lamp.
3. Explain briefly the term stroboscopic effect.
4. Describe briefly the inverse square law applied to lighting.

Example

If the above pendant light produces 2000 candela and is suspended 2 m above a horizontal surface, calculate the illuminance on the surface both directly beneath the lamp (point A) and 3 m away from the lamp (point B).

Point A: $E = \dfrac{I}{d^2} = \dfrac{2000}{2^2} = \dfrac{2000}{4} = 500$ lux

Point B: $E = \dfrac{I \times \cos\theta}{d^2}$ but we don't know $\cos\theta$

We therefore need to use some logic and trigonometry.

Effectively, we have been given the dimensions of two sides of a right-angled triangle:

- the distance from the pendant to point A (2 m)
- the distance between points A and B (3 m).

The right angle 'points' at the hypotenuse (side H) and this is the measurement that we are trying to find as it is the distance from the pendant to point B.

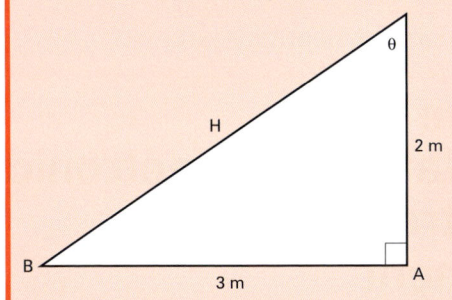

Therefore using Pythagoras' Theorem:

$H^2 = A^2 + B^2$
$H^2 = 2^2 + 3^2$
$H^2 = 4 + 9$

Therefore $H = \sqrt{4 + 9} = 3.6$

In other words, the distance from the pendant to Point B (side H) is 3.6 m

For our illuminance formula we are trying to find $\cos\theta$, and a cosine is the ratio between the adjacent side and the hypotenuse. We now know the hypotenuse, and the adjacent side is the one 'next to' the angle we are working with (θ), which in this case has a measurement of 2 m.

The cosine formula is:

$\cos\theta = \dfrac{\text{Adjacent}}{\text{Hypotenuse}} = \dfrac{2}{3.6} = 0.56$

We can now use this value in our formula

$E = \dfrac{I \times \cos\theta}{d^2} = \dfrac{2000 \times 0.56}{12.96} = \dfrac{1120}{12.96} = 86.4$

Therefore the illuminance at point B = **86.4 lux**.

K11. Understand the principles and applications of electrical heating

The two main types of heating used in buildings are electrical space and electrical water heating. Both of these have been explored in Unit ELTK 04 in Book A. Please refer to these sections for more information.

These sections also looked at some of operating principles, types, limitations and applications of electrical space and water heating appliances and components, namely:

- immersion heaters
- storage heaters
- convector heaters
- underfloor heaters
- controls, timers and programmers for heating systems.

K12. Understand the types, application and limitations of electronic components in electrotechnical systems and equipment

There are several electrotechnical components that use the following electrical components. More information about the following components can be found in Unit ELTK 04 in Book A:

- security alarms
- telephones
- dimmer switches
- heating/boiler controls
- motor control.

Resistors

There are two basic types of resistor: fixed and variable. The resistance value of a fixed resistor cannot be changed by mechanical means (though its normal value can be affected by temperature or other effects). Variable resistors have some means of adjustment (usually a spindle or slider). The method of construction, specifications and features of both fixed and variable resistor types vary, depending on what they are to be used for.

Fixed resistors

Making a resistor simply consists of taking some material of a known resistivity and making the dimensions (CSA and length) of a piece of that material such that the resistance between the two points at which leads are attached (for connecting into a circuit) is the value required.

Most of the very earliest resistors were made by taking a length of resistance wire (wire made from a metal with a relatively high resistivity, such as brass) and winding this on to a support rod of insulating material. The resistance value of the resulting resistor depended on the length of the wire used and its cross-sectional area.

This method is still used today, though it has been somewhat refined. For example, the resistance wire is usually covered with some form of enamel glazing or ceramic material to protect it from the atmosphere and mechanical damage. The external and internal view of a typical wire wound resistor is shown in Figure 8.167.

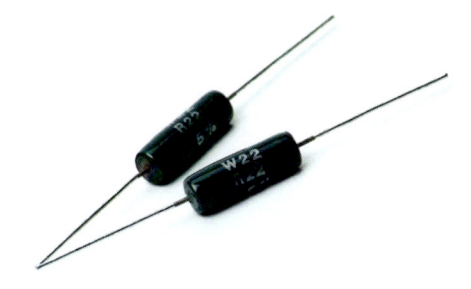

Figure 8.167 Wire wound resistor

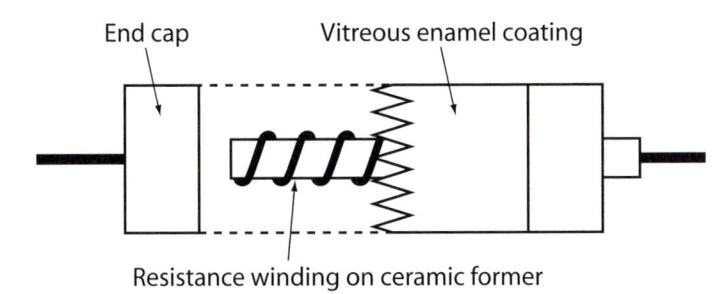

Figure 8.168 Typical wire wound resistor

Most wire wound resistors can operate at fairly high temperatures without suffering damage, so they are useful in applications where some power may be dissipated. They are, however, relatively difficult to mass-produce, which makes them expensive.

Techniques for making resistors from materials other than wire have now been developed for low power applications.

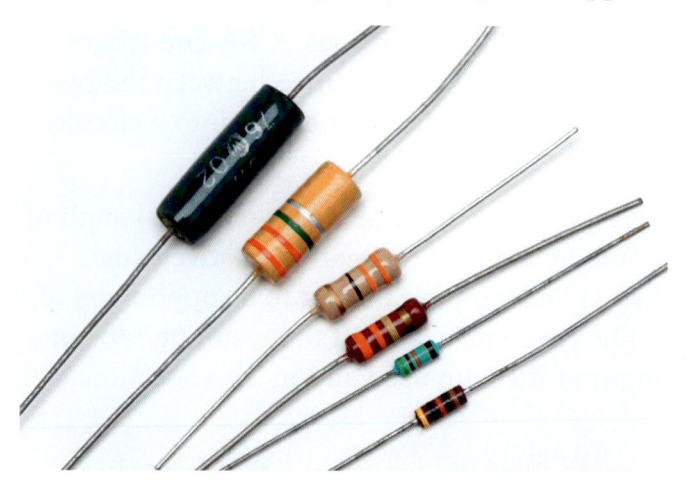

Figure 8.169 Metal oxide and carbon-composition resistors

Resistor manufacture advanced considerably when techniques were developed for coating an insulating rod (usually ceramic or glass) with a thin film of resistive material (see Figure 8.170). The resistive materials in common use today are carbon and metal oxides. Metal end caps fitted with leads are pushed over the ends of the coated rod and the whole assembly is coated with several layers of very tough varnish or similar material to protect the film from the atmosphere and from knocks during handling. These resistors can be mass-produced with great precision at very low cost.

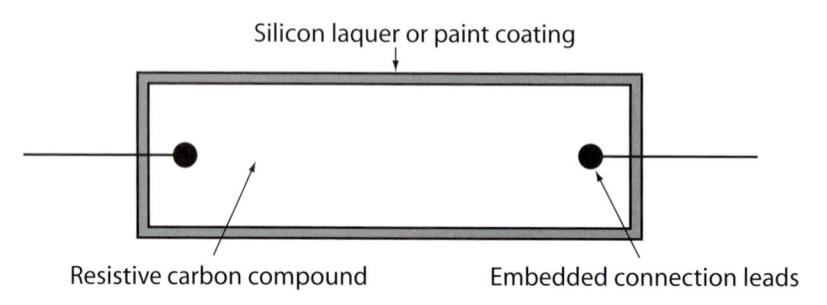

Figure 8.170 Resistor construction

Variable resistors

The development of the techniques for manufacturing variable resistors followed fairly closely that of fixed resistors, though they required some sort of sliding contact together with a fixed resistor element.

Wire wound variable resistors are often made by winding resistance wire onto a flat strip of insulating material, which is then wrapped into a nearly complete circle. A sliding contact arm is made to run in contact with the turns of wire as they wrap over the edge of the wire strip as in Figure 8.171 below. Straight versions are also possible. A straight former is used and the wiper travels in a straight line along it as shown in Figure 8.172.

While wire wound resistors are ideal for certain applications, there are many others where their size, cost and other disadvantages make them unattractive, and as a consequence alternative types have been developed.

The early alternative to the wire wound construction was to make the resistive element (on which the wiper rubs) out of a carbon composition, deposited or moulded as a track and shaped as a nearly completed circle on an insulating support plate. Alternative materials for the track are carbon films, or metal alloys of a metal oxide and a ceramic (cermet) and again straight versions are possible.

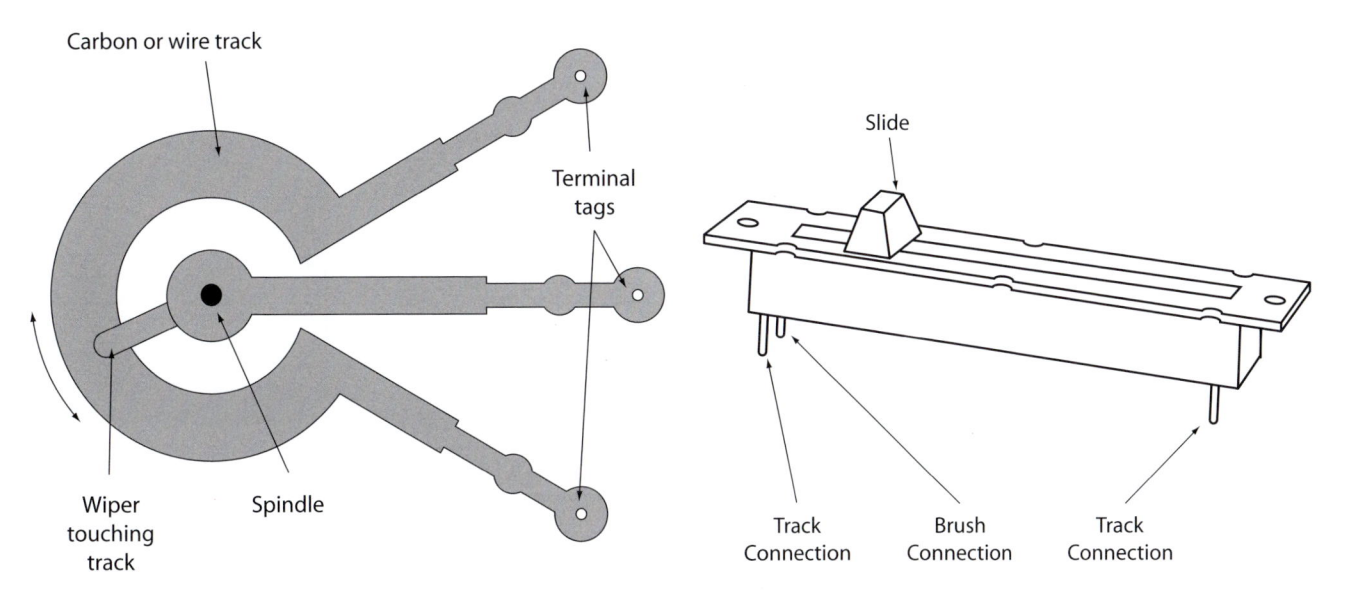

Figure 8.171 Layout of internal track of rotary variable resistor

Figure 8.172 Linear variable resistor

Preferred values

There is a series of preferred values for each tolerance level as shown in Table 8.11, so that every possible numerical value is covered.

E6 series 20% Tol	E12 series 10% Tol	E24 series 5% Tol
10	10	10
		11
	12	12
		13
15	15	15
		16
	18	18
		20
22	22	22
		24
	27	27
		30
33	33	33
		36
	39	39
		43
47	47	47
		51
	56	56
		62
68	68	68
		75
	82	82
		91

Table 8.11 Table of preferred values

In theory, there's no reason why you couldn't have resistors in every imaginable resistance value; from zero to, say, tens or hundreds of megohms. In reality, however, such an enormous range would be totally impractical to manufacture and store, and from the point of view of the circuit designer it's not usually necessary.

So, rather than an overwhelming number of individual resistance values, what manufacturers do is make a limited range of preferred resistance values. In electronics, we use the preferred value closest to the actual value we need.

A resistor with a preferred value of 1000 Ω and a 10% tolerance can have any value between 900 Ω and 1100 Ω. The next largest preferred value, which would give the maximum possible range of resistance values without too much overlap, is 1200 Ω. This can have a value between 1080 Ω and 1320 Ω.

Together, these two preferred value resistors cover all possible resistance values between 900 Ω and 1320 Ω. The next preferred values would be 1460 Ω, 1785 Ω etc.

Resistance markings

There is obviously the need for the resistor manufacturer to provide some sort of markings on each resistor so that it can be identified.

The user should be able to tell, by looking at the resistor, what its nominal resistance value is and its tolerance. Various methods of marking this information on each resistor have been used and sometimes a resistor code will use numbers and letters rather than colours.

Where physical size permits, putting the actual value on the resistor in figures and letters has an obvious advantage in terms of easy interpretation. However, again because of size restrictions, we don't use the actual words and instead use a code system. This code is necessary because when using small text on a small object, certain symbols and the decimal point become very hard to see. This code system is also commonly used to represent resistance values on circuit diagrams for the same reason.

In reality resistance values are generally given in either Ω, kΩ or MΩ using numbers from 1–999 as a prefix (e.g. 10 Ω, 567 kΩ).

In the code system we replace Ω, kΩ and MΩ and represent them instead by using the following letters:

- Ω = R
- kΩ = K
- MΩ = M

These letters are now inserted wherever the decimal point would have been in the value. So for example a resistor of value 10 Ω resistor would now be shown as 10 R, and a resistor of value 567 kΩ resistor would become 567 K.

Table 8.12 gives some more examples of this code system. Table 8.13 shows the letters that are then commonly used to represent the tolerance values. These letters are added at the end of the resistor marking so that, for example, a resistor of value 2.7 MΩ with a tolerance of ±10% would be shown as 2M7K.

> **Remember**
> Whole numbers could have a decimal point at the end (e.g. 10.0 or 567.0), but we normally miss them out when we write the numbers down (e.g. 10 or 567).

0.1 Ω	is coded	R10
0.22 Ω	is coded	R22
1.0 Ω	is coded	1R0
3.3 Ω	is coded	3R3
15 Ω	is coded	15R
390 Ω	is coded	390R
1.8 Ω	is coded	1R8
47 Ω	is coded	47R
820 kΩ	is coded	820K
2.7 MΩ	is coded	2M7

Table 8.12 Examples of resistance value codes

F	=	± 1%
G	=	± 2%
J	=	± 5%
K	=	± 10%
M	=	± 20%
N	=	± 30%

Table 8.13 Codes for common tolerance values

Resistor coding

Standard colour code

Many resistors are so small that it is impractical to print their value on them. Instead, they are marked with a code that uses bands of colour. Located at one end of the component, it is these bands that identify the resistor's value and tolerance.

Most general resistors have four bands of colour, but high-precision resistors are often marked with a five-colour band system. No matter which system is being used, the value of the colours is the same.

> **Remember**
>
> Before you read a resistor, turn it so that the end with bands is on the left-hand side. Now you read the bands from left to right (as shown in Figures 8.173 to 8.176).

Resistor colour code

Band colour	Value
Black	0
Brown	1
Red	2
Orange	3
Yellow	4
Green	5
Blue	6
Violet	7
Grey	8
White	9
Gold	0.1
Silver	0.01

Figure 8.173 Resistor colour code

Tolerance colour code

Band colour	±%
Brown	1
Red	2
Gold	5
Silver	10
None	20

Figure 8.174 Tolerance colour code

What this means

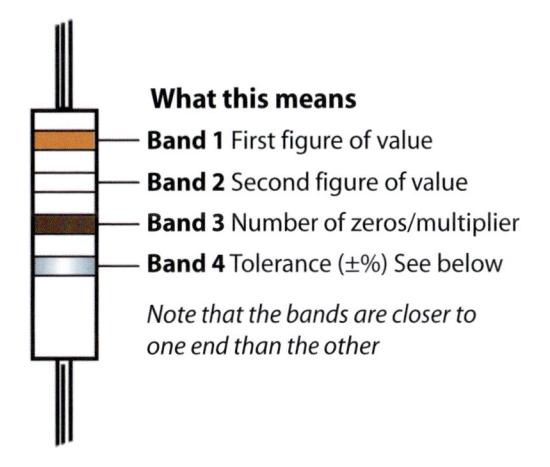

- **Band 1** First figure of value
- **Band 2** Second figure of value
- **Band 3** Number of zeros/multiplier
- **Band 4** Tolerance (±%) See below

Note that the bands are closer to one end than the other

Figure 8.175 What this means

Brown	Green	Orange	Gold
1	5	000	5%

Resistor is 15 000Ω or 15K ± 5%

Yellow	Violet	Silver	Red
4	7	×0.01	2%

Resistor is 47 × 0.01Ω or 0.47Ω ± 2%

Red	Red	Green	None
2	2	00000	20%

Resistor is 2 200 000Ω or 2.2M ± 20%

Brown	Green	Red	Gold
1	5	00	5%

Resistor is 1500Ω or 1.5K ± 5%

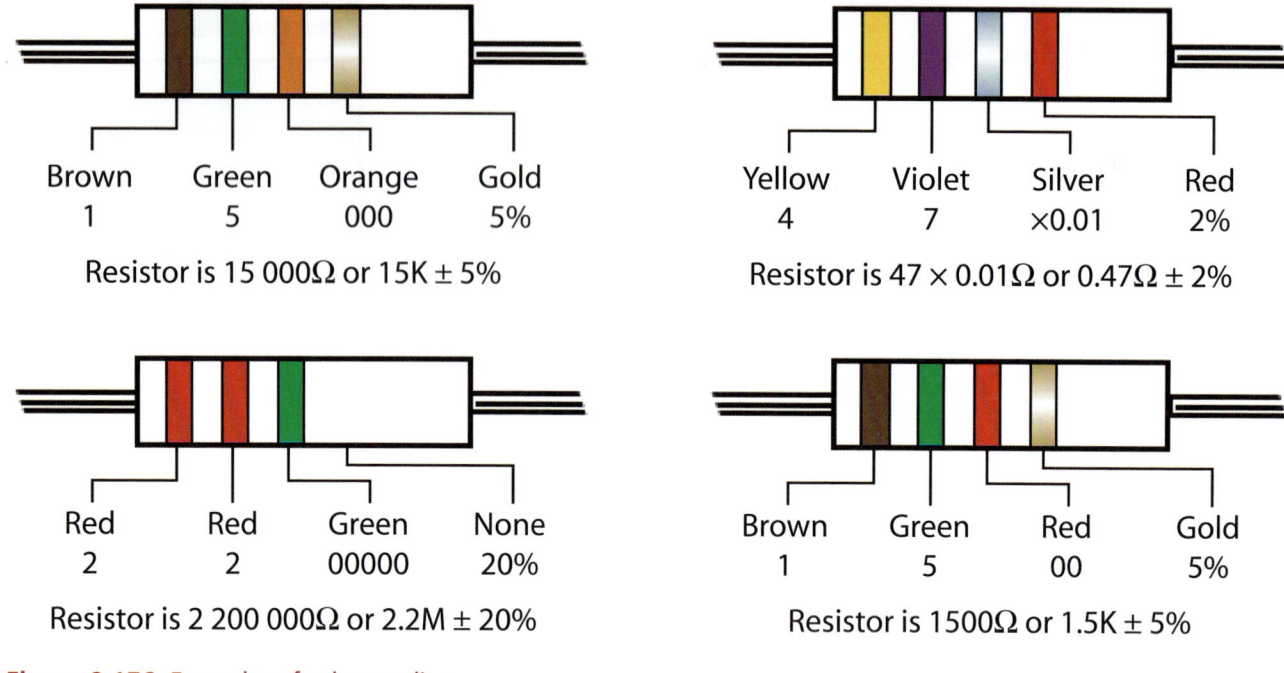

Figure 8.176 Examples of colour coding

> **Example 1**
>
> A resistor is colour-coded red, yellow, orange, gold. Determine the value of the resistor.
>
> - first band red (First digit) 2
> - second band yellow (Second digit) 4
> - third band orange (No of zeros) 3
> - fourth band gold (Tolerance) 5%.
>
> The value is **24 000 Ω ±5%**.

Example 2

A resistor is colour-coded yellow, yellow, blue, silver. Determine the value of the resistor.

- first band yellow (First digit) 4
- second band yellow (Second digit) 4
- third band blue (No of zeros) 6
- fourth band silver (Tolerance) 10%.

The value is **44 000 000 Ω ±10%**.

Example 3

A resistor is colour-coded violet, orange, brown, gold. Determine the value of the resistor.

- first band violet (First digit) 7
- second band orange (Second digit) 3
- third band brown (No of zeros) 1
- fourth band gold (Tolerance) 5%.

The value is **730 Ω ± 5%**.

Example 4

A resistor is colour-coded green, red, yellow, silver. Determine the value of the resistor.

- first band green (First digit) 5
- second band red (Second digit) 2
- third band yellow (No of zeros) 4
- fourth band silver (Tolerance) 10%.

The value is **520 000 Ω ±10%**.

Remember

To help you remember the resistor colour, learn this rhyme:

Barbara	Black	0
Brown	Brown	1
Runs	Red	2
Over	Orange	3
Your	Yellow	4
Garden	Green	5
But	Blue	6
Violet	Violet	7
Grey	Grey	8
Won't	White	9

Testing resistors

Resistors must be removed from a circuit before testing, otherwise readings will be false. To measure the resistance, the leads of a suitable ohmmeter should be connected to each resistor connection lead and a reading obtained which should be close to the preferred value and within the tolerance stated.

Resistors as current limiters

A resistor is often provided in a circuit to limit, restrict or reduce the current flowing in the circuit to some level that better suits the ratings of some other component in the circuit. For example, consider the problem of operating a solenoid valve from a 36 V d.c. supply, given the information that the energising current of the coil fitted to the valve is 100 mA and its resistance is 240 Ω.

Note that the coil, being a wound component, is actually an inductor. However, we are concerned here with the steady d.c. current through the coil and not the variation in coil current at the instant the supply is connected, so we can ignore the effects of its inductance and consider only the effects of its resistance.

If the solenoid valve were connected directly across the 36 V supply, as shown in Figure 8.177, then from Ohm's Law the steady current through its coil would be:

$$I = \frac{V}{R}$$
$$= \frac{36}{240}$$
$$= 0.15 \text{ A or } 150 \text{ mA}$$

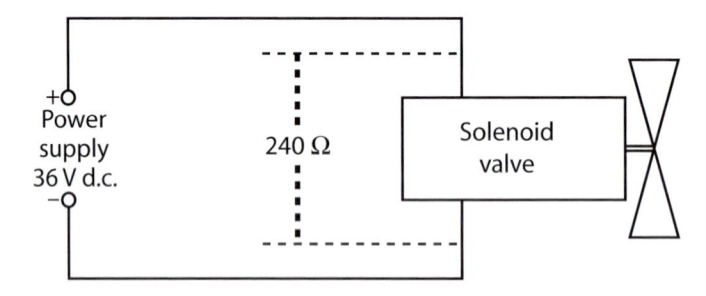

Figure 8.177 Solenoid valve connected across 36 V supply

As the coil was designed to produce an adequate magnetic 'pull' when energised at 100 mA, any increase in the energising current is unnecessary and may in fact be highly undesirable due to the resulting increase in the power, which would be dissipated (as heat) within the coils.

Note: The power dissipated in the coil of the solenoid valve, when energised at the recommended current of 100 mA is:

$$P = I^2 \times R$$
$$= 0.1^2 \text{ A} \times 240 \text{ Ω} = 2.4 \text{ watts}$$

Now if the current were to be 0.15 A on the 36 V supply it would be:

P = V × I
= 36 × 0.15
= 5.4 watts

Thus connecting the coil directly across a 36 V supply would result in the power dissipation in it being more than doubled. If the valve is required to be energised for more than very brief periods of time, the coil could be damaged by overheating.

Some extra resistance must therefore be introduced into the circuit so that the current through the coil is limited to 100 mA even though the supply is 36 V.

For a current of 100 mA to flow from a 36 V supply, the total resistance R_t connected across the 36 V must, from Ohm's Law, be:

$$R_t = \frac{36}{0.1}$$
$$= 360 \, \Omega$$

of which there is already 240 Ω in the coil.

A resistor of value 120 Ω must therefore be fitted in series with the coil to bring the value of R_t to 360 Ω. This limits the current through the coil to 100 mA when the series combination of coil and resistor is connected across the 36 V supply as shown in Figure 8.178.

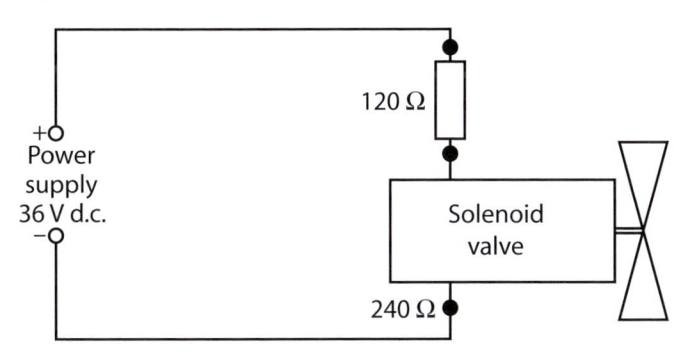

Figure 8.178 Series combination of coil and resistor connected across 36 V supply

Resistors for voltage control

Within a circuit it is often necessary to have different voltages at different stages and we can achieve this by using resistors.

For example, if we physically opened up a resistor and connected its ends across a supply, we would find that, if we then measured the voltage at different points along the resistor, the values would vary along its length. In doing so, we are effectively imitating the 'tapping' technique that we used in the transformers section of this book.

However, reality will stop us from doing this as resistors are sealed components. But we can create the same tapping effect by combining two resistors in series as shown in Figure 8.179, and then our tapping becomes a connection point made between the two resistors.

If we look at Figure 8.179, we can see that the series combination of resistors R_1 and R_2 is connected across a supply that is provided by two rails. One is shown as V+ (the positive supply rail or in other words our input) and the other as 0V (or common rail of the circuit).

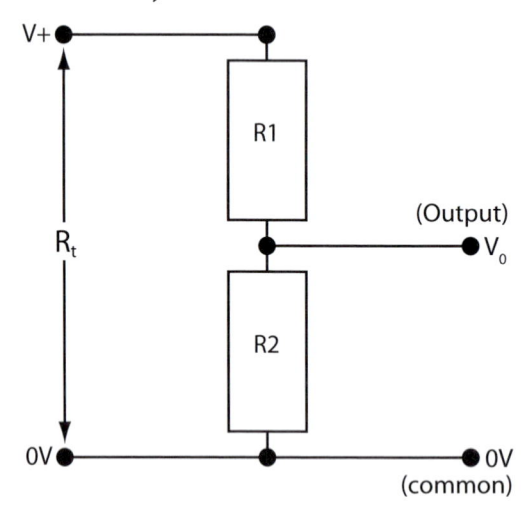

Figure 8.179 Series circuit for voltage control

The total resistance of our network (R_t) will be:

$$R_t = R_1 + R_2$$

We know, using Ohm's Law, that $V = I \times R$ and the same current flows through both resistors. Therefore for this network, we can see that $V+ = I \times R_t$ and the voltage dropped across resistor R2, $V_0 = I \times R_2$.

We now have two expressions, one for V+ and one for V_0. We can find out what fraction V_0 is of V+ by putting V_0 over V+ on the left-hand side of an equation and then putting what we said each one is equal to in the corresponding positions on the right-hand side. This gives us the following formula:

$$\frac{V_0}{V+} = \frac{I \times R_2}{I \times R_t}$$

As current is common on the right hand side of our formula, they cancel each other out. This leaves us with:

$$\frac{V_0}{V+} = \frac{R_2}{R_t}$$

To establish what V_0 (our output voltage) actually is, we can transpose again, which would give us:

$$V_0 = \frac{R_2 V+}{R_t}$$

Finally, we can replace R_t by what it is actually equal to, and this will give us the means of establishing the value of the individual resistors needed to give a desired output voltage (V_0). By transposition, our final formula now becomes:

$$V_0 = \frac{R_2}{R_1 + R_2} \times V$$

This equation is normally referred to as the potential divider rule.

In reality R_1 and R_2 could each be a combination (series or parallel) of many resistors.

However, as long as each combination is replaced by its equivalent resistance so that the simplified circuit looks like Figure 8.180, then the potential divider rule can be applied.

The potential divider circuit is very useful where the full voltage available is not required at some point in a circuit and, as we have seen, by a suitable choice of resistors in the potential divider, the desired fraction of the input voltage can be produced.

In applications where the fraction produced needs to be varied from time to time, the two resistors are replaced by a variable resistor (also known as a potentiometer, which is often abbreviated to the word 'pot'), which would be connected as shown in Figure 8.180.

$$V_0 = \frac{R_2}{R_1 + R_2} \times V+$$

The output V_0 will be dependent upon the position of the arrow. The output V_0 can be anything from V+ to 0V.

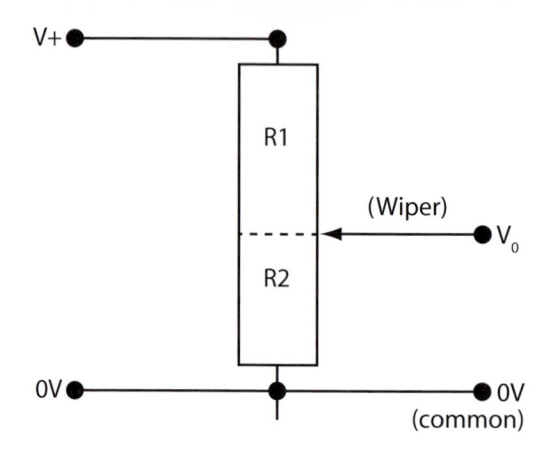

Figure 8.180 Circuit diagram for voltage applied across potentiometer

The potentiometer has a resistor manufactured in the form of a track, the ends of which effectively form our V+ and 0V connections. Our output voltage (V_o) is achieved by means of a movable contact that can touch the track anywhere on its length and this is called the wiper. We have therefore effectively created a variable tapping point.

To compare this with our potential divider, we can say that the part of the track above the wiper can be regarded as R_1 and that part below the wiper as R_2. The fraction of the input voltage appearing at the output can therefore be calculated for any setting of the wiper position by using our potential divider equation:

$$V_0 = \frac{R_2}{R_1 + R_2} \times V+$$

Obviously, when the wiper is at the top of the track, R_1 becomes zero and the equation would give the result that V_0 is equal to V+. Equally, with the wiper right at the bottom of the track, R_2 now becomes zero and therefore V_0 also becomes zero, which is not too surprising as the wiper is now more or less directly connected to the 0V rail.

This sort of circuit finds practical application in a wide variety of control functions such as volume or tone controls on audio equipment, brightness and contrast controls on televisions and shift controls on oscilloscopes.

Power ratings

Resistors often have to carry comparatively large values of current so they must be capable of doing this without overheating and causing damage. As the current has to be related to the voltage, it is the power rating of the resistor that needs to be identified.

The power rating of a resistor is thus really a convenient way of stating the maximum temperature at which the resistor is designed to operate without damage to itself. In general the more power a resistor is designed to be capable of dissipating, the larger physically the resistor is; the resulting larger surface area aids heat dissipation.

Resistors with high power ratings may even be jacketed in a metal casing provided with cooling ribs and designed to be bolted flat to a metal surface – all to improve the radiation and conduction of heat away from the resistance element.

Power is calculated by:

$P = V \times I$

Instead of V we can substitute $I \times R$ for V and $\dfrac{V}{R}$ for I. We can then use the following equations to calculate power:

$P = I^2 \times R$

Or:

$P = \dfrac{V^2}{R}$

What would the power rating of the 50 Ω resistor in Figure 8.181 be?

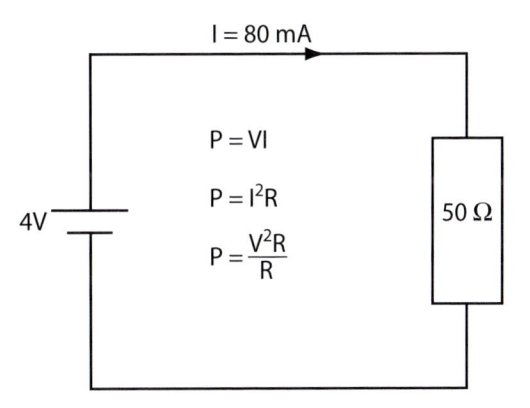

$P = V \cdot I \quad = 4 \cdot 0.08 \quad = 0.32$ watts

$P = I^2 \cdot R \quad = 0.08^2 \cdot 50 \quad = 0.32$ watts

$P = \dfrac{V^2}{R} \quad = \dfrac{4 \cdot 4}{50} \quad = 0.32$ watts

Figure 8.181 Typical power ratings for resistors

Normally only one calculation is required. Typical power ratings for resistors are shown in Table 8.14.

Carbon resistors	0 to 0.5 watts
Ceramic resistors	0 to 6 watts
Wire wound resistors	0 to 25 watts

Table 8.14 Typical power ratings for resistors

Manufacturers also always quote a maximum voltage rating for their resistors on their data sheets. The maximum voltage rating is basically a statement about the electrical insulation properties of those parts of the resistor, which are supposed to be insulators (e.g. the ceramic or glass rod which supports the resistance element or the surface coating over the resistance element).

If the maximum voltage rating is exceeded there is a danger that a flashover may occur from one end of the resistor to the other. This flashover usually has disastrous results. If it occurs down the outside of the resistor it can destroy not only the protective coating but, on film resistors, the resistor film as well.

If it occurs down the inside of the resistor the ceramic or glass rod is frequently cracked (if not shattered) and, of course, this mechanical damage to the support for the resistance element results in the element itself being damaged as well.

Light-dependent resistors

These resistors are sensitive to light. They consist of a clear window with a cadmium sulphide film under it. When light shines onto the film its resistance varies, with the resistance reducing as the light increases.

These resistors are commonly found in street lighting. You may sometimes observe street lights switching on during a thunderstorm in the daytime. This is because the sunlight is obscured by the dark thunderclouds, thus increasing the resistance, which in turn controls the light 'on' circuit.

Thermistors

A thermistor is a resistor which is temperature sensitive. The general appearance is shown in Figure 8.183. They can be supplied in various shapes and are used for the measurement and control of temperature up to their maximum useful temperature limit of about 300°C. They are very sensitive and because of their small construction they are useful for measuring temperatures in inaccessible places.

Thermistors are used for measuring the temperature of motor windings and sensing overloads. The thermistor can be wired into the control circuit so that it automatically cuts the supply to the motor when the motor windings overheat, thus preventing damage to the windings.

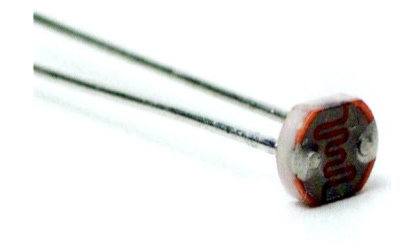

Figure 8.182 Light-dependent resistor

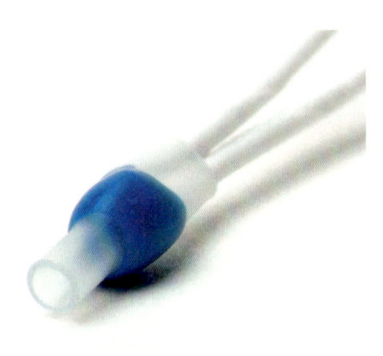

Figure 8.183 Thermistors

Did you know?

Thermistors are used for monitoring the temperature of the water in a car engine.

Thermistors can have a temperature coefficient that may be positive (PTC) or negative (NTC).

- A PTC thermistoris is made from barium titanate and its resistance of the thermistor increases as the surrounding temperature increases.
- A NTC thermistor is made from oxides of nickel, manganese, copper and cobalt and its resistance decreases as the temperature increases.

The rated resistance of a thermistor may be identified by a standard colour code or by a single body colour used only for thermistors. Typical values are shown in Table 8.15.

Colour	Resistance
Red	3000 Ω
Orange	5000 Ω
Yellow	10 000 Ω
Green	30 000 Ω
Violet	100 000 Ω

Table 8.15 Colour coding for rated resistance of thermistor

Thermocouples are two different metals bonded together and each has a lead. When the bonded metals are heated a voltage appears across the two leads. The hotter the metals become, the larger the voltage (mV). An example of their use is for measuring the temperature of furnaces within the steel industry.

Figure 8.184 Thermocouples

Capacitors

We use components known as capacitors to introduce capacitance into a circuit. Capacitance always exists in circuits – though, as you'll see when we have discussed the subject in more detail, capacitance exists between conductors whereas resistance exists in conductors.

Safety tip

Never pick a capacitor up by the terminals as it may still be charged and you will receive a shock. Always ensure the capacitor has been discharged before handling. Some capacitors have a discharge resistor connected in the circuit for this reason.

Basic principles

A capacitor is basically two metallic surfaces usually referred to as plates, separated by an insulator commonly known as the **dielectric**. The plates are usually, though not necessarily, metal and the dielectric is any insulating material. Air, glass, ceramic, mica, paper, oils and waxes are some of the many materials commonly used. The common symbols used for capacitors are identified in Figure 8.185.

> **Key term**
>
> **Dielectric** – an electrical insulator that can be polarised by an applied electric field.

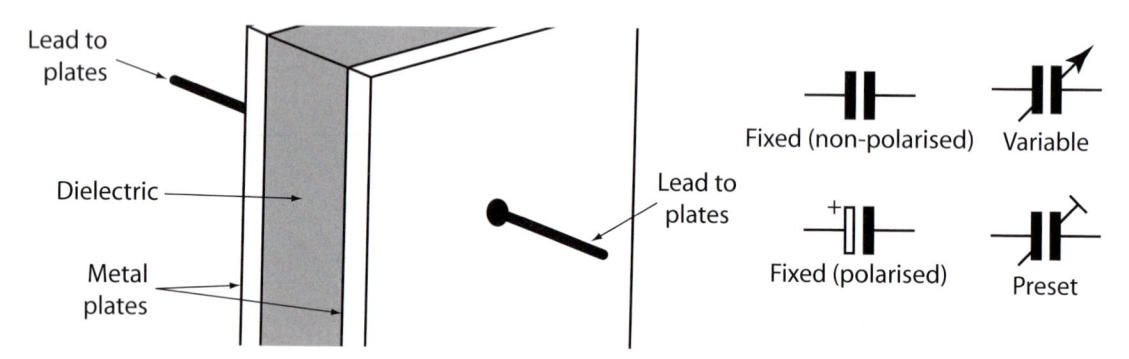

Figure 8.185 Basic construction of a capacitor and circuit symbols

These two plates are not in contact with each other and so they do not form a circuit in the same way that conductors with resistors do. However, the capacitor stores a small amount of electric charge and it can be thought of as a small rechargeable battery, which can be quickly recharged.

The capacitance of any capacitor depends on three factors:

1. The working area of the plates, i.e. the area of the conducting surfaces facing each other.

   We can think of the degree of crowding of excess electrons near the surface of one plate of a capacitor (and the corresponding sparseness of electrons near the surface of the other) as being directly related to the potential difference (p.d.) applied across the capacitor, for example connecting it directly across a battery. If we increase the area of the plates, then more electrons can flow onto one of the plates before the same degree of crowding is reached. The battery voltage determines this level of crowdedness. There is of course a similar increased loss of electrons from the other plate. The working area of the plates is directly proportional to the capacitance. If we double the area of the plates we double the capacitance of the capacitor.

2. The thickness of the dielectric between the plates.

   As mentioned earlier, the capacitance effect depends on the forces of repulsion or attraction caused by an electron surplus or shortage on the plates on either side of the dielectric. The further apart the plates are, the weaker these factors become. As a result, the degree of crowding of electrons on one plate (and the shortage of electrons on the other) produced by a given p.d across the capacitor decreases.

3. The nature of the dielectric or spacing material used.

   This fundamental principle of capacitors and the time constant of capacitor resistor circuits will be looked at later under the heading of electrostatics.

> **Remember**
>
> In any application the capacitor to be used should meet or preferably exceed the capacitor voltage rating. The voltage rating is often called 'the working voltage' and refers to d.c. voltage values. When applied to a.c. circuits the peak voltage value must be used as a comparison to the d.c. working voltage of a capacitor.

Capacitor types

There are two major types of capacitor, fixed and variable, both of which are used in a wide range of electronic devices. Fixed capacitors can be further subdivided into electrolytic and non-electrolytic types and together they represent the majority of the market.

All capacitors possess some resistance and inductance because of the nature of their construction. These undesirable properties result in limitations, which often determine their applications.

Fixed capacitors

Electrolytic capacitors

These capacitors have a much higher capacitance, volume for volume, than any other type. This is achieved by making the plate separation extremely small by using a very thin dielectric (insulator). The dielectric is often mica or paper.

They are constructed on the Swiss roll principle as are the paper dielectric capacitors used for power factor correction in electrical installation circuits, for example fluorescent lighting circuits.

The main disadvantage of an electrolytic capacitor is that it is polarised and must be connected to the correct polarity in a circuit, otherwise a short circuit and destruction of the capacitor will result.

Figure 8.186 illustrates a newer type of electrolytic capacitor using tantalum and tantalum oxide to give a further capacitance/size advantage. It looks like a raindrop with two leads protruding from the bottom. The polarity and values may be marked on the capacitor or the colour code (see Figure 8.193 on page 455) can be used.

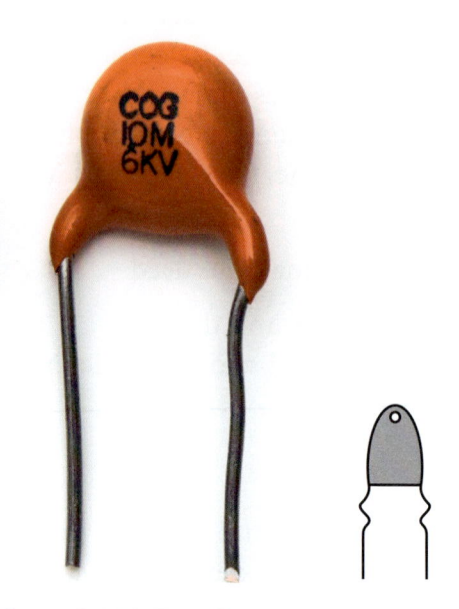

Figure 8.186 Tantalum capacitor

Non-electrolytic capacitors

There are many different types of non-electrolytic capacitor. However, only mica, ceramic and polyester are of any significance. Older types using glass and vitreous enamel are expected to disappear over the next few years and even mica will be replaced by film types.

Mica

Mica is a naturally occurring dielectric and has a very high resistance; this gives excellent stability and allows the capacitors to be accurate within a value of ±1 per cent of the marked value. Since costs usually increase with increased accuracy, they tend to be more expensive than plastic film capacitors. They are used where high stability is required, for example in tuned circuits and filters required in radio transmission. Figure 8.187 illustrates a typical mica capacitor.

Figure 8.187 Mica capacitor

> **Did you know?**
>
> Mica is a common rock-forming mineral. You find it in rocks such as granite and some sandstones and mudstones.

Ceramic capacitors

These consist of small rectangular pieces of ceramic with metal electrodes on opposite surfaces. Figure 8.188 illustrates a typical ceramic capacitor. These capacitors are mainly used in high-frequency circuits subjected to wide temperature variations. They have high stability and low Equivalent Series Resistance (ESR) loss, which is the sum loss caused by the dielectric and the metallic element of the capacitor. This is useful as ceramic capacitors with high losses can drain, and therefore waste, power.

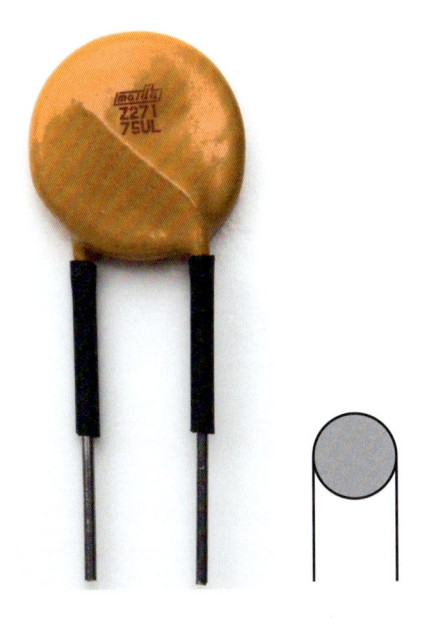

Figure 8.188 Ceramic capacitor

Polyester capacitors

These are an example of a plastic film capacitor. Polypropylene, polycarbonate and polystyrene capacitors are other types of plastic film capacitors. They are widely used in the electronics industry due to their good reliability and relative low cost but are not suitable for high-frequency circuits. Figure 8.189 illustrates a typical polyester capacitor; however, they can also be a tubular shape (see Figure 8.190).

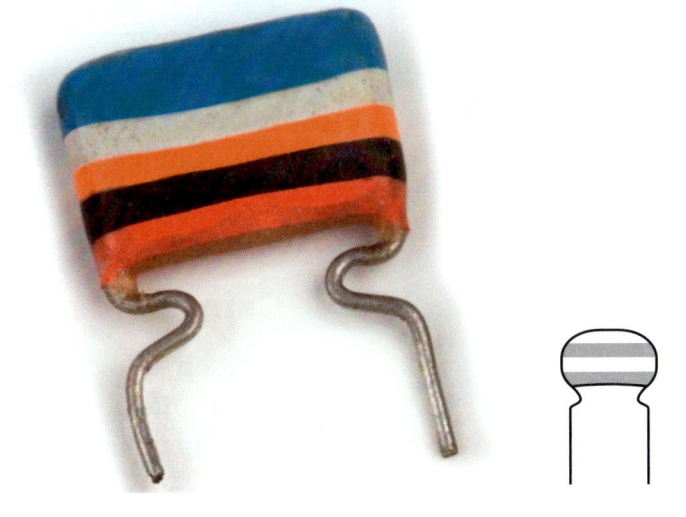

Figure 8.189 Polyester capacitor

Figure 8.190 Tubular capacitor

Variable capacitors

Variable capacitors generally have air or a vacuum as the dielectric, although ceramics are sometimes used. The two main sub-groups are tuning and trimmer capacitors.

Tuning capacitors

These are so called because they are used in radio tuning circuits and consist of two sets of parallel metal plates, one isolated from the mounting frame by ceramic supports while the other is fixed to a shaft which allows one set to be rotated into or out of the first set. The rows of plates interlock like fingers, but do not quite touch each other.

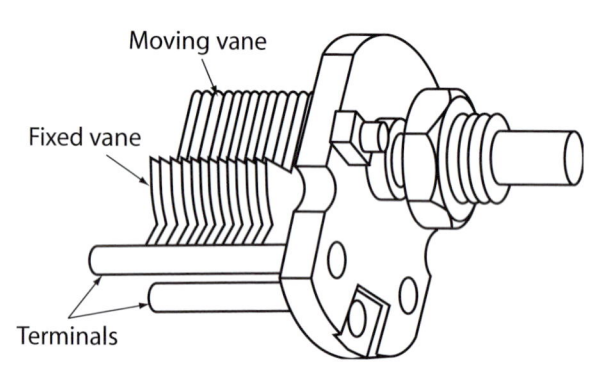

Figure 8.191 A typical variable capacitor of the tuning type

Trimmer capacitors

These are constructed of flat metal leaves separated by a plastic film; these can be screwed towards each other. They have a smaller range of variation than tuning capacitors, and so are only used where a slight change in value is needed.

Capacitor coding

To identify a capacitor the following details must be known: the capacitance, working voltage, type of construction and polarity (if any). The identification of capacitors is not easy because of the wide variation in shapes and sizes. In the majority of cases the capacitance will be printed on the body of the capacitor, which often gives a positive identification that the component is a capacitor.

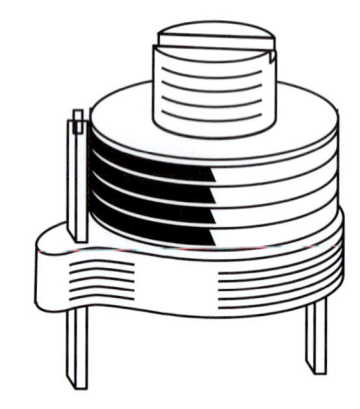

Figure 8.192 A typical capacitor used as a trimmer

The capacitance value is the farad (symbol F); this was named after the English scientist Michael Faraday. However, for practical purposes the farad is much too large and in electrical installation work and electronics we use fractions of a farad as follows:

- 1 microfarad = 1 μF = 1 × 10^{-6} F
- 1 nanofarad = 1 nF = 1 × 10^{-9} F
- 1 picofarad = 1 pF = 1 × 10^{-12} F.

The power factor correction capacitor found in fluorescent luminaires would have a value typically of 8 μF at a working voltage of 400 V. One microfarad is one million times greater than one picofarad.

The working voltage of a capacitor is the maximum voltage that can be applied between the plates of the capacitor without breaking down the dielectric insulating material.

It was quite common for capacitors to be marked with colour codes but today relatively few capacitors are colour coded. At one time nearly all plastic foil type capacitors were colour coded, as in Figure 8.193, but this method of marking is rarely encountered. However, it is a useful skill to know and be able to use the colour-coding method as shown in Table 8.16.

This method is based on the standard four-band resistor colour coding. The first three bands indicate the value in normal resistor fashion, but the value is in picofarads. To convert this into a value in nanofarads it is merely necessary to divide by 1000. Divide the marked value by 1 000 000 if a value in microfarads is required. The fourth band indicates the tolerance, but the colour coding is different from the resistor equivalent. The fifth band shows the maximum working voltage of the component. Details of this colour coding are shown in Figure 8.193 and Table 8.16.

> **Did you know?**
>
> An ideal capacitor, which is isolated, will remain charged forever. But in practice no dielectric insulating material is perfect, and therefore the charge will slowly leak between the plates, gradually discharging the capacitor. The loss of charge by leakage should be very small for a practical capacitor.

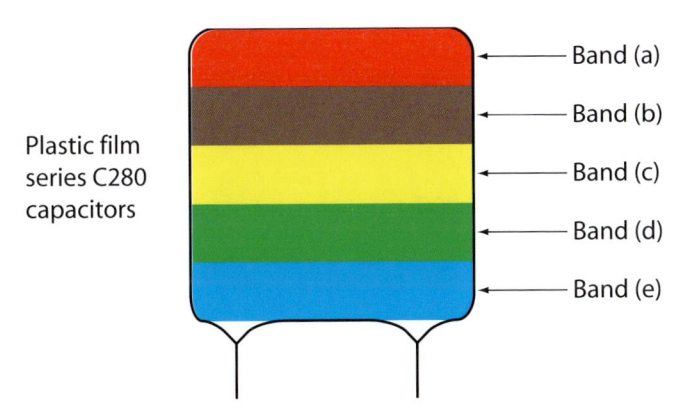

Plastic film series C280 capacitors — Band (a), Band (b), Band (c), Band (d), Band (e)

Figure 8.193 Capacitor colour bands

Standard capacitor colour coding

Colour	1st Digit	2nd Digit	3rd Digit	Tol. Band	Max. Voltage
Black		0	None	20%	
Brown	1	1	1		100 V
Red	2	2	2		250 V
Orange	3	3	3		
Yellow	4	4	4		400 V
Green	5	5	5	5%	
Blue	6	6	6		630 V
Violet	7	7	7		
Grey	8	8	8		
White	9	9	9	10%	

Table 8.16 Standard capacitor colour coding

Bands are then read from top to bottom. Digit 1 gives the first number of the component value; the second digit gives the second number. The third band gives the number of zeros to be added after the first two numbers and the fourth band indicates the capacitor tolerance, which is normally black 20%, white 10% and green 5%.

Example 1

A plastic film capacitor is colour-coded from top to bottom brown, red, yellow, black, red. Determine the value of the capacitor, its tolerance and working voltage.

- band (a) – brown = 1
- band (b) – red = 2
- band (c) – yellow = 4 multiply by 10 000
- band (d) – black = 20% tolerance
- band (e) – red = 250 volts.

The capacitor has a value of 120 000 pF or 0.12 μF with a tolerance of 20% and a maximum working voltage of 250 volts.

Example 2

A plastic film capacitor is colour-coded from top to bottom orange, orange, yellow, green, yellow. Determine the value of the capacitor, its tolerance and working voltage.

- band (a) – orange = 3
- band (b) – orange = 3
- band (c) – yellow = 4 multiply by 10 000
- band (d) – green = 5%
- band (e) – yellow = 400 volts.

The capacitor has a value of 330 000 pF or 0.33 μF with a tolerance of 5% and a maximum working voltage of 400 volts.

Example 3

A plastic film capacitor is colour-coded from top to bottom violet, blue, orange, black, and brown. Determine the value of the capacitor, its tolerance and working voltage.

- band (a) – violet = 7
- band (b) – blue = 6
- band (c) – orange = 3 multiply by 1 000
- band (d) – black = 20%
- band (e) – brown = 100 volts.

The capacitor has a value of 76 000 pF or 0.076 μF with a tolerance of 20% and a maximum working voltage of 100 volts.

Often the value of a capacitor is simply written on its body, possibly together with the tolerance and/or its maximum operating voltage. The tolerance rating may be omitted, and it is generally higher for capacitors than resistors. Most modern resistors have tolerances of 5% or better, but for capacitors the tolerance rating is generally 10% or 20%. The tolerance figure is more likely to be marked on a close tolerance capacitor than a normal 10% or 20% type.

The most popular form of value marking on modern capacitors is for the value to be written on the components in some slightly cryptic form. Small ceramic capacitors generally have the value marked in much the same way that the value is written on a circuit diagram.

Where the value includes a decimal point, it is standard practice to use the prefix for the multiplication factor in place of the decimal point. This is the same practice as was used for resistors.

The abbreviation μ means microfarad; n means nanofarad; p means picofarad. Therefore:

- 3.5 pF capacitor would be abbreviated to 3 p5
- 12 pF capacitor would be abbreviated to 12 p
- 300 pF capacitor would be abbreviated to 300 p or n 30
- 4500 pF capacitor would be abbreviated to 4 n5
- 1000 pF = 1 nF = 0.001 μF.

Polarity

Once the size, type and d.c. voltage rating of a capacitor have been determined it now remains to ensure that its polarity is known. Some capacitors are constructed in such a way that if the component is operated with the wrong polarity its properties as a capacitor will be destroyed, especially electrolytic capacitors. Polarity may be indicated by a + or – as appropriate. Electrolytic capacitors that are contained within metal cans will have the can casing as the negative connection. If there are no markings a slight indentation in the case will indicate the positive end.

Tantalum capacitors have a spot on one side as shown in Figure 8.195.

When this spot is facing you the right-hand lead will indicate the positive connection.

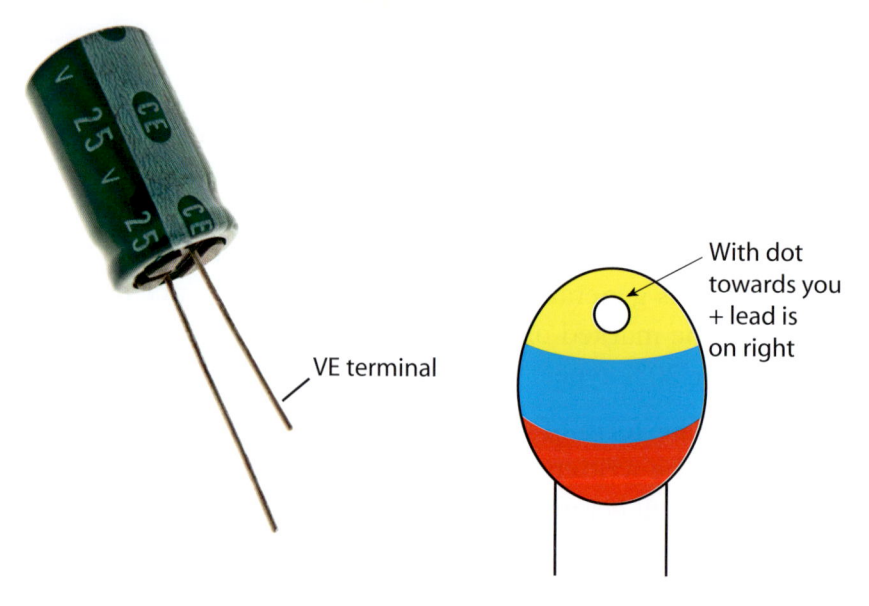

Figure 8.194 Capacitor showing polarity

Figure 8.195 Tantalum capacitor

Electrostatics and calculations with capacitors

The charge stored on a capacitor is dependent on three main factors: the area of the facing plates; the distance between the plates; and the nature of the dielectric. The charge stored by a capacitor is measured in coulombs (Q) and is related to the value of capacitance and the voltage applied to the capacitor:

Charge (coulombs) = Capacitance (farads) × Voltage (volts)
$Q = C \times V$

The formula for energy stored in a capacitor can be calculated by using the formula:

$$W = \frac{1}{2}CV^2$$

Capacitors in combination

Capacitors, like resistors, may be joined together in various combinations of series or parallel connections. Figures 8.196 and 8.197 illustrate the equivalent capacitance C_t of a number of capacitors. C_t can be found by applying similar formulae as for resistors. However, these formulae are the opposite way round to series and parallel resistors.

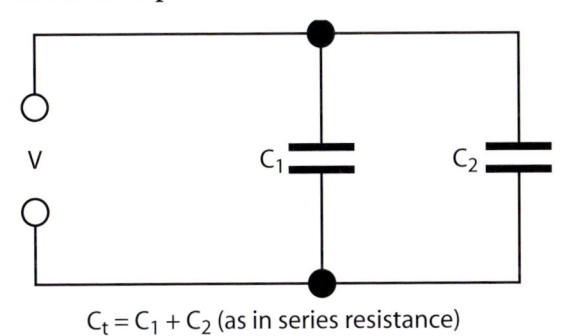

$C_t = C_1 + C_2$ (as in series resistance)

Figure 8.196 Capacitors connected in parallel

$\frac{1}{C_t} = \frac{1}{C_1} + \frac{1}{C_2}$ (as in parallel resistance)

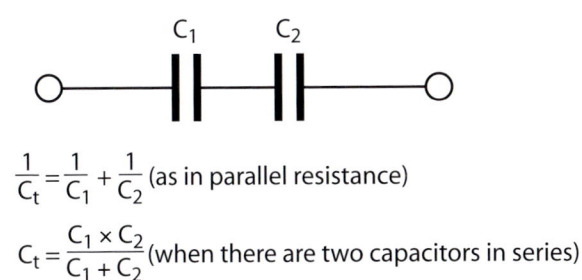

$C_t = \frac{C_1 \times C_2}{C_1 + C_2}$ (when there are two capacitors in series)

Figure 8.197 Capacitors connected in series

Example 1

Capacitors of 10 µF and 40 µF are connected in series and then in parallel. Calculate the effective capacitance for each connection.

Series:

$$\frac{1}{C_t} = \frac{1}{C_1} + \frac{1}{C_2}$$

$$\frac{1}{C_t} = \frac{1}{10\mu F} + \frac{1}{40\mu F}$$

$$\frac{1}{C_t} = \frac{4\mu F + 1\mu F}{40\mu F}$$

$$\frac{1}{C_t} = \frac{5\mu F}{40\mu F}$$

Parallel:

$$C_t = C_1 + C_2$$

$$C_t = 10\mu F + 40\mu F$$

$$C_t = 50\mu F$$

Therefore:

$$\frac{C_t}{1} = \frac{40\mu F}{5\mu F}$$

$$C_t = 8\mu F$$

Example 2

Three capacitors of 30 µF, 20 µF and 15 µF are connected in series across a 400 V d.c. supply. Calculate the total capacitance and the charge on each capacitor.

$$\frac{1}{C_t} = \frac{1}{C_1} + \frac{1}{C_2} + \frac{1}{C_3}$$

$$= \frac{1}{30\mu F} + \frac{1}{20\mu F} + \frac{1}{15\mu F}$$

$$= \frac{9\mu F}{60\mu F}$$

$$\therefore C_t = 6.66\mu F$$

Q, the charge, is common to each capacitor. Therefore:

Q = C × V
Q = 6.66 × 10⁻⁶ × 400
Q = 2.664 mC

Example 3

Three capacitors of 30 μF, 20 μF and 15 μF are connected in parallel across a 400 V d.c. supply. Calculate the total capacitance, the total charge and the charge on each capacitor.

$C_t = C_1 + C_2 + C_3$
$C_t = 30 + 20 + 15$
$C_t = 65 \mu F$

Total charge $Q = C \times V$

$Q_t = C_1 \times V$
$Q_t = 65 \times 10^{-6} \times 400$
$Q_t = \mathbf{26\,mC}$

$Q_1 = C_1 \times V$
$Q_1 = 30 \times 10^{-6} \times 400$
$Q_1 = \mathbf{12\,mC}$

$Q_2 = C_2 \times V$
$Q_2 = 20 \times 10^{-6} \times 400$
$Q_2 = \mathbf{8\,mC}$

$Q_3 = C_3 \times V$
$Q_3 = 15 \times 10^{-6} \times 400$
$Q_3 = \mathbf{6\,mC}$

Charging and discharging capacitors

Figure 8.198 shows a typical charge and discharge circuit for an uncharged capacitor (C) connected via a three-position switch (S1) to a 6 V supply (V_s), with a voltmeter (V) connected across the capacitor.

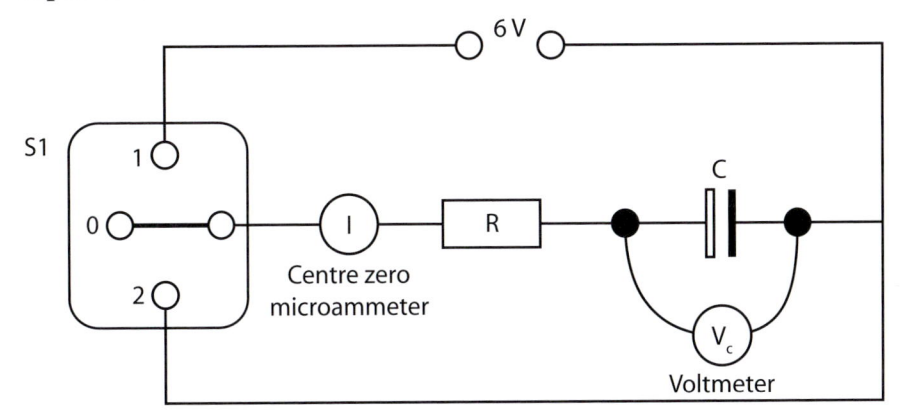

Figure 8.198 A typical charge and discharge circuit

The charging phase

With the switch in position (0), we can see that the circuit is open; no voltage exists across the capacitor (V_C as measured by the voltmeter) and therefore no electrostatic field can exist between the plates.

If we now close switch S1 to position (1), the capacitor will be connected to the supply; current will flow through the resistor (R) and both positive and negative charges will be deposited on the capacitor plates. This results in an increasing potential difference (V_C) being created that will rise exponentially from zero to a maximum value that will be the same value as the supply, in this case 6 V.

Once the voltage at the terminals of the capacitor (V_C) is equal to the supply voltage of 6 V (V_S), then no further current can flow and the capacitor can be said to be charged.

We can look at this in another way. From Ohm's Law we know that the supply voltage (V_S) is equal to the sum of the p.d.s across each individual component (V = V1 + V2 etc.).

We can therefore use Ohm's Law to find the charging current, where the voltage across the resistor (V_R) will be ($V_S - V_C$).

$$\text{Charging current (I)} = \frac{(V_S - V_C)}{R}$$

When we first switch to position (1), V_C is zero and therefore from Ohm's Law, our initial charging current will be: $(I) = \dfrac{V_S}{R}$

We also said that V_C increases exponentially as a charge starts to build up on the capacitor. In turn this will reduce the voltage across the resistor ($V_R = V_S - V_C$) and therefore reduce the charging current. This means that the rate of charging becomes progressively slower as V_C increases.

As we have discussed, the charging and discharging of a capacitor is never instantaneous. The time taken for the capacitor to charge/discharge to within a certain percentage of its maximum supply value is the time constant and is taken as the product of the circuit resistance and the capacitance. Therefore:

Time constant (T) = Resistance (Ω) × Capacitance (μF)

In other words: T = RC

As the capacitor charges, the potential difference across its plates increases with the actual time taken for the charge on the capacitor to reach 63% of its maximum possible voltage (V_S). This is known as the time constant or 1RC.

In theory our capacitor (C) never fully charges to the supply voltage (V_S). In the first time constant, (C) charges to 63% of V_S and in the second time constant (C) charges to 86% of V_S. This is also 63% of the remaining voltage difference between V_S and V_C.

This continues indefinitely, with V_C always approaching, but never quite reaching, the full value of V_S.

However, at the end of five time constants (5T or 5RC), V_C will reach 99% of the value of V_S and to all intents and purposes the capacitor is deemed to be fully charged. We can show these relationships in Figures 8.199 and 8.200.

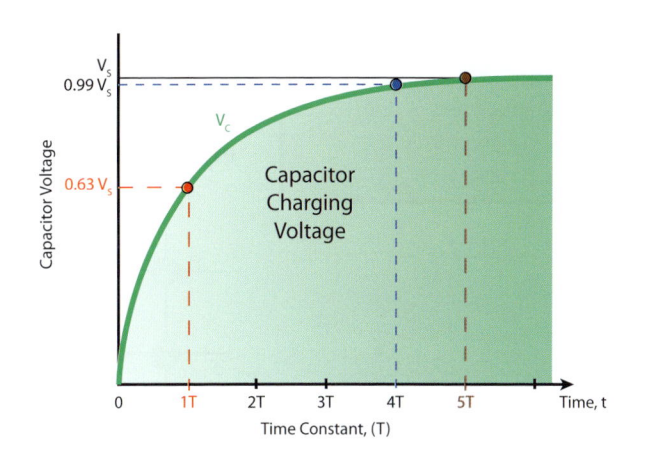

Figure 8.199

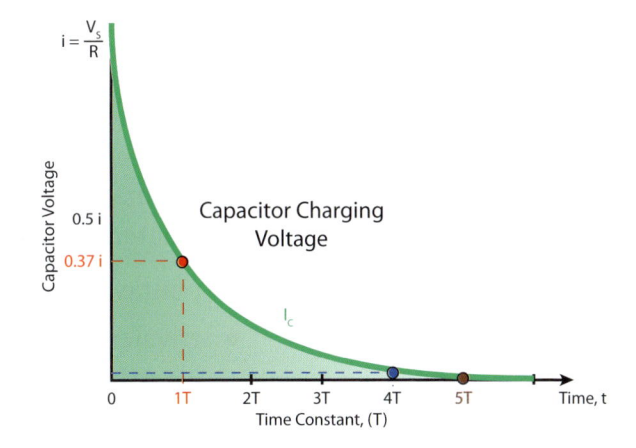

Figure 8.200

The discharging phase

To discharge the capacitor, we simply move the switch (S1) to position (2) where it will discharge through the resistor, again falling exponentially over five time constants and again not fully discharging, but instead reaching a value just above zero.

Semiconductor devices

The crystal radio that can be constructed from modern electronic game sets depends on the detector action produced by a 'cat's whisker' and a crystal (a cat's whisker was a piece of wire, the point of which was pressed firmly into contact with a suitably mounted piece of natural crystal).

This crystal detector was in fact a diode. It is the use of diodes, semiconductors and semiconductor devices that we will investigate in this section.

Semiconductor basics

Try to think of a semiconductor as being a material that has an electrical quality somewhere between a conductor and an insulator, in that it is neither a good conductor nor a good insulator. Typically, we use semiconducting materials such as silicon or germanium, where the atoms of these materials are

> **Key term**
>
> **Valence electrons** – the electrons in an atom's outermost orbit

arranged in a 'lattice' structure. The lattice has atoms at regular distances from each other, with each atom 'linked' or 'bonded' to the four atoms surrounding it. Each atom then has four **valence electrons**.

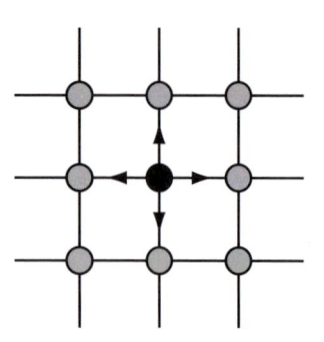

Figure 8.201 Lattice structure of semiconducting material

However, we have a problem in that, with atoms of pure silicon or germanium no conduction is possible because we have no free electrons. To allow conduction to take place we add an impurity to the material via a process known as **doping**. When we dope the material we can add two types of impurity:

- pentavalent – e.g. arsenic which contains five valence electrons
- trivalent – e.g. aluminium that contains three valence electrons.

As we can see by the number of valence electrons in each, adding a pentavalent (five) material introduces an extra electron to the semiconductor and adding a trivalent (three) material to the semiconductor 'removes' an electron (also known as creating a hole).

When we have an extra electron, we have a surplus of negative charge and call this type of material 'n-type'. When we have 'removed' an electron we have a surplus of positive charge and call this material 'p-type'. It is the use of these two materials that will allow us to introduce the component responsible for rectification, the diode.

The p–n junction

A semiconductor diode is basically created when we bring together an 'n-type' material and a 'p-type' material to form a p–n junction. The two materials form a barrier where they meet which we call the **depletion** layer. In this barrier, the coming together of unlike charges causes a small internal p.d. to exist.

We now need to connect a battery across the ends of the two materials, where we call the end of the p-type material the anode and the end of the n-type material the cathode.

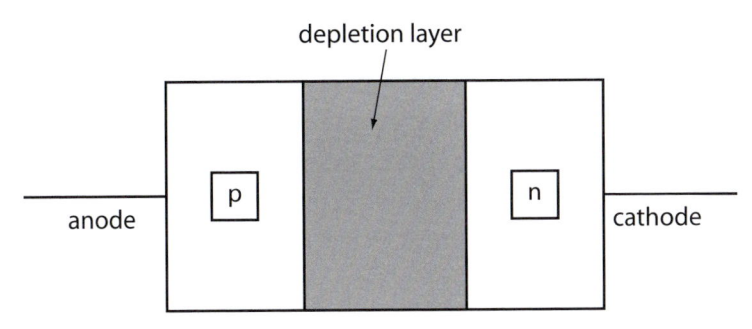

Figure 8.202 p–n junction

If the anode is positive and the battery voltage is big enough, it will overcome the effect of the internal p.d. and push charges (both positive and negative) over the junction. In other words, the junction has a low enough resistance for current to flow. This type of connection is known as being forward biased.

Reverse the battery connections so that the anode is now negative and the junction becomes high resistance and no current can flow. This type of connection is known as being reverse biased.

When the junction is forward biased, it only takes a small voltage (0.7 V for silicon) to overcome the internal barrier p.d.

When reversed biased, it takes a large voltage (1200 V for silicon) to overcome the barrier and thus destroy the diode, effectively allowing current to flow in both directions. As a general summary of its actions, we can therefore say that a diode allows current to flow through it in one direction only.

We normally use the symbol in Figure 8.204 to represent a diode.

In this symbol, the direction of the arrow can be taken to represent the direction of current flow.

Figure 8.203 Diodes

Figure 8.204 Symbol representing a diode

Zener diode

We have just established that a conventional diode will not allow current flow if reverse biased and below its reverse breakdown voltage. We also said that when forward biased (in the direction of the arrow), the diode exhibits a voltage drop of roughly 0.6 V for a typical silicon diode. If we were to exceed the breakdown voltage, the internal barrier of the diode would be destroyed, thus allowing current flow in both directions. However, this normally results in the total destruction of the device.

Zener diodes are p–n junction devices that are specifically designed to operate in the reverse breakdown region without completely destroying the device. The breakdown voltage of a zener diode V_z, (known as the zener voltage, named after the American physicist Clarence Zener who first discovered the effect) is set by carefully controlling the doping level during manufacture. The breakdown voltage can be controlled quite accurately in the doping process and tolerances to within 0.05% are available, although the most widely used tolerances are 5% and 10%.

Figure 8.205 Zener diode characteristics

Therefore, a reverse biased zener diode will exhibit a controlled breakdown, allow current to flow and thus keep the voltage across the zener diode at the predetermined zener voltage. Because of this characteristic, the zener diode is commonly used as a form of voltage limiting/regulation when connected in parallel across a load.

When connected so that it is reverse biased in parallel with a variable voltage source, a zener diode acts as a short circuit when the voltage reaches the diode's reverse breakdown voltage and therefore limits the voltage to a known value.

A zener diode used in this way is known as a shunt voltage regulator (shunt meaning connected in parallel and voltage regulator being a class of circuit that produces a fixed voltage).

For a low current power supply, a simple voltage regulator could be made with a resistor (to limit the operating current) and a reverse biased zener diode as shown in Figure 8.206. Here, V_s is the supply voltage, remembering that V_z is our zener breakdown voltage.

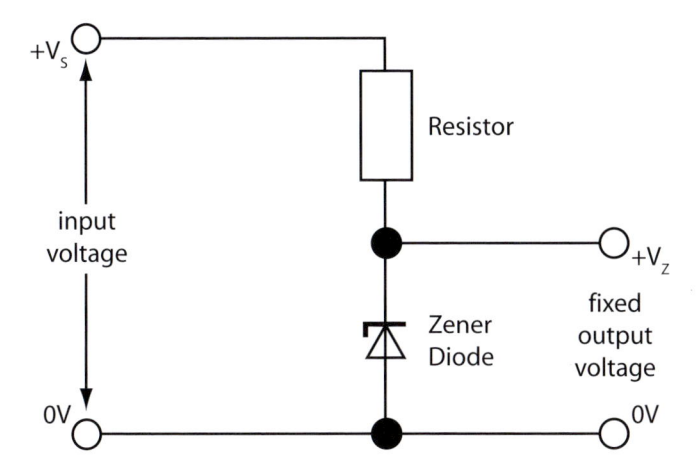

Figure 8.206 A simple voltage regulator made with a resistor and a reverse biased zener diode

As a summary, we can therefore say that a zener's properties are as follows:

- When forward biased (although not normally used for a zener) the behaviour is like an ordinary semiconductor diode.
- When reverse biased, at voltages below V_z the device essentially doesn't conduct, and it behaves just like an ordinary diode.
- When reverse biased, any attempt to apply a voltage greater than V_z causes the device to be prepared to conduct a very large current. This has the effect of limiting the voltage we can apply to around V_z.

As with any characteristic curve, the voltage at any given current, or the current at any given voltage, can be found from the curve of a zener diode (see Figure 8.206).

Light emitting diodes (LEDs)

The light emitting diode is a p–n junction especially manufactured from a semi-conducting material, which emits light when a current of about 10 mA flows through the junction. No light is emitted if the diode is reverse biased and if the voltage exceeds 5 volts then the diode may be damaged. If the voltage exceeds 2 volts then a series connected resistor may be required.

Figure 8.207 illustrates the general appearance of a LED.

Did you know?

Zener diodes are readily available with power ratings ranging from a few hundred milliwatts to tens of watts. Low power types are usually encapsulated in glass or plastic packages, and heat transfer away from the junction is mainly by conducting along the wires. High power types like power rectifiers are packaged in metal cases designed to be fitted to heat sinks so that heat can be dissipated by conduction, convection and radiation.

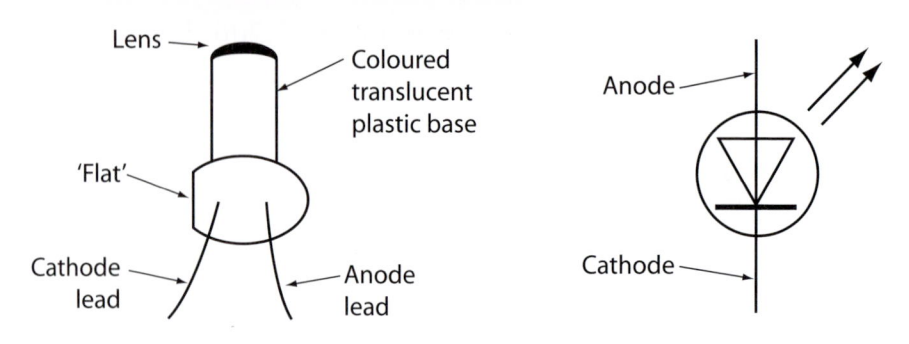

Figure 8.207 Light emitting diode

Photo cell and light-dependent resistor

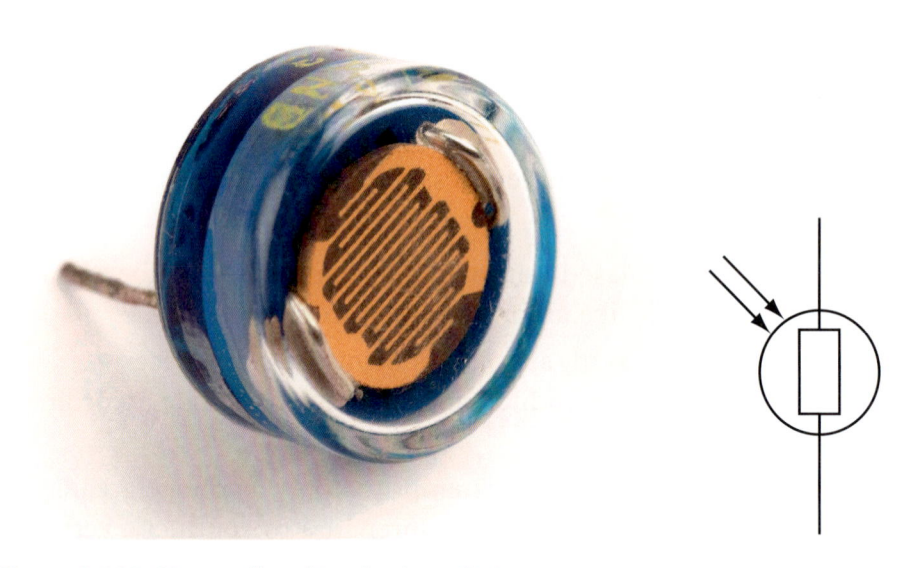

Figure 8.208 Photo cell and its circuit symbol

The photo cell shown in Figure 8.208 changes light (also infrared and ultraviolet radiation) into electrical signals and is useful in burglar and fire alarms as well as in counting and automatic control systems. Photoconductive cells or light-dependent resistors make use of the semiconductors in which resistance decreases as the intensity of light falling on them increases. The effect is due to the energy of the light setting free electrons from donor atoms in the semiconductor, making it more conductive. The main use of this type of device is for outside lights along the streets, roads and motorways. There are also smaller versions for domestic use within homes and businesses.

Photodiode

A photodiode is a p–n junction designed to be responsive to optical input. As a result, they are provided with either a window or optical fibre connection that allows light to fall on the sensitive part of the device.

Photodiodes can be used in either zero bias or reverse bias. When zero biased, light falling on the diode causes a voltage to develop across the device, leading to a current in the forward bias direction. This is called the photovoltaic effect and is the principle of operation of the solar cell, a solar cell being a large number of photodiodes. The use of the solar cell can then be seen in providing power to equipment such as calculators, solar panels and satellites orbiting the earth.

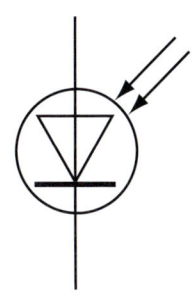

Figure 8.209 Circuit symbol for photodiode

When reverse biased, diodes usually have extremely high resistance. This resistance is reduced when light of an appropriate frequency shines on the junction. When light falls on the junction, the energy from the light breaks down bonds in the 'lattice' structure of the semiconductor material, thus producing electrons and allowing current to flow. Circuits based on this effect are more sensitive to light than ones based on the photovoltaic effect. Consequently, the photodiode is used as a fast counter or in light meters to measure light intensity.

Opto-coupler

The opto-coupler, also known as an opto-isolator, consists of an LED combined with a photodiode or phototransistor in the same package, as shown in Figure 8.210 and 8.211.

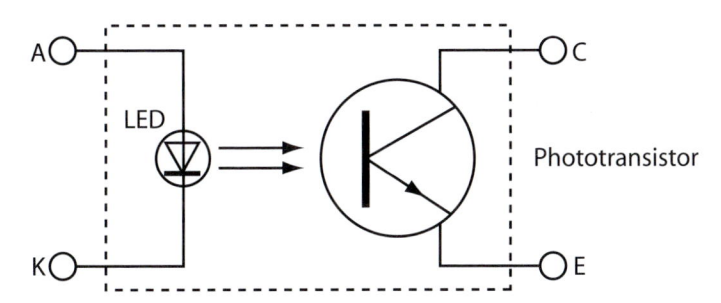

Figure 8.210 Opto-coupler circuit

Figure 8.211 Opto-coupler package

The opto-coupler package allows the transfer of signals, analogue or digital, from one circuit to another in cases where the second circuit cannot be connected electrically to the first, for example, due to different voltages.

Light (or infrared) from the LED falls on the photodiode/transistor which is shielded from outside light. A typical use for one of these is in a VCR to detect the end or start of the tape.

Infrared source and sensor

An infrared beam of light is projected from an LED which is a semiconductor made from gallium arsenide crystal. The light emitted is not visible light, but very close to the white light spectrum. Figure 8.112 shows various housings for both the source output and the sensors within the security alarms industry.

Infrared beams have a receiver, which reacts to the beam in differing ways depending upon its use. Infrared sources/receivers are used for alarm detection and as remote control signals for many applications. The passive infrared (PIR) detector is housed in only one enclosure and uses ceramic infrared detectors. The device does not have a projector but detects the infrared heat radiated from the human body.

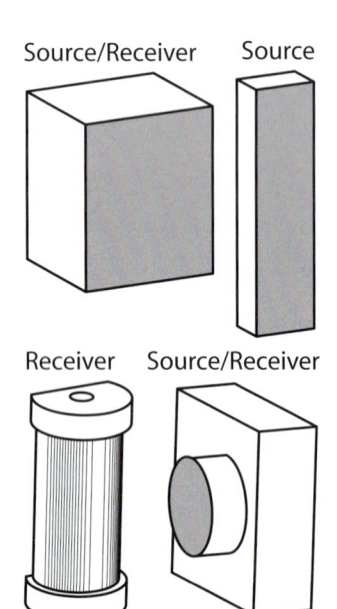

Figure 8.212 Housings for infrared source output and sensors for security alarms

Fibre optic link

The simplified block diagram in Figure 8.213 shows a system of communication, which can be several thousand kilometres in length. On the far left we input information such as speech or visual pictures as electrical signals. They are then pulse code modulated in the coder and changed into equivalent digital light signals by the optical transmitter via a miniature laser or LED at the end of the fibre optic cable.

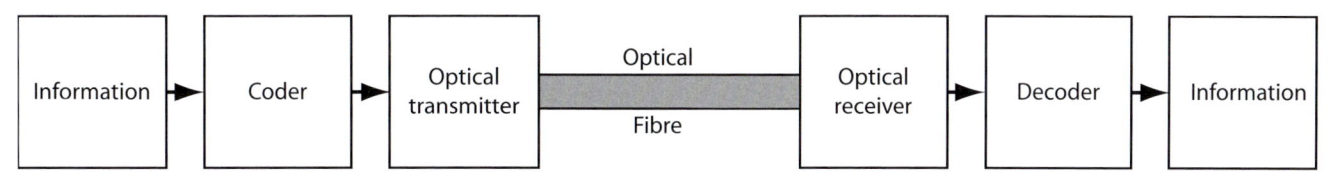

Figure 8.213 Simplified block diagram of fibre optic link

The light is transmitted down the cable to the optical receiver which uses a photodiode or phototransistor which converts the incoming signals back to electrical signals before they are decoded back into legible information.

The advantages of this type of link over a conventional communication system are:

- high information carrying capacity
- free from the noise of electrical interference
- greater distance can be covered, as there is no volt drop
- the cable is lighter, smaller and easier to handle than copper
- crosstalk between adjacent channels is negligible
- it offers greater security to the user.

The fibre optic cable

The fibre optic cable (see Figure 8.214) has a glass core of higher refractive index than the glass cladding around it. This maintains the light beam within the core by total internal reflection at the point where the core and the cladding touch.

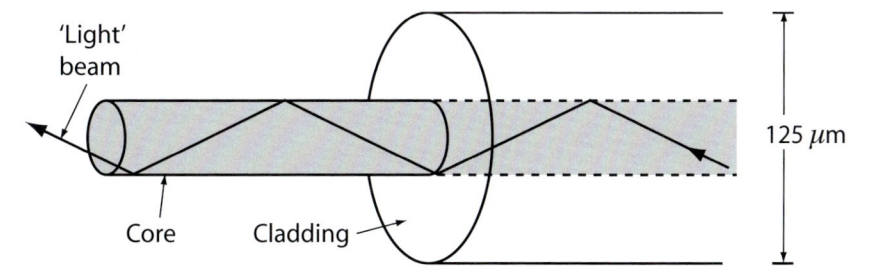

Figure 8.214 Fibre optic cable

This is similar to the insulation of the single core cable preventing the current leaking from the conductor. The beam of light bounces off the outer surface of the core in a zigzag formation along its length.

There are two main types of cable: multimode and singlemode (monomode).

Multimode

The wider core of the multimode fibre (Figure 8.215) allows the infrared to travel by different reflected paths or modes. Paths that cross the core more often are longer and take more time to travel along the fibre. This can sometimes cause errors and loss of information.

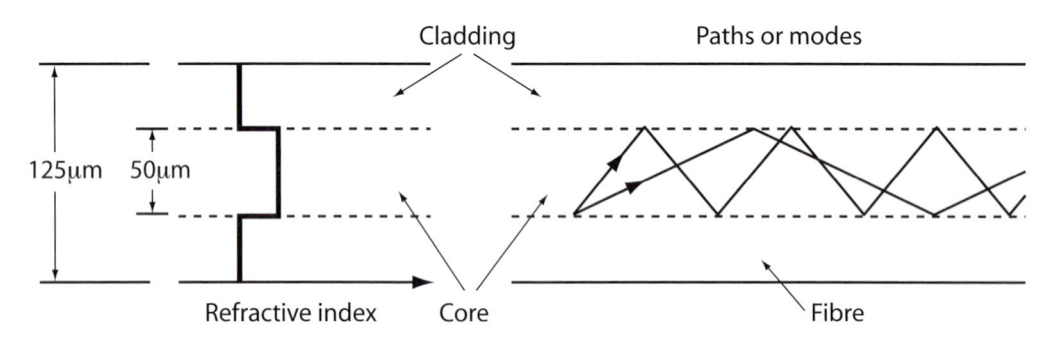

Figure 8.215 Multimode fibre optic cable

Singlemode

The core of the singlemode fibre (Figure 8.216) is about one tenth of the multimode and only straight through transmission is possible.

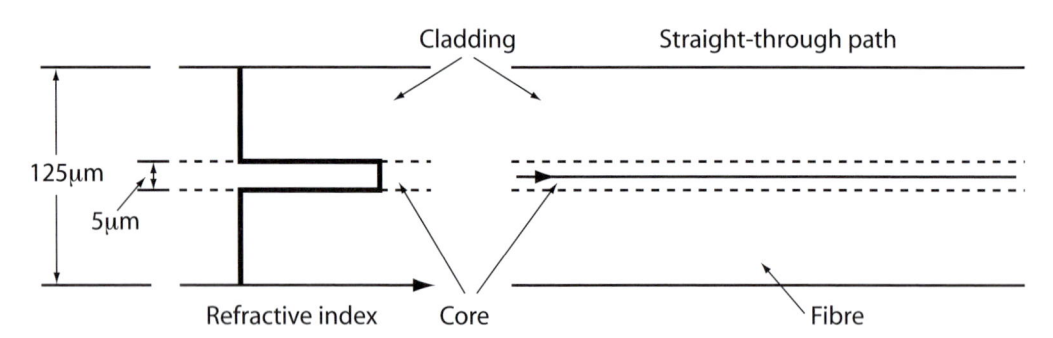

Figure 8.216 Singlemode fibre optic cable

Diode testing

Diodes allow electricity to flow in only one direction. When we test diodes, we are testing not only the condition of the diode but its ability to do this. The arrow in the circuit symbol shows the direction in which the current can flow.

We therefore need to remind ourselves of a diode's connections. Figure 8.217 shows the connections of the diode.

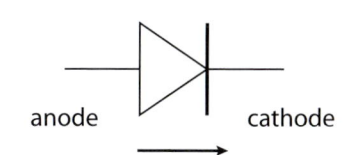

Figure 8.217 Diode connections

We can therefore say that:

- a good diode will show an extremely high resistance with reverse bias and a very low resistance with forward bias
- a defective open diode will show an extremely high resistance for both forward and reverse bias
- a defective shorted or resistive diode will show zero or a low resistance for both forward and reverse bias.

A defective open diode is the most common type of failure.

It is also important to remember that the test method used will depend upon the type of multimeter available, i.e. whether it is an analogue or digital meter.

Using an analogue multimeter

The following list explains the key points you should remember when using an analogue multimeter to carry out tests of diodes.

- Set the multimeter to a low value resistance scale such as ×10.
- **An important point** – when on the resistance settings only, most analogue meters reverse the polarity of the leads, so black become positive.
- Therefore connect the black (+) positive lead to the anode and the red (−) negative lead to the cathode. The diode should conduct and the meter will display a low resistance (the needle will be heading towards zero).
- Reverse the connections so that black (+) positive is connected to the cathode.
- The diode should **not** conduct this way so the meter will show a high resistance (the needle will be heading towards infinity).

Using a digital multimeter

Digital multimeters have a special setting for testing a diode, usually indicated by the diode symbol.

When connected to this diode setting the meter provides an internal voltage sufficient to forward bias and reverse bias a diode.

- Connect the red (+) positive lead to the anode and the black (−) negative lead to the cathode.
- If the diode is good, expect a reading of between approximately 0.5 V and 0.9 V, with 0.7 V being typical for forward bias.
- Reverse the connections.
- The diode should not conduct this way, so if the diode is working properly, expect a voltage reading based on the meter's internal voltage source. The value indicates that the diode has an extremely high reverse resistance with essentially all of the internal voltage therefore appearing across it.

Using a digital multimeter on Ohms function

If the multimeter does not have a special diode setting then the diode can be checked using a low resistance scale, much as with the analogue meter. However, there will be no need to reverse the lead polarity. Red should be positive this time!

For a forward bias check of a good diode, expect a low resistance reading and for the reverse bias check of a good diode expect an extremely high value or a reading of 'OL' as the reverse resistance is too high for the meter to measure. The actual resistance of forward biased diode is typically much less than 100 Ω.

Rectification

Rectification is the conversion of an a.c. supply into a d.c. supply. Despite the common use of a.c. systems in our day-to-day work as an electrician, there are many applications (e.g. electronic circuits and equipment) that require a d.c. supply. The following section looks at the different forms this can take.

Half-wave rectification

A diode will only allow current to flow in one direction and it does this when the anode is more positive than the cathode. In the case of an a.c. circuit, this means that only the positive half cycles are allowed 'through' the diode and, as a result, we end up with a signal that resembles a series of 'pulses'. This tends to be unsuitable for most applications, but can be used in situations such as battery charging. A transformer is also commonly used at the supply side to ensure that the output voltage is to the required level. The waveform for this form of rectification would look as in Figure 8.218.

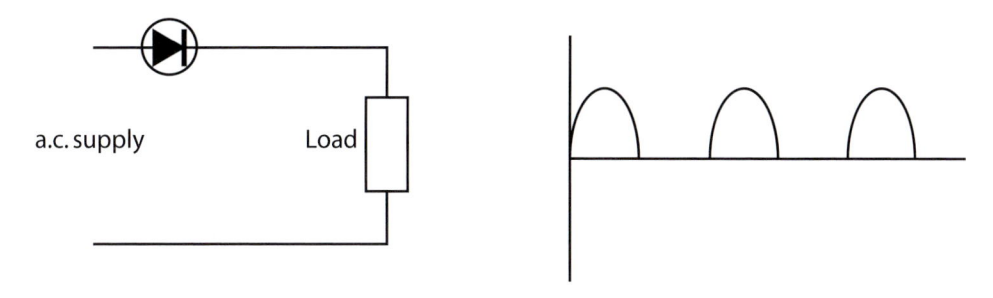

Figure 8.218 Half-wave rectification

Full-wave rectification

We have seen that half-wave rectification occurs when one diode allows the positive half cycles to pass through it. However, we can connect two diodes together to give a more even supply. We call this type of circuit **biphase**. In this method, we connect the anodes of the diodes to the opposite ends of the secondary winding of a centre-tapped transformer. As the anode voltages will be 180° out of phase with each other, one diode will effectively rectify the positive half cycle and one will rectify the negative half cycle. The output current will still appear to be a series of pulses, but they will be much closer together, with the waveform shown in Figure 8.219.

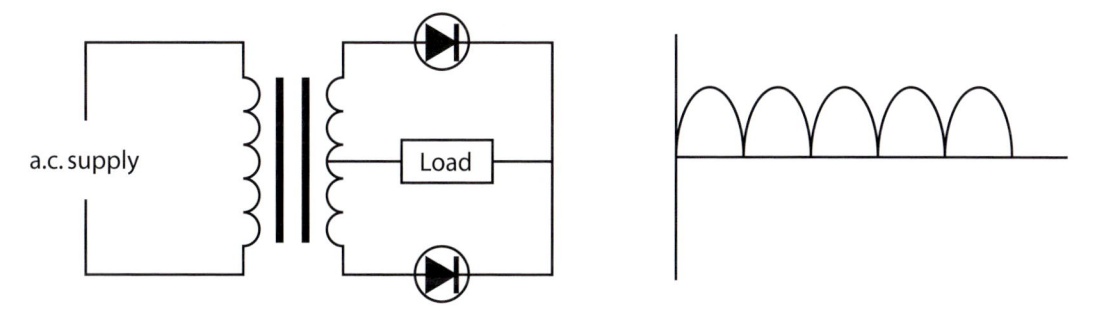

Figure 8.219 Full-wave rectification

The full-wave bridge rectifier

This method of rectification doesn't have to rely on the use of a centre-tapped transformer, but the output waveform will be the same as that of the biphase circuit previously described. In this system, we use four diodes, connected in such a way that at any instant in time two of the four will be conducting. The connections would be as in Figure 8.220, where we have shown two drawings to represent the route through the network for each half cycle.

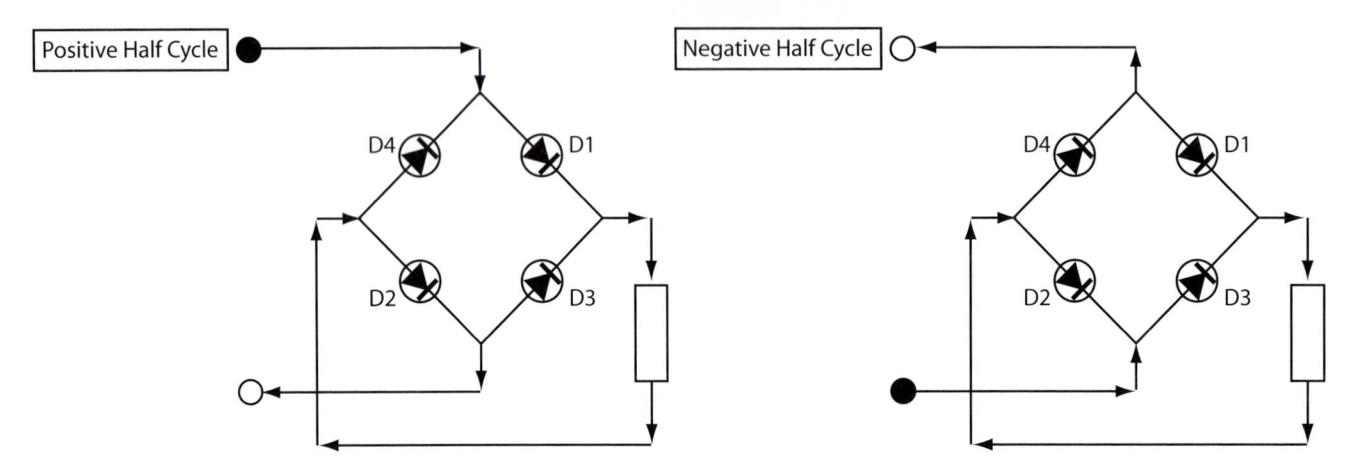

Figure 8.220 Full-wave bridge rectifier

The circuits that we have looked at so far convert a.c. into a supply which, although never going negative, is still not a true d.c. supply. This brings us to the next stage of the story – smoothing.

Smoothing

We have seen that the waveform produced by our circuits so far could best be described as having the appearance of a rough sea. The output current is not at a constant value, but constantly changing. This, as we have said before, is acceptable for battery charging, but useless for electronic circuits where a smooth supply voltage is required.

To make it useful for electronic circuits, we need to smooth out the waveform by creating a situation that is sometimes referred to as **ripple-free** and in essence there are three ways to achieve this, namely capacitor smoothing, choke smoothing and filter circuits.

Capacitor smoothing

If we connect a capacitor in parallel across the load, then the capacitor will charge up when the rectifier allows a flow of current and discharge when the rectifier voltage is less than the capacitor. However, the most effective smoothing comes under no-load conditions. The heavier the load current, the heavier the ripple. This means that the capacitor is only useful as a smoothing device for small output currents.

Choke smoothing

If we connect an inductor in series with our load, then the changing current through the inductor will induce an e.m.f. in opposition to the current that produced it. This means that the e.m.f. will try to maintain a steady current. Unlike the capacitor, this means that the heavier the ripple (rate of change of current) the more that ripple will be smoothed. This effectively means that the choke is more useful in heavy current circuits.

Filter circuits

This is the name given to a circuit that removes the ripple and is basically a combination of the two previous methods. The most effective of these is the capacitor input filter, shown in Figure 8.221.

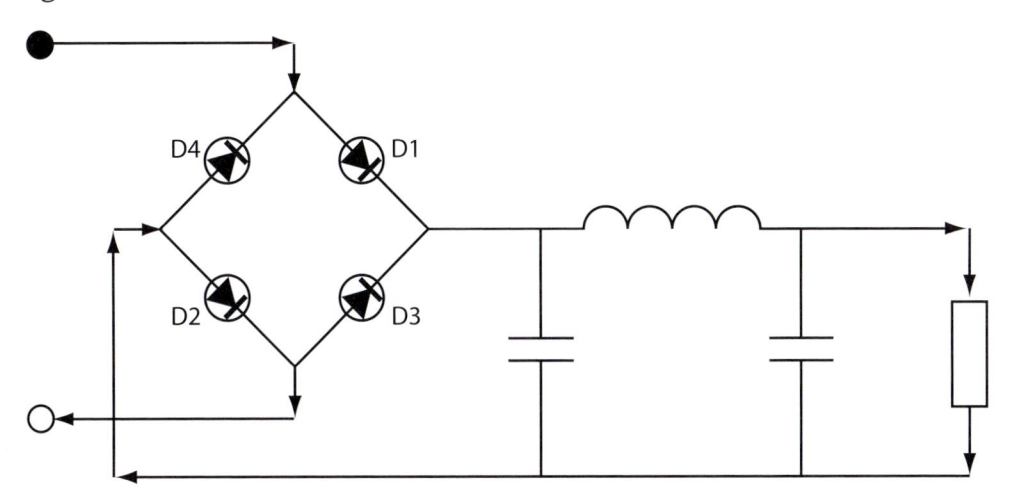

Figure 8.221 Capacitor input filter

The waveform for the filter circuit that we have just spoken about is shown in Figure 8.222, where the dotted line indicates the waveform before smoothing.

Three-phase rectifier circuits

Whereas Figure 8.222 indicates that a reasonably smooth waveform can be obtained from a single-phase system, we can get a much smoother wave from the three-phase supply mains. To do so we use six diodes connected as a three-phase bridge circuit (see Figure 8.223). These types of rectifier are used to provide high-powered d.c. supplies.

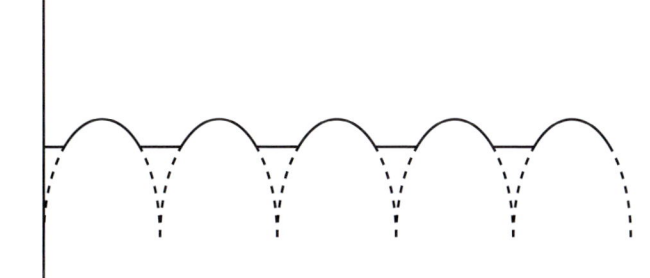

Figure 8.222 Waveform for capacitor input filter

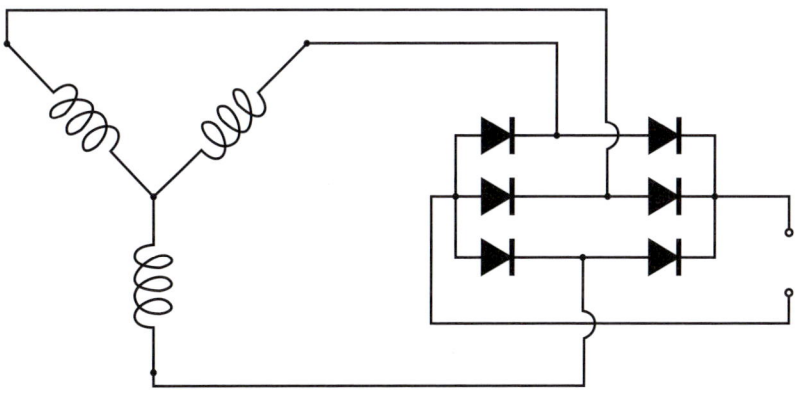

Figure 8.223 Waveform for capacitor input filter

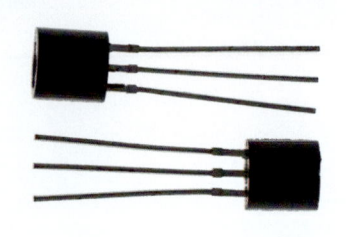

Figure 8.224 Thyristor

Thyristors, diacs and triacs

Sometimes referred to as a 'silicon controlled rectifier' (SCR), the thyristor is a four-layer semiconducting device, with each layer consisting of an alternating n or p type material as shown in Figure 8.223.

The main terminals (the anode and cathode) are across the full four layers, while the control terminal (the gate) is attached to one of the middle layers. The circuit symbol for a thyristor is shown in Figure 8.226.

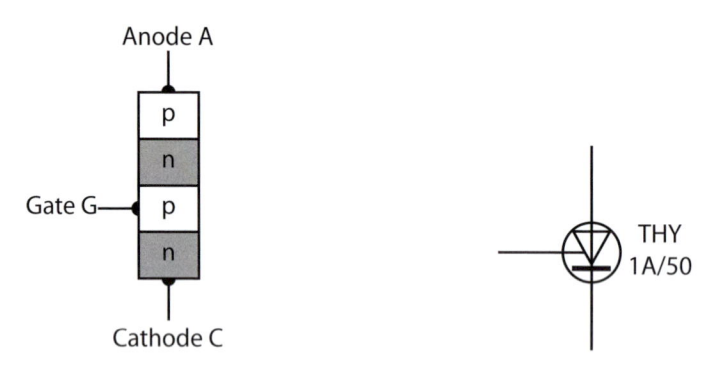

Figure 8.225 Construction of a thyristor   **Figure 8.226** Thyristor circuit symbol

Effectively acting as a high speed switch, thyristors are available that can switch large amounts of power (as high as MW) and can therefore be seen in use within high voltage direct current (HVDC) systems. These can be used to interconnect two a.c. regions of a power-distribution grid, albeit the equipment needed to convert between a.c. and d.c. can add considerable cost. That said, above a certain distance (about 35 miles for undersea cables and 500 miles for overhead cables), the lower cost of the HVDC electrical conductors can outweigh the cost of the electronics required.

Principle of operation

A thyristor acts like a semiconductor version of a mechanical switch, having two states; in other words it is either 'on' or 'off' with nothing in between. This is how they gained their name from the Greek word thyra (which means door), the inference being something that is either open or closed.

The thyristor is very similar to a diode, with the exception that it has an extra terminal (the gate) which is used to activate it. Effectively in its normal or 'forward biased' state, the thyristor acts as an open-circuit between anode and cathode, thus

> **Did you know?**
>
> The conversion from a.c. to d.c. is known as rectification and from d.c. to a.c. as inversion.

preventing current flow through the device. This is known as the 'forward blocking' state.

However, the thyristor can allow current to flow through it by the application of a control (gate) current to the gate terminal. It is this concept that allows a small signal at the gate to control the switching of a higher power load. In this respect the thyristor is performing in a similar way to a relay (pages 292–94).

Once activated, a thyristor doesn't require a control (gate) current to continue operating and will therefore continue to conduct until either the supply voltage is turned off, reversed or when a minimum 'holding' current is no longer maintained between the anode and cathode.

These concepts are shown in the following diagram:

In Figure 8.227, switch S_1 acts as a master isolator and no supply is present at either the thyristor or at switch S_2. Closing switch S_1 will allow a supply to be present at the thyristor, but there is no signal at the gate terminal as switch S_2 is open and therefore no current will flow to the indicator lamp. However, if we now close switch S_2, the gate will be energised and the thyristor will operate, thus allowing current to flow through it to the indicator lamp. The current at the anode would be large enough in this situation to allow the thyristor to continue operating, even if we opened switch S_2.

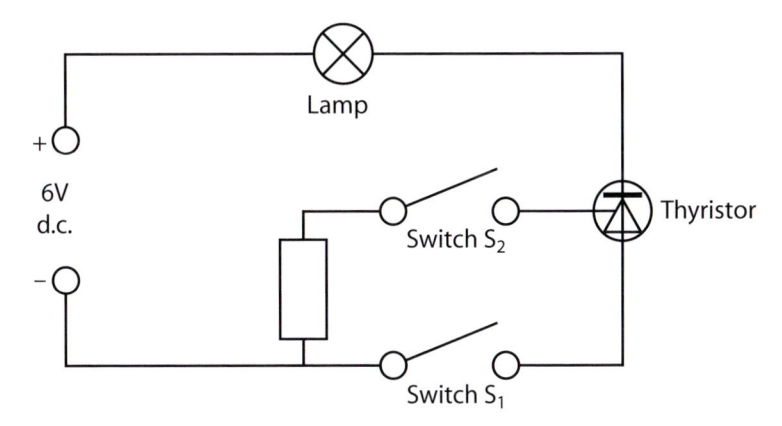

Figure 8.227 Circuit diagram of a thyristor

The control of a.c. power can also be achieved with the thyristor by allowing current to be supplied to the load during part of each half cycle. If a gate pulse is applied automatically at a certain time during each positive half cycle of the input, then the thyristor will conduct during that period until it falls to zero for the negative half cycle.

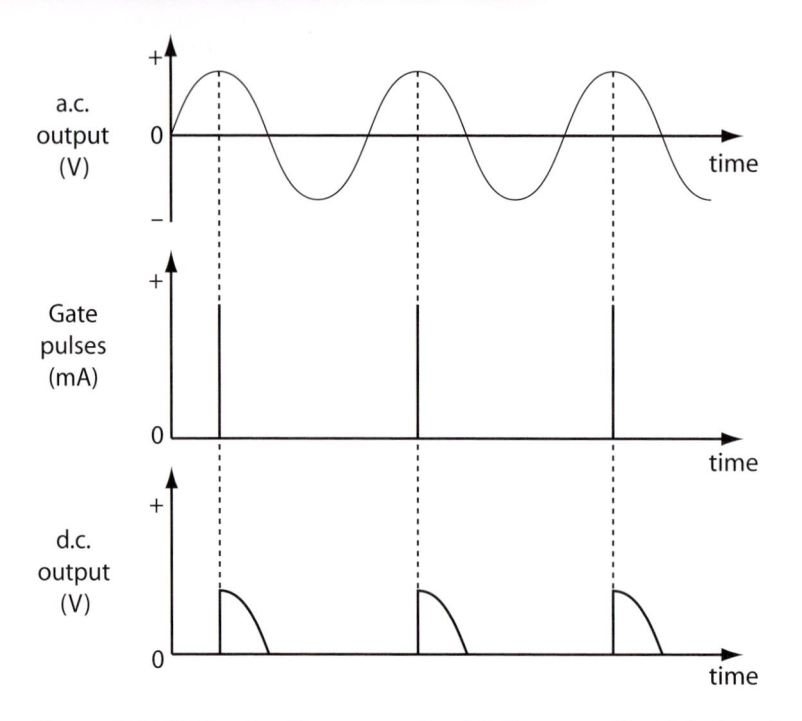

Figure 8.228 Supply of current to load during part of each half cycle

You will see from Figure 8.228 that the gate pulse (mA) occurs at the peak of the a.c. input (V). During negative half cycles the thyristor is reverse biased and will not conduct and will not conduct again until half way through the next positive half cycle. Current actually flows for only a quarter of the cycle, but by changing the timing of the gate pulses, this can be decreased further or increased. The power supplied to the load can be varied from zero to half wave rectified d.c.

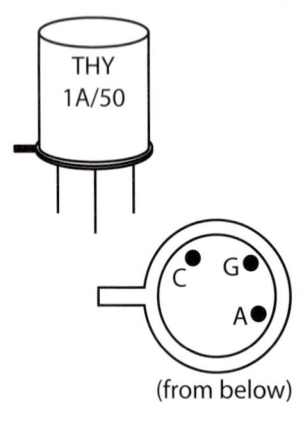

Figure 8.229 Typical small value thyristor

Thyristor testing

To test thyristors a simple circuit needs to be constructed as shown in Figure 8.230. When switch B only is closed the lamp will not light, but when switch A is closed the lamp lights to full brilliance. The lamp will remain illuminated even when switch A is opened. This shows that the thyristor is operating correctly. Once a voltage has been applied to the gate the thyristor becomes forward conducting like a diode and the gate loses control.

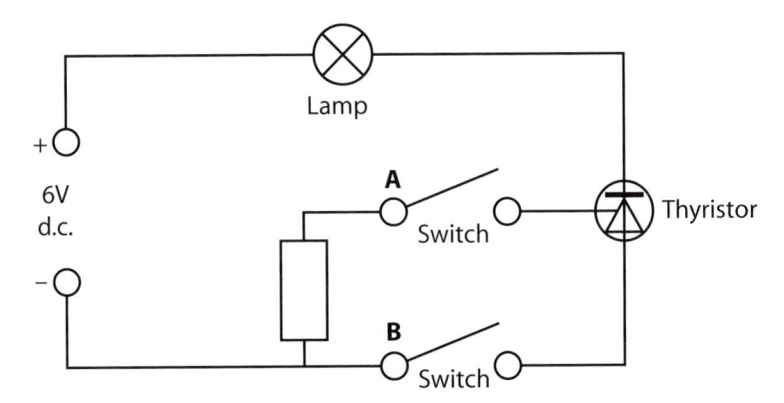

Figure 8.230 Circuit for testing thyristor

The triac

The triac is a three terminal semiconductor for controlling current in either direction and the schematic symbol for a triac is shown in Figure 8.231.

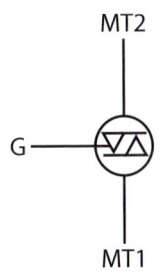

Figure 8.231 Triac symbol

If we look at Figure 8.231 more closely, we can see that the symbol looks like two thyristors that have been connected in parallel, albeit in opposite facing directions and with only one gate terminal. We refer to this type of arrangement as an inverse parallel connection.

The main power terminals on a triac are designated as MT1 (Main Terminal 1) and MT2. When the voltage on terminal MT2 is positive with regard to MT1, if we were to apply a positive voltage to the gate terminal the left thyristor would conduct. When the voltage is reversed and a negative voltage is applied to the gate, the right thyristor would conduct.

As with the thyristor generally, a minimum holding current must be maintained in order to keep a triac conducting. A triac

> **Did you know?**
>
> Triacs were originally developed for, and used extensively in, the consumer market. They are used in many low power control applications such as food mixers, electric drills and lamp dimmers, etc.

therefore generally operates in the same way as the thyristor, but operating in both a forward and reverse direction. It is therefore sometimes referred to as a bidirectional thyristor, in that it can conduct electricity in both directions. One disadvantage of this is that triacs can require a fairly high current pulse to turn them on.

The diac

Before consideration is given to practical triac applications and circuits, it is necessary to examine the diac. This device is often used in triac triggering circuits because it, along with a resistor-capacitor network, produces an ideal pulse-style waveform. It does this without any sophisticated additional circuitry, due to its electrical characteristics. Also it provides a degree of protection against spurious triggering from electrical noise (voltage spikes).

The symbol is shown in Figure 8.232. The device operates like two breakdown (zener) diodes connected in series, back to back. It acts as an open switch until the applied voltage reaches about 32–35 volts, when it will conduct.

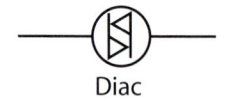

Figure 8.232 Diac symbol

Lamp dimmer circuit

Figure 8.233 shows a typical 230 V GLS lamp dimmer circuit.

The GLS lamp has a tungsten filament, which allows it to operate at about 2500°C and is wired in series with the triac. The variable resistor is part of a trigger network providing a variable voltage into the gate circuit, which contains a diac connected in series. Increasing the value of the resistor increases the time taken for the capacitor to reach its charge level to pass current into the diac circuit. Reducing the resistance allows the triac to switch on faster in each half cycle. By this adjustment the light output of the lamp can be controlled from zero to full brightness.

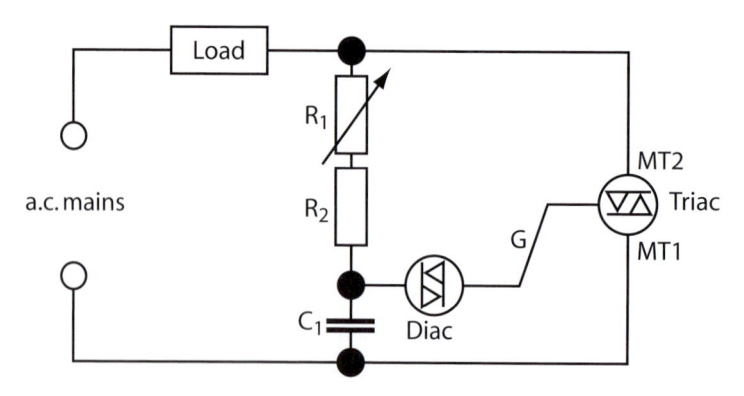

Figure 8.233 GLS lamp dimmer circuit

The capacitor is connected in series with the variable resistor. This combination is designed to produce a variable phase shift into the gate circuit of the diac. When the p.d. across the capacitor rises, enough current flows into the diac to switch on the triac.

The diac is a triggering device having a relatively high switch on voltage (32–35 volts) and acts as an open switch until the capacitor p.d. reaches the required voltage level.

The triac is a two-directional thyristor, which is triggered on both halves of each cycle. This allows it to conduct current in either direction of the a.c. supply. Its gate is in series with the diac, allowing it to receive positive and negative pulses.

A relatively high resistive value resistor R2 (100 Ω) is placed in series with a capacitor to reduce false triggering of the triac by mains voltage interference. The capacitor is of a low value (0.1 mF). This combination is known as the **snubber circuit**.

> ### Working life
>
> You are asked to look at a problem at a small private residential care home. When you arrive the warden explains that there is a problem with the nurse call system, in that it has very recently been installed but, as the electrical contractor that fitted it has gone into receivership, no one is quite sure of its operation. Additionally, they require another 'patient call' button to be installed in a further bedroom. The warden has a circuit diagram of the system, which is shown below.
>
> All components, with the exception of the patient call buttons, are located inside the nurse call panel, which is located at the nurses' station. Looking at the diagram, identify the components and prepare a written report for the warden, explaining in writing how the system operates. Now produce a revised circuit diagram to show how an additional patient could be added to the system. (Please assume that the values and ratings of any components for this exercise will be acceptable.) Your new drawing and report will then be held by the warden for future reference.

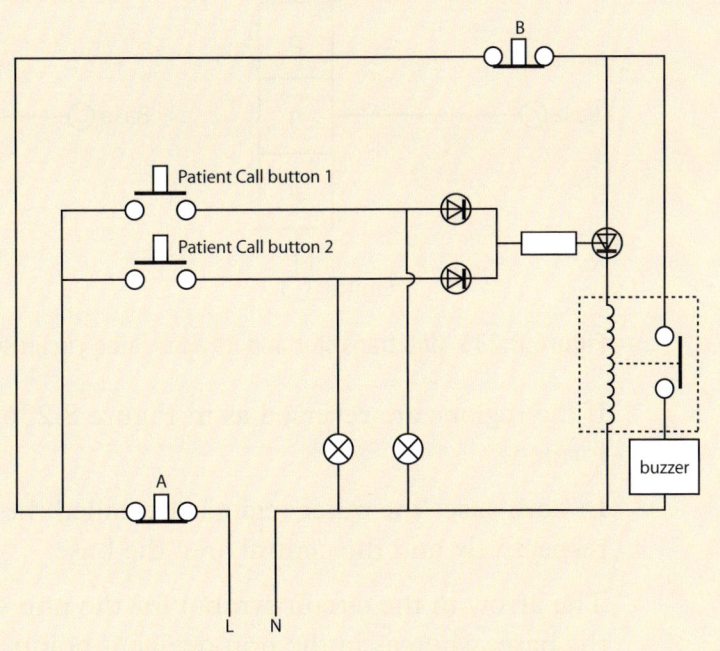

Transistors

> **Did you know?**
>
> The term 'transistor' is derived from the two words 'transfer-resistor'. This is because in a transistor approximately the same current is transferred from a low to a high resistance region. The term 'bipolar' means both electrons and holes are involved in the action of this type of transistor.

What is usually referred to as simply a transistor, but is more accurately described as a bipolar transistor, is a semiconductor device, which has two p–n junctions. It is capable of producing current amplification and, with an added load resistor, both a load and voltage power gain can be achieved.

Figure 8.234 Transistor

Transistor basics

A bipolar transistor consists of three separate regions or areas of doped semiconductor material and, depending on the configuration of these regions, it is possible to manufacture two basic types of device.

When the construction is such that a central n-type region is sandwiched between two p-type outer regions, a pnp transistor is formed as in Figure 8.235.

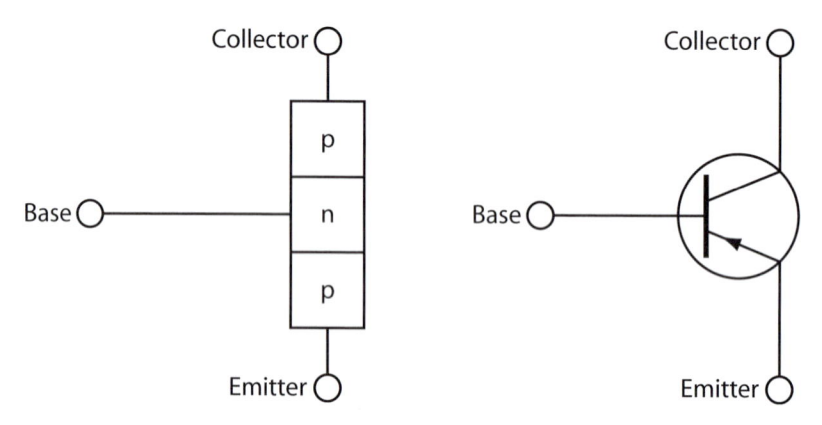

Figure 8.235 pnp transistor and its associated circuit symbols

If the regions are reversed as in Figure 8.236, an npn transistor is formed.

In both cases the outer regions are called the emitter and collector respectively and the central area the base.

The arrow in the circuit symbol for the pnp device points towards the base, whereas in the npn device it points away from it. The arrow indicates the direction in which conventional current would normally flow through the device.

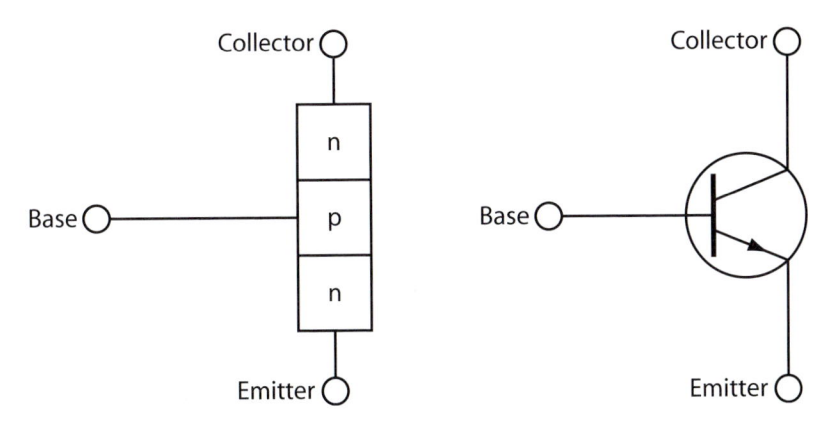

Figure 8.236 npn transistor and its associated circuit symbols

Electron flow is of course in the opposite direction as explained on pages 264–81. Note that in these idealised diagrams, the collector and emitter regions are shown to be the same size. This is not so in practice; the collector region is made physically larger since it normally has to dissipate the greater power during operation. Further, the base region is physically very thin, typically only a fraction of a micron (a micron is one millionth of a metre).

Hard-wire connections are made to the three regions internally; wires are then brought out through the casing to provide an external means of connection to each region. Either silicon or germanium semiconductor materials may be used in the fabrication of the transistor but silicon is preferred for reasons of temperature stability.

Transistor operation

For transistors to operate three conditions must be met:

1. The base must be very thin.
2. Majority carriers in the base must be very few.
3. The base-emitter junction must be forward biased and the base-collector junction reverse biased.

Electrons from the emitter enter the base and diffuse through it. Due to the shape of the base most electrons reach the base-collector junction and are swept into the collector by the strong positive potential. A few electrons stay in the base long enough to meet the indigenous holes present and recombination takes place.

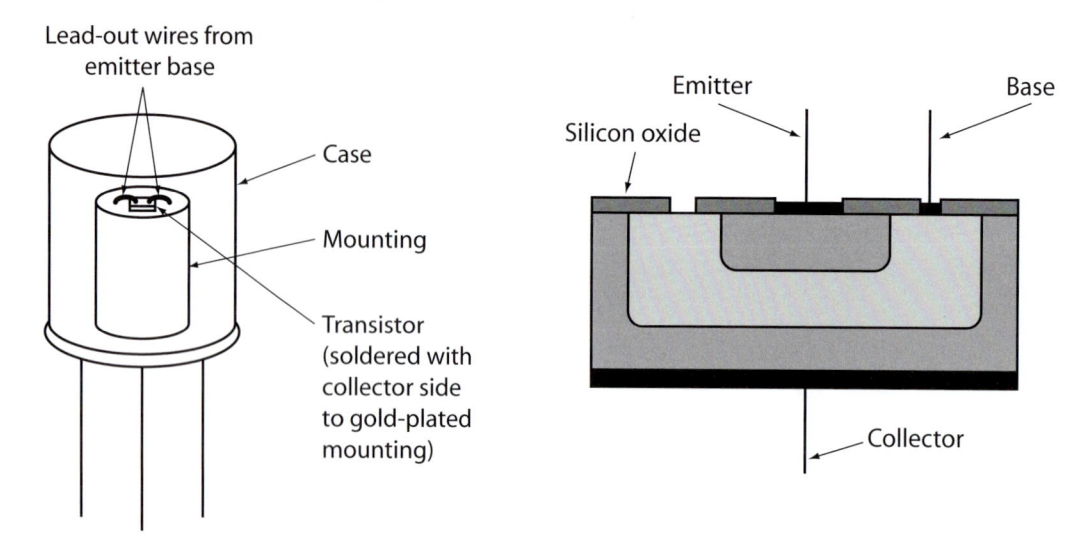

Figure 8.237 Transistor operation

To maintain the forward bias on the base-emitter junction, holes enter the base from the base bias battery. It is this base current which maintains the base-emitter forward bias and therefore controls the size of the emitter current entering the base. The greater the forward bias on the base-emitter junction, the greater the number of emitter current carriers entering the base.

The collector current is always a fixed proportion of the emitter current set by the thinness of the base and the amount of doping. Holes from the emitter enter the base and diffuse through it – see Figure 8.238. Due to the shape of the base most holes reach the base-collector junction and are swept into the collector by the strong negative potential.

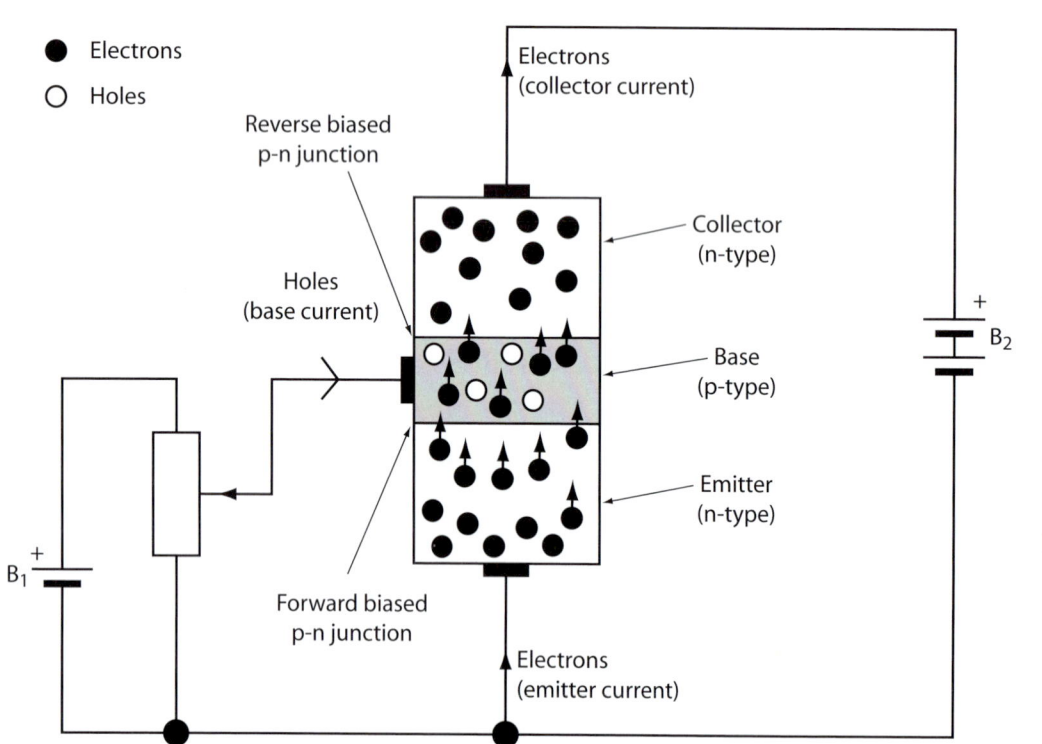

Figure 8.238 Circuit diagram for operation of transistor

Current amplification

Consider, for example, as in Figure 8.239, a base bias of some 630 mV has caused a base current of 0.5 mA to flow but more importantly has initiated a collector current of 50 mA.

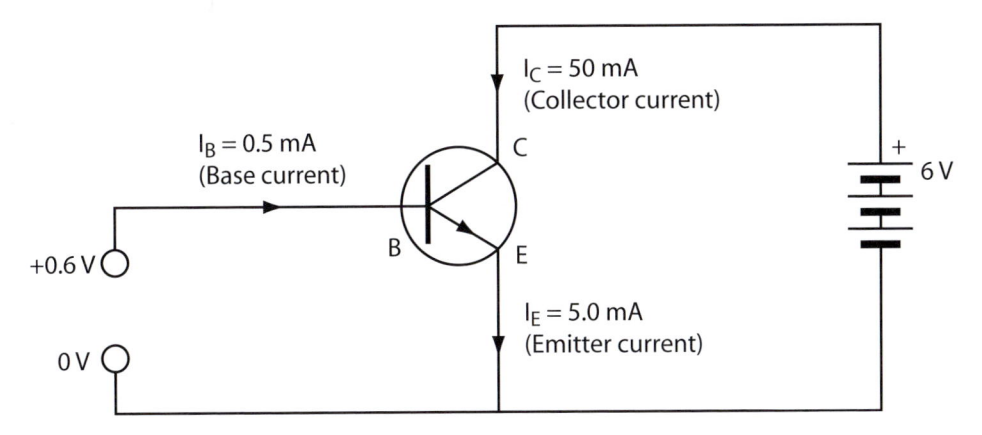

Figure 8.239 Current flow in transistor

Although these currents may seem insignificant, it's the comparison between the base and collector currents which is of interest. This relationship between I_B and I_C is termed the 'static value of the short-circuit forward current transfer' – we normally just call it the gain of the transistor, and it is simply a measure of how much amplification we would get. The symbol that we use for this is h_{FE}. This is the ratio between the continuous output current (collector current) and the continuous input current (base current). Thus when I_B is 0.5 mA and I_C is 50 mA the ratio is:

$$h_{FE} = \frac{I_C}{I_B}$$
$$= \frac{50 \text{ mA}}{0.5 \text{ mA}}$$

> **Remember**
> There are no units since this is a ratio.

i.e. approximately equal to 100.

It can therefore be said that a small base current initiated by the controlling forward base-bias voltage produces a significantly higher value of collector current to flow dependent on the value of h_{FE} for the transistor. Thus current amplification has been achieved.

Voltage amplification

Mention was previously made of the derivation of the word 'transistor', and quoted the device as transferring current from a low resistive circuit to approximately the same current in a high resistive circuit. This is an npn transistor so the low resistive reference is the emitter circuit and the high resistive reference, the collector circuit, the current in both being almost identical.

The reason the emitter circuit is classed as having a low resistance is because it contains the forward biased (pn) base-emitter junction. Conversely, the collector circuit contains the reverse biased (np) base-collector junction, which is of course in the order of tens of thousands of ohms (it varies with I_C). In order to produce a voltage output from the collector, a load resistor (R_B) is added to the collector circuit as indicated in the circuit diagram in Figure 8.240.

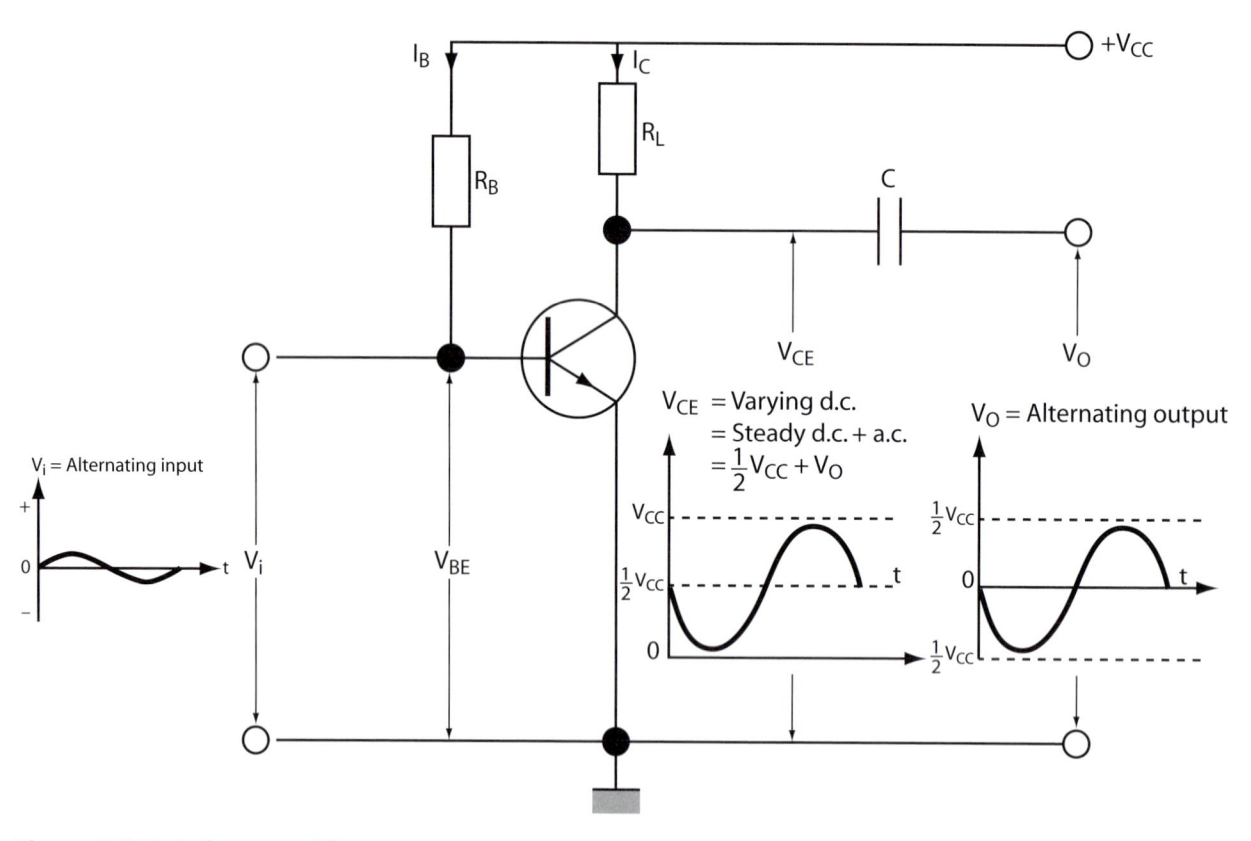

Figure 8.240 Voltage amplification

This shows the simplest circuit for a voltage amplifier. To see how voltage amplification occurs we have to consider that there is no input across V_i, which is called the quiescent (quiet) state. For transistor action to take place the base emitter junction V_{BE} must be forward biased (and has to remain so when V_i goes positive and negative due to the a.c. signal input).

By introducing resistor R_B between collector and base, a small current I_B will flow from V_{CC} through R_B into the base and down to 0V via the emitter, thus keeping the transistor running (ticking over).

Component resistor values R_B and R_L are chosen so that the steady base current I_B makes the quiescent collector-emitter voltage V_{CE} about half the power supply voltage V_{CC}. This allows V_O to replicate the input signal V_i at an amplified voltage with a 180° phase shift. When an a.c. signal is applied to the input V_i and goes positive it increases V_{BE} slightly to around 0.61 V. When V_i swings negative, V_{BE} drops slightly to 0.59 V. As a result a small alternating current is superimposed on the quiescent base current I_B, which in effect is a varying d.c. current.

The collector emitter voltage (V_{CE}) is a varying d.c. voltage, or an alternating voltage superimposed on a normal steady d.c. voltage. The capacitor C is there to block the d.c. voltage, but allow the alternating voltage to pass on to the next stage. So in summing up, a bipolar transistor will act as a voltage amplifier if:

- it has a suitable collector load R_L
- it is biased so that so that the quiescent value V_{CE} is around half the value of V_{CC}, which is known as the class A condition
- the transistor and load together bring about voltage amplification
- the output is 180° out of phase with the input signal as Figure 8.240 indicates
- the emitter is common to the input, output and power supply circuits and is usually taken as the reference point for all voltages, i.e. 0 V. It is called 'common', 'ground' or 'earth' if connected to earth.

Transistor as a switch

We have looked at the transistor as an amplifier of current and voltage. If we connect the transistor as in Figure 8.241, we can operate it as a switch. Compared with other electrically operated switches, transistors have many advantages, whether in discrete or integrated circuit (IC) form. They are small, cheap, reliable, have no moving parts and can switch millions of times per second – the perfect switch that has infinite resistance when 'off', no resistance when 'on' and changes instantaneously from one state to another, using up no power.

Figure 8.241 shows the basic circuit for an npn common emitter as in previous diagrams with a load resistor R_L connected in series with the supply (V_{CC}) and the collector.

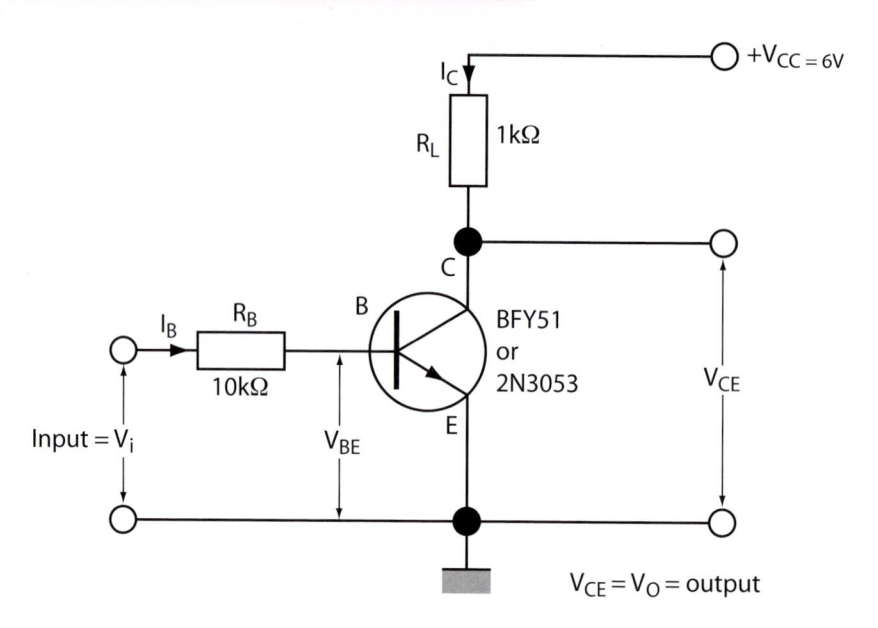

Figure 8.241 Circuit diagram for transistor used as a switch

R_B prevents excessive base currents, which would seriously damage the transistor when forward biased. With no input across V_I, the transistor is basically turned off. This means then that there will be no current (I_C) through R_L, therefore there will be no volt drop across R_L so the $+V_{CC}$ voltage (6 V) will appear across the output V_{CE}.

If we now connect a supply of between 2 V–6 V across V_i input, the transistor will switch on, current will flow through the collector load resistor R_L and down to common, making the output V_{CE} around 0 V. From this we can state that:

- when the input V_i = 0 V, the output V_{CE} = 6 V
- when the input V_i = 2 V–6 V, the output V_{CE} = 0 V.

From this we can see that the transistor is either High (6 V) or Low (0 V), or we can confirm, like a switch i that it is 'On' (6 V) or 'Off' (0 V).

This circuit can be used in alarms and switch relays for all types of processes and is the basic stage for programmable logic control (plc) which uses logic gates with either one or zero to represent what the output is from a possible input.

In Figure 8.242 are identified basic logic gate circuits with their inputs/outputs, 'truth table' and symbols.

There is also an Exclusive NOR which gives an output as indicated in Figure 8.243.

Unit ELTK 08

Circuit NOT Gate

Symbol
British
Input — 1 — Output
American
Input — Output

Truth table
NOT gate

Input	Output
0	1
1	0

Circuit NOR Gate

Symbol
British
A, B — ≥1 — F
American
A, B — F

Truth table
NOR gate
(2-input)

Input		Output
A	B	
0	0	1
0	1	0
1	0	0
1	1	0

Circuit OR Gate

British: A, B — NOR — C — NOT — F
American: A, B — NOR — C — NOT — F

Symbol
British
A, B — ≥1 — F
American
A, B — F

Truth table
OR gate
(2-input)

Input		Output
A	B	
0	0	0
0	1	1
1	0	1
1	1	1

Circuit NAND Gate

Symbol
British
A, B — & — F
American
A, B — F

Truth table
NAND gate
(2-input)

A	B	F
0	0	1
0	1	1
1	0	1
1	1	0

Circuit AND Gate

British: A, B — & — C — 1 — F
American: A, B — C — F

Symbol
British
A, B — & — F
American
A, B — F

Truth table
AND gate
(2-input)

A	B	C	F
0	0	1	0
0	1	1	0
1	0	1	0
1	1	0	1

Symbol
British
A, B — =1 — F
American
A, B — F

Truth table
Exclusive-OR gate

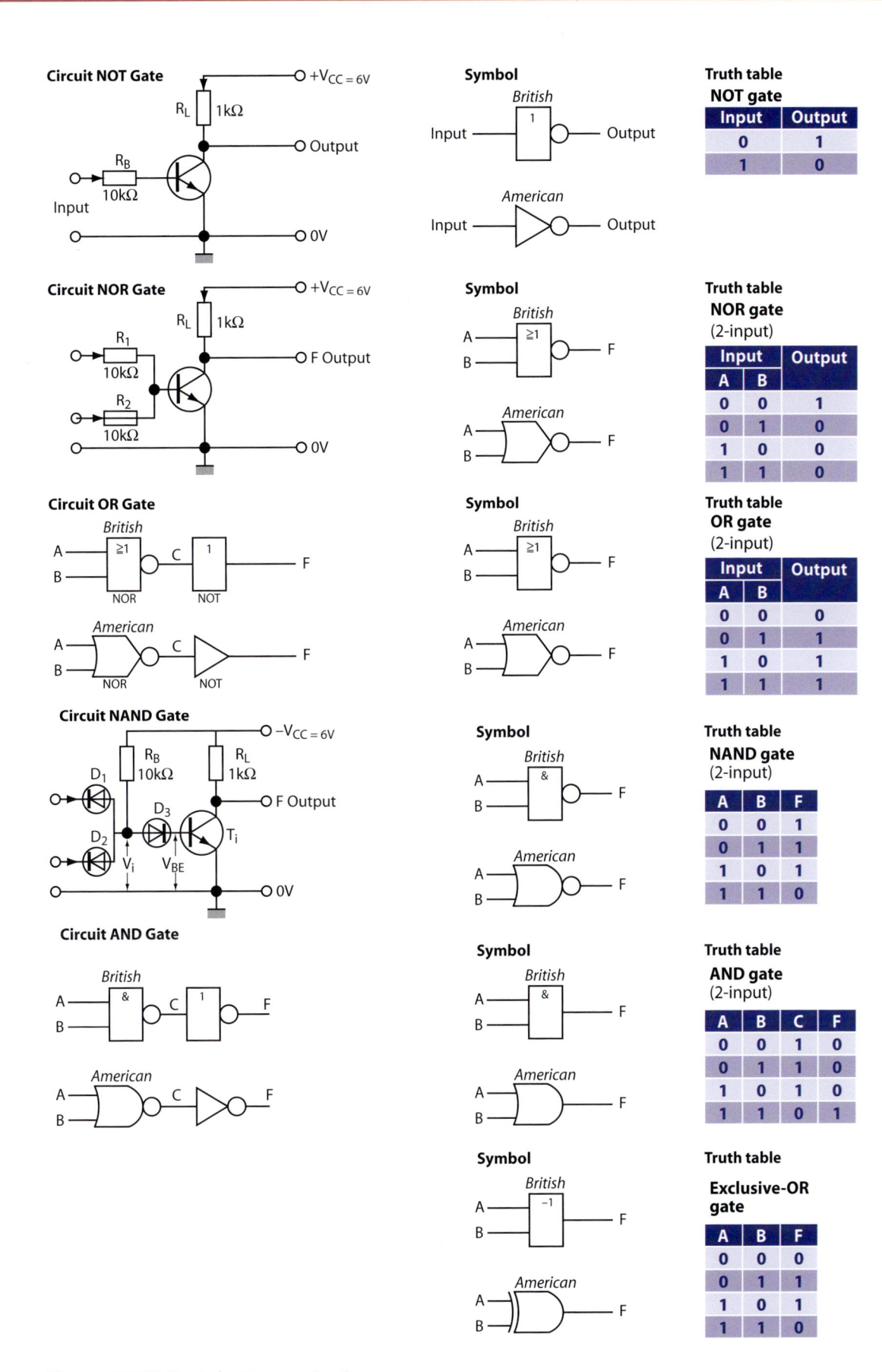

A	B	F
0	0	0
0	1	1
1	0	1
1	1	0

Figure 8.242 Basic logic gate circuits

Symbol

British

American

Truth table

Exclusive NOR gate

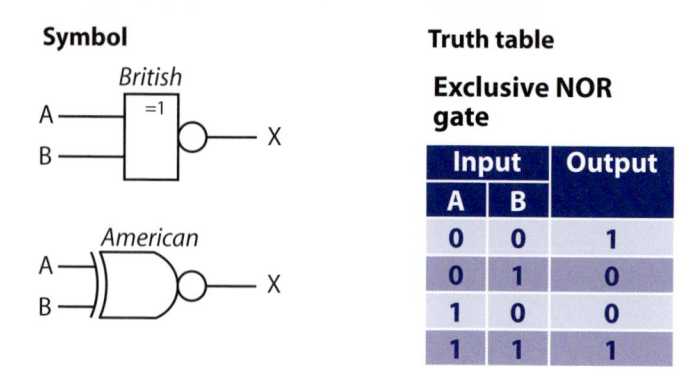

Input		Output
A	B	
0	0	1
0	1	0
1	0	0
1	1	1

Figure 8.243 Exclusive NOR gate

Testing transistors

As all transistors consist of either a npn or a pnp construction the testing of them is similar to diodes. Special meters with three terminals for testing transistors are available and many testing instruments have this facility. However, an ohmmeter can be used for testing a transistor to check if it is conducting correctly. The following results should be obtained from a transistor assuming that the red lead of an ohmmeter is positive.

Note: This is not always the case. With some older analogue meters, the battery connections internally are the opposite way round, so it is always good to check both ways across base and emitter as shown in Figures 8.244 and 8.245.

A good npn transistor will give the following readings:

- Red to base and black to collector or emitter will give a low resistance.
- However, if the connections are reversed it will result in a high resistance reading.
- Connections of any polarity between the collector and emitter will also give a high reading.

A good pnp transistor will give the following readings:

- Black to base and red to collector or the emitter will give a low resistance reading.
- However, if the connections are reversed a high resistance reading will be observed.
- Connections of either polarity between the collector and emitter will give a high resistance reading.

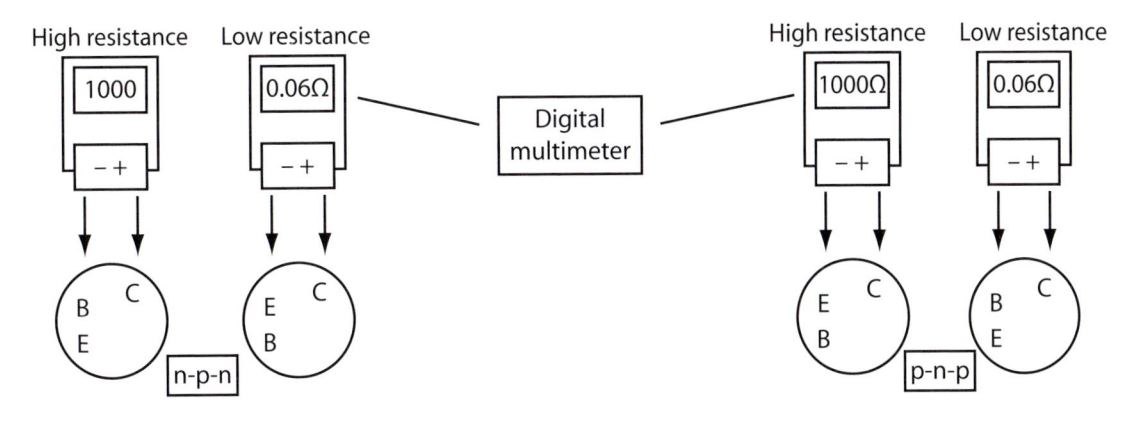

Figure 8.244 Testing transistors with digital multimeter

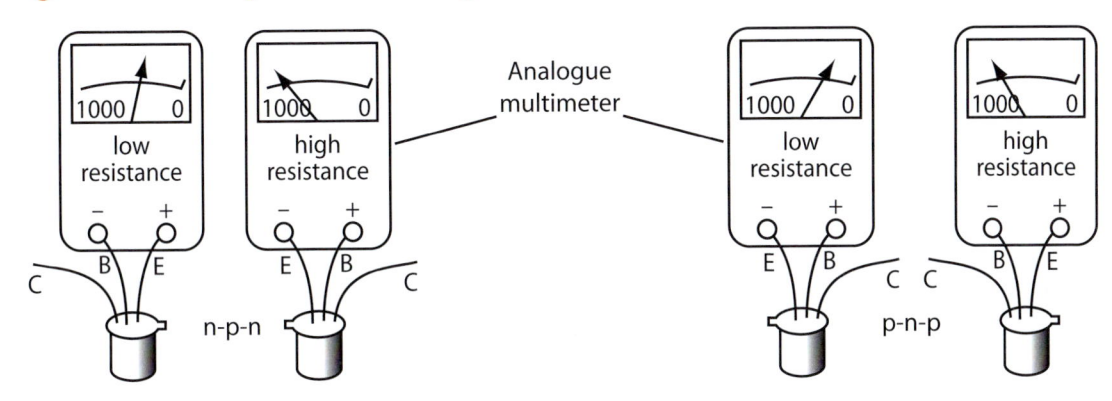

Figure 8.245 Testing transistors with analogue multimeter

Field effect transistors (FETs)

Field effect transistor devices first appeared as separate (or discrete) transistors, but now the field effect concept is employed in the fabrication of large-scale integration arrays such as semiconductor memories, microprocessors, calculators and digital watches.

There are two types of field effect transistor: the junction gate field effect transistor, which is usually abbreviated to JUGFET, JFET or FET, and the metal oxide semiconductor field effect transistor known as the MOSFET. They differ significantly from the bipolar transistor in their characteristics, operation and construction.

The main advantages of an FET over a bipolar transistor are:

- Its operation depends on the flow of majority current carriers only. It is, therefore, often described as a unipolar transistor.
- It is simpler to fabricate and occupies less space in integrated form.
- Its input resistance is extremely high, typically above 10 MΩ especially for MOSFET devices. In practice, this is why

voltage measuring devices such as oscilloscopes and digital voltmeters employ the FET in their input circuitry, so that the voltage being measured is not altered by the connection of the instrument.

- Electrical noise is the production of random minute voltages caused by the movement of current carriers through the transistor structure. Since the FET does not employ minority carriers, it therefore has the advantage of producing much lower noise levels compared with the bipolar transistor.
- Also due to its unipolar nature it is more stable during changes of temperature.

The main disadvantages of an FET over its bipolar counterpart are:

- Its very high input impedance renders it susceptible to internal damage from static electricity.
- Its voltage gain for a given bandwidth is lower. Although this may be a disadvantage at low frequencies (below 10 MHz), at high frequencies the low noise amplification that an FET achieves is highly desirable. This facet of FET operation, though, is usually only exploited in radio and TV applications, where very small high frequency signals need to be amplified.
- The FET cannot switch from its fully on to its fully off condition as fast as a bipolar transistor. It is for this reason that digital logic circuits employing MOSFET technology are slower than bipolar equivalents, although even faster switching speeds are being achieved as FET production technology continues to advance.

> **Remember**
>
> The main reason why the FET has such a differing characteristic from the bipolar transistor is because current flow through the device is controlled by an electric field, which is not the case with the bipolar transistor. It is for this reason that the FET is considered to be a voltage operated device rather than current operated.

Figure 8.246 illustrates the basic construction of the FET, which consists of a channel of n-type semiconductor material with two connections, source (S) and drain (D). A third connection is made at the gate (G), which is made of p-type material to control the n-channel current. The symbol is shown in Figure 8.247.

Figure 8.246 Basic construction of field effect transistor (FET)

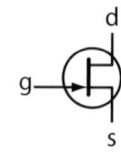

Figure 8.247 Field effect transistor (FET) symbol

In theory, the drain connection is made positive with respect to the source, and electrons are attracted towards the D terminal. If the gate is made negative there will be reverse bias between G and S, which will limit the number of electrons passing from S to D.

The gate and source are connected to a variable voltage supply, such as a potentiometer, and increasing or lowering the voltage makes G more or less negative, which in turn reduces or increases the drain current.

Component positional reference

As electronic diagrams become more complex a system called the component positional reference system is used. This system uses a simple grid reference to identify holes on a board on which components are installed. This is done by counting along the columns at the top of the board starting from the left and numbering them as you count. Then starting from the left and counting down rows, each row is given a letter in turn from the alphabet starting from A. For example the position reference point 7:J would be 7 holes from the left and 10 holes down.

Inverters

Earlier we looked at the conversion of a.c. to d.c. by the process of rectification. The opposite process, namely converting d.c. into a.c. is called inversion and a device called an inverter is used to do so.

We know them as inverters because of their origin. Early devices were actually mechanical devices that used an a.c. motor connected to a generator to produce d.c. and were known as converters. By running the converter in reverse, d.c. could be applied to the field and commutator windings to turn the machine, thus producing a.c. and leading to them being called an 'inverted rotary converter'.

Nowadays, the inverter uses electronic components such as thyristors and mosfets to switch the d.c. to create pulses of a.c. As we saw earlier in this unit, alternating current produces a sine wave that is smooth with continuous curves from positive to negative and back again, and the modern power inverter creates an approximation of a sine wave that meets all equipment needs and is commonly referred to as a pure sine wave inverter.

Solid-state inverters have no moving parts and can be found from switching power supplies in computers to large high-voltage direct current systems (HVDC) that transport large amounts of power. More commonly, inverters are used to supply a.c. from a d.c. source such as solar panels or batteries.

Equally, in terms of emergency supplies, an 'uninterruptible power supply' (UPS) will use batteries and an inverter to supply a.c. to equipment when there has been a mains failure. Additionally, as has also been mentioned earlier in this unit, variable-frequency drives control the operating speed of an a.c. motor by controlling the frequency and voltage of the power supplied to the motor via an inverter.

Integrated circuits

Integrated circuits are complete electronic circuits within a plastic case (known as the **black box**). The chip contains all the components required, which may include diodes, resistors, capacitors, transistors etc.

There are several categories, which include analogue, digital and memories. The basic layout is shown in Figure 8.248, which is an operational amplifier (**dual in-line IC**).

> **Key term**
>
> **Dual in-line ICs** – the types of IC with the pins lined up down each side

The plastic case has a notch at the end and, if you look at the back of the case with the notch at the top, Pin 1 is always the first one on the left hand side, sometimes noted with a small dot. The other pin numbers follow down the left hand side, 2, 3 and 4 and then back up the right hand side from the bottom right to the top 5, 6, 7 and 8. This is an 8-pin chip but you can get some chips with 32 pins or more.

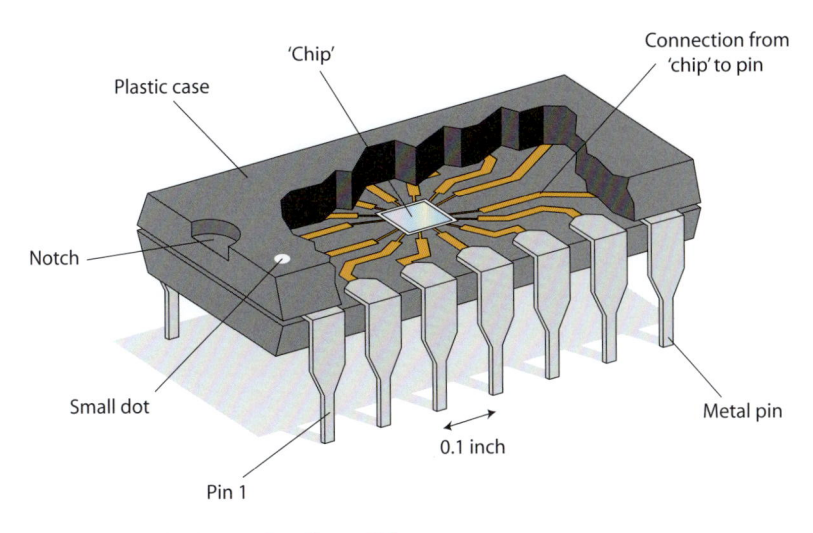

Figure 8.248 Operational amplifier

Progress check
1 Explain briefly what a semiconductor diode is.
2 How does a light emitting diode work?
3 Give a common example of the use of an inverter.

Getting ready for assessment

This unit focuses on some of the key scientific concepts that underpin all your work as an electrician, and relate directly to all the practical tasks you will undertake as part of your course and your career.

For this unit you will need to be familiar with:

- mathematical principles which are appropriate to electrical installation, maintenance and design work
- standard units of measurement used in electrical installation, maintenance and design work
- basic mechanics and the relationship between force, work, energy and power
- relationship between resistance, resistivity, voltage, current and power
- fundamental principles which underpin the relationship between magnetism and electricity
- electrical supply and distribution systems
- how different electrical properties can effect electrical circuits, systems and equipment
- operating principles and application of d.c. machines and a.c. motors
- operating principles of different electrical components
- principles and applications of electrical lighting systems
- principles and applications of electrical heating
- types, applications and limitations of electronic components in electrotechnical systems and equipment

For each learning outcome, there are several skills you will need to acquire, so you must make sure you are familiar with the assessment criteria for each outcome. For example for Learning Outcome 6 you will need to be able to describe how electricity is generated and transmitted for both domestic and commercial use, as well as specifying the features and characteristics of the generation and transmission systems, as well as explaining how electricity is generated from other sources. You will need to describe the main characteristics of a number of supply systems and the operating principles, application and limitations of transformers and the different types used. You should also be able to determine by calculation and measurements primary and secondary voltages and current and the kVA rating of a transformer.

It is important to read each question carefully and take your time. Try to complete both progress checks and multiple choice questions, without assistance, to see how much you have understood. Refer to the relevant pages in the book for subsequent checks. Always use correct terminology as used in BS 7671. There are some simple tips to follow when writing answers to exam questions.

- **Explain briefly** – usually a sentence or two to cover the topic. The word to note is 'briefly' meaning do not ramble on. Keep to the point.
- **Identify** – refer to reference material, showing which the correct answers are.
- **List** – a simple bullet list is all that is required. An example could include, listing the installation tests required in the correct order.
- **Describe** – a reasonably detailed explanation to cover the subject in the question.

Much of the knowledge in this unit is theoretical rather than practical. As such you will need to make extensive use of your functional skills in order to complete the unit effectively. In particular, this unit will require extensive use of your maths skills, making calculations and working with equations. Just as with practical work, it is important to regularly check any calculations you are making to ensure that they are correct. Mistakes in sums can be as costly and time-consuming as errors when installing components.

Good luck!

CHECK YOUR KNOWLEDGE

1. The SI unit for weight is known as a:
   a) Mole
   b) Joule
   c) Newton
   d) Kilogram

2. Which one of the following prefixes represents a multiplier of 1 000 000?
   a) Tera
   b) Kilo
   c) Giga
   d) Mega

3. What is the result of the following calculation 10 + 2 × 4 – 2?
   a) 14
   b) 16
   c) 24
   d) 46

4. What is the formula for calculating Force?
   a) Force = Mass + Acceleration
   b) Force = Mass – Acceleration
   c) Force = Mass ÷ Acceleration
   d) Force = Mass × Acceleration

5. What is the total resistance for a parallel combination consisting of resistors valued at 20 Ω and 30 Ω?
   a) 10 Ω
   b) 12 Ω
   c) 25 Ω
   d) 50 Ω

6. A 2 kW electric fire is used 6 hours a day in winter. What will be the total consumption over a 13 week quarter?
   a) 546 kWh
   b) 54.6 kWh
   c) 156 kWh
   d) 15.6 kWh

7. Which of the following materials **cannot** be magnetised?
   a) Iron
   b) Steel
   c) Cobalt
   d) Aluminium

8. When selecting a relay, which of the following would **not** be considered?
   a) Coil voltage
   b) Contact ratings
   c) Coil material
   d) Contact arrangements

9. What is the force on a conductor 20 m long, lying at right angles to a magnetic field of 8 tesla, when 10 A flows in the coil?
   a) 38 T
   b) 38 N
   c) 1600 N
   d) 1600 T

10. Which of these is a suitable transmission voltage?
    a) 400 kV
    b) 11 kV
    c) 400 V
    d) 230 V

11. What is the effective r.m.s. value of an a.c. sine wave given by?
    a) $0.637 \times I_{max}$
    b) $0.707 \times I_{max}$
    c) $0.637 \times I_{peak}$
    d) $0.707 \times I_{peak}$

12. Complete this sentence: The current in an inductive circuit…?
    a) is in phase with the voltage
    b) leads the voltage
    c) lags the voltage
    d) is resistive only

CHECK YOUR KNOWLEDGE

13. Which of the following is used to improve power factor?
    a) Resistor
    b) Inductor
    c) Capacitor
    d) Transformer

14. What is the impedance of a circuit with 30 Ω resistance and a 40 Ω inductive reactance?
    a) 10 Ω
    b) 30 Ω
    c) 40 Ω
    d) 50 Ω

15. What would the balanced three-phase system in a neutral current be?
    a) The same as the phase current
    b) The same as the line current
    c) Zero
    d) Low

16. How many windings does an autotransformer have?
    a) None
    b) One
    c) Two
    d) Three

17. What is the turns ratio of a safety isolating transformer?
    a) 1:1
    b) 2:1
    c) 10:1
    d) 20:1

18. Which of the following motors would be used on an a.c. supply?
    a) Long shunt
    b) Short shunt
    c) Universal
    d) Compound

19. What is the major advantage a star-delta starter has compared with a direct on line starter?
    a) Reduced starting current
    b) Increased starting current
    c) Increased voltage
    d) Faster motor speed

20. What is stroboscopic effect associated with?
    a) Tungsten filament lamp
    b) Tungsten halogen lamp
    c) Incandescent lighting
    d) Fluorescent discharge lighting

21. What would be the colour coding of a 630 kOhm resistor with a tolerance of 5%?
    a) Blue, red, orange and silver
    b) Blue, orange, yellow and silver
    c) Blue, orange, yellow and gold
    d) Blue, yellow, orange and gold

22. What device is used to measure the temperature of a motor winding?
    a) Resistor
    b) Thermistor
    c) Transistor
    d) Thyristor

Index

a.c. machines
   principles of 370
      series-wound (universal)
         motors 370–1
      single-phase induction
         motors 376–85
      speed and slip calculation 385–8
      synchronised motors 388–9
      three phase induction
         motors 372–7
   windings 390
accessories, faults with 193–7
accuracy of instruments 141
adiabatic equation 60–3
algebra 229
'all in one' instruments 145
alternative sources of
      electricity 309–10
ambient temperature 51
ammeters 357, 358
angles 234–5
antistatic precautions 198–9
arcing 183
armature resistance control 401
assessment preparation
   electrical principles 498
   fault diagnosing and correcting 209
   inspection and testing 161
   planning and selection for
      installation 73
atomic theory of matter
   atoms 260–2
   batteries 265–7
   and circuits 264–8
   conductors 263
   current, electric 264–8
   electron flow 265–6
   insulators 262
   potential difference 266–8
   states of matter 259–60
atoms 260–2
Automatic Disconnection of Supply
     (ADS) 20–2
   testing 114
automatic star-delta starters 397–8
autotransformers 351
auto-transformer starters 398–9
availability of replacement 203
average value 312

bar charts 242
base quantities 213
batteries 201, 265–7, 310
bayonet caps 404
BODMAS 216–17
bonding and earthing 13–14
breaking capacity of fuses 38
BS 1361/1362 cartridge fuses 31–2
BS 88 high breaking capacity
     fuses 32–3
BS 3036 rewireable fuses 30–1
building fabric restoration 207

cables
   adiabatic equation 60–3
   conduit and trunking, size
      of 66–71
   design current 47
   diversity 63–6
   external influences on 46
   fibre optic 198, 471–2
   inspection of 85, 95–7, 132
   installation and reference
      methods 48–9, 50
   interconnections 188–90
   rating factors 49–54
   rating of the protective
      device 47–8
   shock protection 56, 58–60
   symbols and definitions 45
   termination of 190–1
   thermal constraints 60–3, 64
   voltage drop 55–6, 57
calibration of instruments 141–2
capacitance 320, 325–30
capacitors
   calculations with 459–61
   ceramic 453
   charging/discharging 461–3
   coding of 454–8
   in combination 459–61
   electrolyte 451–2
   fixed 451–3
   mica 452
   non-electrolyte 452–3
   polarity 458
   polyester 453
   principles of 450–1
   smoothing 476
   trimmer 454
   values of 455–8
   variable 454
capacitor start-capacitor run
     motors 382–3

capacitor-start motor 379–81, 385
cartridge fuses 31–2
cells 310
ceramic capacitors 453
certificates
   Electrical Installation
      Certificate 146–50, 151
   Minor Electrical Installation
      Works Certificate 150,
      152–4, 157
charts 241–2
checklist for inspection 89–97
chemical effects of electric
     current 268–71
choke smoothing 476–7
circuit breakers
   miniature (MCBs) 34–7
   testing 139
circuit protective conductors
     (CPC) 21
circuits
   and atomic theory of matter 264–8
   basic logic gate 491–2
   controlling 268
   and electron theory 264–8
   filter 477
   integrated 496–7
   lamp dimmer 482–3
   open 184
   parallel 278–9, 282–7
   resistance and resistivity 271–5
   series 276–7, 282–7
   series/parallel 280–1
   voltage drop 282
coding
   capacitors 454–8
   resistors 439–41
colour coding
   capacitors 455–7
   resistors 439–41
commutator 366–7
compact fluorescent lamps 418–19
competence of inspectors 77
compound motors 369
condition reports 140, 155, 157–60
conductors
   atomic structure 262
   inspection 85, 132
   magnetism and electricity 294–7
   testing continuity of 100–5,
      123–4, 137–8, 172–3
   uses of 263

conduit
  inspection checklist 93–4
  size of and cable selection 66–70
connections and joints, inspection of 131
consumer supply systems
  earthing 2–18
  isolation and switching 10–11
contactors, faults with 194–5
continuity meters 123
continuity of conductors, testing 100–5, 123–4, 137–8, 172–3
control equipment
  faults with 193–4
  inspection checklist 90–1
cooker control unit, inspection checklist for 93
core-type transformers 348–50
correction of faults. See fault diagnosing and correcting
cost of replacement 202–3
coulombs 264
current, electric
  amplification 487
  causes of 268–9
  effects of 269–71
  electron theory 264–8
  magnetic effects of 288–303
  measurement 357
  parallel circuits 278–9
  prospective fault, testing 118–19
  resistors as limiters of 442–3
  series circuits 276–7
  series/parallel circuits 280–1
current transformers 351

d.c. generators 301–3
d.c. machines 365–9, 400–2
delta connection 336
design current 47
design of installations 187–8
diacs 482
diagnosis of faults. See fault diagnosing and correcting
diazed (D-type) fuses 33
diodes. See semiconductor devices
Direct-On-Line' (DOL) starters 391–5
discharge lighting 410–18
disconnection times 44
discrimination 42–3
distribution of electricity 309
diversity and selection of cables 63–6
doping 464
downtime due to faults 203–4

earthing
  and bonding 13–14
  classification of arrangements 2–3
  earth electrode resistance 111–14, 144
  earth fault loop 14–16, 114–18, 126, 139, 143–4, 174–5
  internal/external arrangements 13–18
  IT system 8
  lightning protection 17–18
  protection of electrical systems 19–26
  protective multiple earthing (PME) 7–8
  supply systems 8–10
  TN-C-S system (PME) 6–7
  TN-S system 4–5
  TT system 3–4
  unearthed equipment, danger of 16–17
Edison screw cap 404–5
effective value of a wave form 313–14
efficiency 256–8, 287
Electrical Installation Certificate 146–50, 151
electrical separation 22–3
Electricity at Work Regulations (EaWR) 1989 76
electrolyte capacitors 451–2
electromagnetic induction 298–301
electromagnets 291–2
electronic components
  capacitors 449–63
  diacs 482
  integrated circuits 496–7
  inverters 495–6
  lamp dimmer circuits 482–3
  rectification 474–7
  resistors 432–49
  semiconductor devices 463–74
  thermistors 448–9
  thyristors 478–81
  transistors 484–95
  triacs 481–2
electron theory
  batteries 265–7
  conductors 263
  current, electric 264–8
  and electrical circuits 264–8
  electron flow 265–6
  insulators 262
  molecules and atoms 260–2

potential difference 266–8
  states of matter 259–60
electrostatic discharge 199–200
emergency/stand-by supply 208
emergency switching 11
enclosures, inspection of 133
energy 253–5
energy saving lamps 418–19
equations. See mathematical concepts
equipment
  faults with 193–7, 197–8
  instrumentation and metering 197
  IT, shut down of 200
  ohmmeters 142–3
  replacement of 202–3
  testing instruments 140–5, 176–7
  unearthed, danger of 16–17
exploratory surveys before inspection 129–30
extra low voltage 23

fault current 27, 41–2
fault diagnosing and correcting
  accessories 193–7
  access to system 206–7
  antistatic precautions 198–9
  arcing 183
  assessment preparation 209
  building fabric restoration 207
  common symptoms of faults 178–87
  contactors 194–5
  control equipment 193–4
  customer relations 167–8
  design of installations 187–8
  downtime due to faults 203–4
  electrostatic discharge 199–200
  emergency/stand-by supply 208
  factors affecting correction 202–8
  fibre optic cabling 198
  halt-split technique 170
  health and safety 198–201
  high frequency on high capacitative circuits 201
  information on systems and faults 165–7
  instrumentation and metering 197
  insulation failure 184
  interconnections, cable 188–90
  isolation 164, 207–8
  legal responsibilities 204–5

logical approach to 170–1
loss of supply 181
maintenance to prevent faults 187
misuse, faults due to 185
open circuits 184
overload 181–3
over voltage 200
portable appliances 197
position of faults 178–80
preparatory work 169–71
protection against faults 178–9
protective devices 195–6
replacement of equipment 202–3
reporting and recording 165–8
resource availability 204
seals and entries 191–2
shut down of IT equipment 200
solid state and electronic
 devices 195
special precautions 198–201
storage batteries 201
termination of cables 190–1
testing 171–7
transient voltages 183–4
types of faults 187–98
wear and tear, faults due to 185
fibre optic cabling 198, 471–2
field control 401
field effect transistors
 (FETs) 493–5
filter circuits 477
fire, protection against 25–6
firefighters' switches 11–12
fixed capacitors 451–3
fixed resistors 432–4
fluorescent lamps 410–11
force 252–3
fractions 223–5, 227
frequency 315, 363
frequency distribution 242–3
frequency of inspection and
 testing 127–9
full-wave rectification 475–6
functional switching 11–12
functional testing 119–20
fused connection unit, inspection
 checklist for 93
fuses 30–3, 37–44

gears 248
generation of electricity 307,
 309–11
 See also supply systems
geothermal generation 311
glow-type starter circuit 412

GLS (General Lighting Service)
 lamps 407–8
grouping of cables 51–2

halogen lamp caps 405–6
halt-split technique 170
hand-operated star-delta
 starters 396–7
health and safety
 fault diagnosing and
  correcting 198–201
 isolation 76–7
heating, electrical 432
high breaking capacity (HBC)
 fuses 32–3
high pressure sodium vapour 417–18
histograms 244

illuminance 424
illumination, principles of 423–31
impedance 321–2, 363–4
incandescent lamps 406–10
inclined planes 249
indices 230–2
inductance 317–19, 323–5,
 327–30, 347–8
infrared sources/sensors 470
inspection
 assessment preparation 161
 cables 132
 checklist 89–97
 competence and
  responsibility 77
 condition reports 140, 157–60
 conductors 132
 enclosures 133
 exploratory surveys 129–30
 frequency 127–9
 general procedures 129–31
 initial 79–80
 items to be checked 83–7
 joints and connections 131
 legislation and regulations 80–1
 marking and labelling 134–5
 need for 127
 periodic 79–80, 127–35
 protection of electrical systems 133
 protective devices 133
 routine checks 129, 130
 sampling 130–1
 schedule 88
 scope of 131
 switches 132
 verification, initial 78, 81–2
 See also certificates

installation and reference
 methods 48–9, 50
instantaneous value 312
instruments
 ammeters 357, 358
 choosing 360
 correct functioning of 360
 loading errors 358–9
 meter displays 359–60
 ohmmeters 358
 voltmeters 357, 358
instrument transformers 351
insulation
 double/reinforced 22
 failure of 184
 resistance testing 105–10, 125,
  138, 143, 173–4
 thermal, and selection of
  cables 52
insulation resistance testers 123
insulators 262
integrated circuits 496–7
inverse square law 427–9
inverters 495–6
IP Codes 108–10
isolation
 fault diagnosing and correcting 164
 health and safety 76–7
 inspection 86
 regulations 76–7
 and switching 10–11
 testing of devices 139–40
IT equipment, shut down of 200
IT system 8

joint boxes, inspection checklist
 for 93
joints and connections, inspection
 of 131

kilowatt hours 286–7
kinetic energy 255

labelling, inspection of 134–5
Lambert's cosine law 429–30
lamp dimmer circuits 482–3
LED lighting 419–20
legal responsibilities for correcting
 faults 204–5
legislation
 inspection and testing 80–1
 isolation 76
 testing 99
levers 247–8
light-dependent resistors 448, 469

light emitting diodes
    (LEDs) 419–20, 467–8
light fittings. See luminaires
lighting systems
    development of 404
    illuminance 424
    illumination principles 423–31
    inspection checklist 91–2
    inverse square law 427–9
    Lambert's cosine law 429–30
    lumen method 426–7
    luminaires 404–20
    luminous flux 424
    luminous intensity 424
    maintenance factor 424–5
    measuring light 424–31
    regulations 420–2
    utilisation factor 425–6
lightning protection 17–18
load balancing 340–2
loading errors 358–9
loss of supply 181
low pressure mercury caps 406
low-pressure mercury vapour
    lamps 410–11
low pressure sodium vapour 416–17
low-resistance ohmmeters 142
lumen method 426–7
luminaires
    bayonet caps 404
    discharge lighting 410–18
    Edison screw cap 404–5
    GLS (General Lighting Service)
        lamps 407–8
    halogen lamp caps 405–6
    incandescent lamps 406–10
    LED lighting 419–20
    low pressure mercury caps 406
    symbols used in 423
    tungsten halogen lamps 408–10
    See also lighting systems
luminous flux 424
luminous intensity 424

magnetism and electricity
    a.c. supply, generating 304–6
    d.c. generators 301–3
    effects of magnetism 288
    electromagnetic induction 298–301
    electromagnets 291–2
    force between current carrying
        conductors 294–7
    magnetic effects of electrical
        currents 268–71, 288–303
    permanent magnets 288–91

relays 292–4
solenoids 297–8, 300
maintenance factor 424–5
maintenance to prevent faults 187
marking and labelling, inspection
    of 134–5
mass and weight 246
mathematical concepts
    algebra 229
    basic rules 215–17
    BODMAS 216–17
    fractions 223–5, 227
    indices 230–2
    percentages 225–8
    powers 217–22, 230–2
    statistics 241–5
    transposition 232–4
    triangles 234–40
    trigonometry 237–40
matter, states of 259–60
mean 244–5
measurement
    current 357
    of electricity 264
    frequency 363
    impedance 363–4
    loading errors 358–9
    meter displays 359–60
    power 360–3
    resistance 358
    units of 212–15
    voltage 357
mechanical maintenance, switching
    for 11
mechanics
    gears 248
    inclined planes 249
    levers 247–8
    mass and weight 246
    mechanical advantage 250
    pulleys 249–50
    velocity ratio 252
median 245
metal halide lamps 418
metering equipment 197, 357–60
mica capacitors 452
miniature circuit breakers
    (MCBs) 34–7
Minor Electrical Installation Works
    Certificate 150, 152–4, 157
mode 245
molecules 260
motors
    basic concept of 365
    d.c. 365–9

See also a.c. machines
mutual inductance 347–8

neozed (DO-type) fuses 33
neutral currents 339–40
non-electrolyte capacitors 452–3
notices, inspection of 87, 134–5

ohmmeters 142–3, 358
Ohm's Law 272
open circuits 184
opto-couplers 469–70
overload 27
overload faults 181–3

parallel circuits 278–9, 282–7
peak value 312–15
percentages 225–8
periodic inspection and
    testing 127–40
permanent magnets 288–91
permanent spilt capacitor (PSC)
    motors 381–2
phase sequence, testing 119, 144–5
phasor representation 316–17
phasors 321
photo cells 469
photodiodes 469–70
pie charts 241–2
planning and selection for installation
    assessment preparation 73
    See also cables; consumer supply
        systems; protection of
        electrical systems
p-n junction 464–6
polarity 110–11, 125, 138–9, 174,
    458
polyester capacitors 453
portable appliances 197
position of faults 178–80
potential difference 266–8
potential energy 254
power
    in an a.c. circuit 330–3
    measurement of 360–3
    in a mechanical system 255–6
    resistors, ratings of 446–8
    in series and parallel circuits 282–7
    in three-phase supplies 343–5
    triangle 333–4
power factor 315–16, 318–19,
    362–3
power factor correction (PFC) 346
powers
    of 10 217–19

504

of all numbers 219–20
indices 230–2
multiplying and dividing
    with 221–2
prospective fault current,
    testing 118–19
protection of electrical systems
    against faults 178–9
    inspection of 85–6, 133
    IP Codes 108–10
    protection by barriers/enclosures,
        testing 108–10
    SELV, PELV or electrical
        separation 107–8
    shock 19–24
    thermal effects 25–6
    See also protective devices
protective devices
    BS 7671 26–7
    discrimination between 42–3
    fault diagnosing and
        correcting 195–6
    fuses 30–3, 37–44
    inspection 86–7, 133
    miniature circuit breakers
        (MCBs) 34–7
    rating of 47–8
    residual current devices
        (RCDs) 27–30
    and selection of cables 53
protective multiple earthing
    (PME) 7–8
pulleys 249–50
pulse width modulation
    (PWM) 401–2
Pythagoras' Theorem 236–7

rating factors 49–54
rating of the protective device 47–8
RCDs. See residual current devices
    (RCDs)
rectification 474–7
reference methods 48–9
regulations
    isolation 76–7
    lighting systems 420–2
relays 292–4
remote stop/start starters 395–6
replacement of equipment 202–3
residual current devices
    (RCDs) 27–30, 120–2,
    126–7, 139, 144, 175
resistance
    and capacitance 325–30
    and circuits 271–5

and inductance 317–19, 323–5,
    327–30
measurement 358
parallel circuits 278–9
and phasor representation 316–17
series circuits 276–7
series/parallel circuits 280–1
resistivity and circuits 271–5
resistors
    colour coding 439–41
    as current limiters 442–3
    fixed 432–4
    light-dependent 448, 469
    power ratings of 446–8
    preferred values 435–6, 437
    resistance markings 437–8
    testing 441
    thermistors 448–9
    variable 432, 434–5
    voltage control 443–6
resource availability 204
rewireable fuses 30–1
root mean square (r.m.s.) 313–14
rotor-resistance starters 399–400
routine checks 129, 130

safety isolating transformers 352
sampling before inspection 130–1
scope of inspection 131
screw rule 291–2
seals and entries 191–2
selection for installation
    assessment preparation 73
    See also cables; consumer supply
        systems; protection of
        electrical systems
semiconductor devices
    basics of 463–4
    diode testing 473–4
    doping 464
    fibre optic links 471–2
    infrared sources/sensors 470
    light-dependent resistors 469
    light emitting diodes
        (LEDs) 467–8
    opto-couplers 469–70
    photo cells 468
    photodiodes 468–9
    p-n junction 464–6
    Zener diodes 466–7
semi-resonant starting 413
separation, electrical 22–3
series circuits
    current, voltage and resistance
        in 276–7

resistance and capacitance
    in 325–8
resistance and inductance
    in 323–5, 327–8
values of power in 282–7
series motors 367–8
series/parallel circuits 280–1
series-wound (universal)
    motors 370–1
shaded pole motors 383
shell-type transformers 350
shock
    protection of electrical systems
        against 19–24
    selection of cables 56, 58–60
shunt motors 368
shut down of IT equipment 200
signs and notices, inspection of 87,
    134–5
sine waves 304, 312–15
single-phase a.c. synchronous
    motors 389
single-phase induction
    motors 376–85
single phase supply 9, 10
SI units 212–15
slip and speed calculation 385–8
smoothing 476–7
sockets, inspection checklist
    for 92–3
sodium vapour lamps 416–17
soft starters 398
solenoids 297–8, 300
solid state and electronic
    devices 195
speed and slip calculation 385–8
speed control of d.c.
    machines 400–2
spilt-phase motors 377–79
squirrel-cage rotors 374–5, 376
staff availability 204
stand-by supply 208
star connection 336–8
star-delta starters 396–8
starters
    automatic star-delta starters 397–8
    auto-transformer 398–9
    Direct-On-Line' 391–5
    hand-operated star-delta 396–7
    remote stop/start 395–6
    rotor-resistance 399–400
    soft 398
states of matter 259–60
statistics 241–5
stator construction 373

step-up/step-down
    transformers 352
storage batteries 201
stroboscopic effect 414–15
supply systems 8–10
    a.c. supply, generating 304–6
    alternative sources 309–10
    characteristics of
        different 335–40
    d.c. generators 301–3
    delta connection 336
    distribution of electricity 309
    emergency/stand-by supply 208
    generation of electricity 307
    load balancing 340–2
    neutral currents 339–40
    power factor correction
        (PFC) 346
    three-phase supplies 335–9,
        343–5
    transformers 346–56
    transmission of electricity 308–9
surveys before inspection 129–30
switches
    inspection 132
    testing of devices 139–40
    transistors as 489–92
switchgear
    faults with 193–4
    inspection checklist 90–1
switching 10–12, 86
symbols used in luminaires 423
synchronised motors 388–9

tally charts 242–3
termination of cables 190–1
testing
    accuracy of instruments 141
    'all in one' instruments 145
    assessment preparation 161
    calibration of instruments 141–2
    circuit breakers 139
    condition reports 140
    continuity of conductors 100–5,
        123–4, 137–8, 172–3
    earth electrode
        resistance 111–14, 144
    earth fault loop
        impedance 114–18, 126,
        139, 143–4, 174–5
    exploratory surveys 129–30
    fault finding and
        diagnosis 171–7
    frequency 127–9
    functional 119–20, 175

general procedures 129–31
initial 99
instruments for 140–5, 176–7
insulation resistance 105–10,
    125, 138, 173–4
isolation/switching
    devices 139–40
legislation and regulations 80–1
need for 127
ohmmeters 142–3
periodic 127–31, 135–40
phase sequence 119, 144–5
polarity 110–11, 125, 138–9,
    174
prospective fault current 118–19
protection by automatic
    disconnection of supply
    (ADS) 114
protection by barriers/
    enclosures 108–10
protection by SELV, PELV or
    electrical separation 107–8
RCDs 120–2, 139, 144, 175
regulations 99
resistors 441
sampling 130–1
scope of 131
sequence of 100–22
transistors 492–3
unsatisfactory results from 123–7
voltage drop 122
See also certificates
thermal constraints on selection of
    cables 60–3, 64
thermal effects
    of electric current 268–71
    protection against 25–6, 132
thermal insulation and selection of
    cables 52
thermal MCBs 34–5
thermistors 448–9
thermocouples 449
three-phase a.c. synchronous
    motors 388–9
three phase induction motors 372–7
three-phase supply 9–10, 335–9,
    343–5
thyristors 478–81
time current characteristics 39, 44
TN-C-S system (PME) 6–7
TN-S system 4–5
transformers 346–56
transient voltages 183–4
transistors
    basics of 484–5

current amplification 487
field effect (FETs) 493–5
operation of 485–9
as switches 489–92
testing 492–3
voltage amplification 487–9
transmission of electricity 308–9
transposition of equations 232–4
triacs 481–2
triangles 234–40
trimmer capacitors 454
trunking
    inspection checklist 94–5
    size of and cable selection 70–1
TT system 3–4
tungsten halogen lamps 408–10

unearthed equipment, danger
    of 16–17
units of measurement 212–15
universal (series-wound)
    motors 370–1
utilisation factor 425–6

variable capacitors 454
variable resistors 432, 434–5
variable speed drives 402
velocity ratio 252
verification, initial 78, 81–2
voltage
    amplification 487–9
    control of with resistors 443–6
    extra low 23
    indicating devices 176–7
    measurement 357
    over 200
    parallel circuits 278–9
    series circuits 276–7
    series/parallel circuits 280–1
    transient voltages 183–4
voltage drop 55–6, 57, 122, 282
voltage transformers 352
voltmeters 357, 358

wave generation 311
weight and mass 246
windings 390
wiring systems
    external influences on selection
        of 46
    See also cables
work 253, 282–3
wound rotors 374–6

Zener diodes 466–7